Linux Command Line and Shell Scripting Techniques

Master practical aspects of the Linux command line and then use it as a part of the shell scripting process

Vedran Dakic

Jasmin Redzepagic

BIRMINGHAM—MUMBAI

Linux Command Line and Shell Scripting Techniques

Copyright © 2022 Packt Publishing

Group Product Manager: Vijin Boricha

Publishing Product Manager: Shrilekha Malpani

Senior Editor: Shazeen Iqbal

Content Development Editor: Rafiaa Khan

Technical Editor: Shruthi Shetty

Copy Editor: Safis Editing

Project Coordinator: Shagun Saini

Proofreader: Safis Editing

Indexer: Subalakshmi Govindhan

Production Designer: Vijay Kamble

Marketing Coordinator: Hemangi Lotlikar

First published: April 2022

Production reference: 1150222

Published by Packt Publishing Ltd.
Livery Place
35 Livery Street
Birmingham
B3 2PB, UK.

ISBN 978-1-80020-519-2

www.packt.com

Writing this book was quite a journey for me. It was a very busy year, compounded by a lot of weird situations – earthquakes, COVID-19, health issues in the family – basically, this year threw everything that it possibly could at me. Often, I found myself pondering the fact that writing a book is as much therapeutic as it is potentially useful. Writing a book is as much about talking to your inner self as it is about talking to your future audience, especially when knowledge sharing is involved. And, as an ex-IT journalist/editor who used to write IT magazine articles for a living, I was constantly reminded of that fact.

To my co-workers, for offering valuable insight and for being a whiteboard to bounce ideas off (Jasmin, Zlatan, and Andrej). Your coherent and random "in passing" thoughts helped a lot.

To my Packt crew, for putting up with the extreme, parallel universe-level randomness that was my year 2021 – thank you.

To my partner, Sanja, for pushing me forward and kicking my derrière, when it needs to be kicked.

– Vedran Dakic

My first Linux, Slackware, came on 50+ floppy disks. Installing it was a pain, and somewhere in the middle of the process, a floppy disk was faulty so I started my Linux journey on a machine that was unable to install a GUI. Being confined to a text-only terminal for a few weeks got me hooked.

The nice thing about scripting is that I rarely have any need to leave the terminal even now, 25+ years later, while still being able to do almost anything on any machine I encounter.

Having said that, running scripts would not be possible if it weren't for a few special people in my life, so here goes:

To Filip, my son, who is learning to spell while I try to teach him how to type.

To Dinka, my wife, who keeps us all together through the sun and the rain.

And to all the SysAdmins and SysOps I have encountered during this quarter of a century, exchanging ideas and solutions.

Remember, it's usually DNS.

– Jasmin Redzepagic

Contributors

About the authors

Vedran Dakic is a master of electrical engineering and computing and an IT trainer, covering system administration, cloud, automatization, and orchestration courses. He is a certified Red Hat, VMware, and Microsoft trainer. He's currently employed as head of department of operating systems at Algebra University College in Zagreb. As part of this job, he's a lecturer for 3- and 5-year study programs in system engineering, programming, and multimedia tracks. Also, he does a lot of consulting and systems integration for his clients' projects – something he has been doing for the past 25 years. His approach is simple – bring real-world experience to all the courses that he teaches as it brings added value to his students and customers.

Jasmin Redzepagic is a professional master in computer engineering, with a sub-specialization in system engineering, but is a person with many hats worn over the years. Having worked in IT as long as he has, he has been a sound technician, SysOps, DevOps, hardware tester, editor-in-chief of a major magazine, writer, IT support lead, and head of IT at a couple of companies. He is currently employed by Algebra University College in Zagreb. This enables him to have a very wide overview of different technologies while still maintaining his focus – implementing and teaching the implementation of open source technologies in the field.

Right now, his main interest is trying to teach his students and clients to understand the immense expanse of tools available for any conceivable task, with his goal being to get people to understand computers as something that is the most formidable tool we, as humanity, have ever had.

About the reviewers

Sergio Guidi Tabosa Pessoa is a software engineer with more than 30 years of experience with software development and maintenance, from complex enterprise software projects to modern mobile applications. In the early days, he was working primarily with the Microsoft stack, but soon discovered the power of the Unix and Linux operating systems. Even though he has worked with many languages over the years, C and C++ remain his favorite languages on account of their power and speed.

He has a bachelor's degree in computer science and an MBA in IT management, and is always hungry to learn new technologies, break code, and learn from his mistakes. He currently lives in Brazil with his wife, two Yorkshire terriers, and two cockatiels.

First and foremost, I would like to thank all the people involved in this project, including the author, for such a great piece of work, and those from Packt Publishing for giving me this opportunity. I also would like to thank my beautiful wife, Lucia, and my children, Touché and Lion, for their patience and for allowing me the time needed to help with this book.

Nicholas Cross was born in the UK and educated in New Zealand. He is a distinguished engineer at a tier-1 technology company and has 20 years' experience in Linux. He was working with Linux and Bash scripting a long time before the cloud, automation, and DRY were cool.

Nicholas is passionate about infrastructure automation, DevOps culture, SRE, automation, containerization, and security, and everything else associated with these broad topics.

When not at his computer hacking code for work or pleasure, he enjoys running, walking his dogs, and watching his sons play rugby.

Jason Willson has been working in IT for 17 years since his first job at the help desk at his alma mater, Grove City College. He was first introduced to Linux in 2007 at a start-up in Boston and has worked with it professionally and personally ever since. He has used command-line and shell scripting techniques for a variety of tasks relating to data analysis, systems administration, and DevOps. He currently works as a Linux systems administrator at the Software Engineering Institute of Carnegie Mellon University.

I'd like to thank the incredible LinkedIn community for making this connection possible with Packt Publishing. I'd also like to thank all my coworkers, classmates, and mentors (personal, professional, and academic) who have helped to shape me into the person I am today.

Table of Contents

5

Using Commands for File, Directory, and Service Management

6

Shell-Based Software Management

7

Network-Based File Synchronization

8

Using the Command Line to Find, Extract, and Manipulate Text Content

9

An Introduction to Shell Scripting

10

Using Loops

15

Troubleshooting Shell Scripts

16

Shell Script Examples for Server Management, Network Configuration, and Backups

17
Shell Script Examples

Index

Other Books You May Enjoy

Preface

Linux Command-Line and Shell Scripting Techniques is a book that will help you learn how to use the **Command-Line Interface (CLI)** and to further expand your CLI knowledge with the ability to do scripting. It looks at a big collection of CLI commands, shell scripting basics (loops, variables, and functions), and advanced scripting topics – such as troubleshooting. It also includes two chapters with script examples that should get you further ahead in your understanding of scripting while also offering good insight into how the shell scripting process works.

Who this book is for

This book is for beginners and professionals alike, as it doesn't necessarily need a lot of prior Linux knowledge. That's partially what this book is for – to get to grips with using the command line and to further that usage model to shell scripting. For more advanced users, there's a bulk of content about shell scripting and corresponding examples that will help you to organize and improve your knowledge about shell scripting.

What this book covers

Chapter 1, Basics of the Shell and Text Terminal, discusses the concept of the shell and text terminal, configuration of the Bash shell, using some basic shell commands, and using the screen to get access to multiple virtual terminals in text mode.

Chapter 2, Using Text Editors, takes us to the highly subjective world of text editors, where discussions have been happening for the past 30-40 years on the topic of *best editor*. As a part of this chapter, we're going to use vi(m), nano, and some more advanced vi(m) settings.

Chapter 3, Using Commands and Services for Process Management, is about using files, folders, and services, specifically, how to administer them, how to secure them (files and folders), and how to manage them (services). A big chunk of this chapter is related to ACLs and systemctl, essential tools for system administrators.

Chapter 4, Using Shell to Configure and Troubleshoot Network, is all about working with files, folders, and services – working with permissions, manipulating file content, archiving and compressing files, and managing services. Throughout this chapter, there will be a lot of simple commands that we will be using later when we go on to scripting, as well.

Chapter 5, Using Commands for File, Directory, and Service Management, is about making sure that we know the basics of fundamental networking configuration – nmcli and netplan, FirewallD and ufw, DNS resolving, and diagnostics. These are some of the settings that we most commonly re-configure post-deployment, so deep insight into them is a necessity.

Chapter 6, Shell-Based Software Management, takes us through two of the most commonly used packaging systems (dnf/yum and apt), as well as some more advanced concepts, such as using additional repositories, streams, and profiles, creating custom repositories, and third-party software. Every Linux deployment needs us to have knowledge about package management, so this chapter is all about that.

Chapter 7, Network-Based File Synchronization, teaches us about the most commonly used tools to send and receive files and connect to remote destinations via a network – ssh and scp, rsync, and vsftpd. For anything ranging from hosting a Linux distribution mirror all the way to synchronizing files and backups, this is mandatory knowledge.

Chapter 8, Using the Command Line to Find, Extract, and Manipulate Text Content, is all about using basic and more advanced ways of manipulating text files and content. We start off by doing simple things such as paste and dos2unix, and then move on to some of the most used commands in the IT world – cut, (e)grep, and sed.

Chapter 9, An Introduction to Shell Scripting, is the starting point for the second part of this book, which is all about shell scripting and using previously mentioned tools and commands to create shell scripts. This chapter is about the basics of shell scripting and working with general concepts, such as input, output, error, and shell script hygiene.

Chapter 10, Using Loops, goes deep into the concept of loops. We cover all the most used loops here – the for loop, break and continue, the while loop, the test-if loop, the case loop, and logical looping with conditions such as and, or, and not. This will further enhance our ability to do more things in shell scripts.

Chapter 11, Working with Variables, is about using variables in our shell script code – shell variables, quoting and special characters in variable values, assigning external variables via commands, as well as some logical operations on variables. Variables are the spine of shell scripting, and all permanent and temporary data gets stored in them, so, for whatever purpose we're developing a shell script, variables are a must-have.

Chapter 12, Using Arguments and Functions, is about further customizing and modularizing the shell script code, as we can use functions to do that. For that purpose, we are going to use external and shell arguments, to do away with the static nature of most of our previous shell script examples.

Chapter 13, Using Arrays, is about using arrays to store and manipulate data. Arrays are just one of those structures – we need them, we learn to not necessarily like them, but we can't live without them, especially as we venture into the world of working with their many different capabilities, such as indexing, adding and removing members, and working with files as a de facto array source.

Chapter 14, Interacting with Shell Scripts, is about moving from the idea of shell script code as a purely text-driven principle, and going in the opposite direction – to create a TUI-based interface to interact with a script. We are also going to have a play with `expect` script, which makes it easier for us to create a script that's waiting for specific output and then doing something based on that output, which can be useful for the configuration of third-party systems at times.

Chapter 15, Troubleshooting Shell Scripts, deals with shell script troubleshooting – common mistakes, debugging output via echoing values during script execution, Bash -xv, and other concepts. This is the last chapter before we start dealing with the many script examples that we prepared for you to use both as a learning tool and to work with them in production, if you so desire.

Chapter 16, Shell Script Examples for Server Management, Network Configuration, and Backups, sets us off in the direction of simple shell scripts – nine different examples to be exact. Topics vary from simple, modular code that can be implemented in any shell script (for example, how to check if we're executing a script as root), to more complex examples such as dealing with date and time, the interactive configuration of network settings and firewalls, as well as some backup script examples.

Chapter 17, Advanced Shell Script Examples, deals with more complex examples, such as a script to modify web server and security settings, bulk-creating users and groups with random passwords, scripted KVM virtual machine installation, and scripted KVM virtual machine administration (start, stop, getting info, manipulating snapshots, and so on). These are examples that we use in everyday life to drive the point of shell scripting home, which is all about automating boring, repetitive tasks and offloading them to a script that can do all of that for us.

To get the most out of this book

Software/hardware covered in the book	OS requirements
CentOS 8	A virtualization app such as VMware Workstation, Player, Fusion, or equivalent
Ubuntu 20.10	A virtualization app such as VMware Workstation, Player, Fusion, or equivalent

Download the color images

We also provide a PDF file that has color images of the screenshots/diagrams used in this book. You can download it here: `https://static.packt-cdn.com/downloads/9781800205192_ColorImages.pdf`.

Conventions used

There are a number of text conventions used throughout this book.

`Code in text`: Indicates code words in text, database table names, folder names, filenames, file extensions, pathnames, dummy URLs, user input, and Twitter handles. Here is an example: *To configure the host side of the network, you need the tunctl command from the User Mode Linux (UML) project.*

A block of code is set as follows:

```
#include <stdio.h>
#include <stdlib.h>
int main (int argc, char *argv[])
{
    printf ("Hello, world!\n");
    return 0;
}
```

Any command-line input or output is written as follows:

```
$ sudo tunctl -u $(whoami) -t tap0
```

Bold: Indicates a new term, an important word, or words that you see onscreen. For example, words in menus or dialog boxes appear in the text like this. Here is an example: *Click* **Flash** *from Etcher to write the image.*

> **Tips or Important Notes**
> Appear like this.

Get in touch

Feedback from our readers is always welcome.

General feedback: If you have questions about any aspect of this book, mention the book title in the subject of your message and email us at customercare@packtpub.com.

Errata: Although we have taken every care to ensure the accuracy of our content, mistakes do happen. If you have found a mistake in this book, we would be grateful if you would report this to us. Please visit www.packtpub.com/support/errata, selecting your book, clicking on the Errata Submission Form link, and entering the details.

Piracy: If you come across any illegal copies of our works in any form on the Internet, we would be grateful if you would provide us with the location address or website name. Please contact us at copyright@packt.com with a link to the material.

If you are interested in becoming an author: If there is a topic that you have expertise in and you are interested in either writing or contributing to a book, please visit authors.packtpub.com.

Share Your Thoughts

Once you've read *Linux Command Line and Shell Scripting Cookbook*, we'd love to hear your thoughts! Scan the QR code below to go straight to the Amazon review page for this book and share your feedback.

https://packt.link/r/1-800-20519-8

Your review is important to us and the tech community and will help us make sure we're delivering excellent quality content.

1
Basics of Shell and Text Terminal

An ancient Chinese proverb states that *A journey of a thousand miles begins with a single step*. This chapter is going to be that single, first step on our journey to mastering the Linux **command-line interface (CLI)** and **shell scripting**. Specifically, we're going to learn how to use Terminal, the shell, some basic commands, and one very handy utility to work on many things at once, called **screen**.

As you progress further through the book, you'll notice that we will be using these concepts a lot, as they're the basis for what we're going to do in the later chapters. When dealing with systems administration, we can usually do a lot more in some kind of CLI than in any kind of **graphical user interface (GUI)**. This stems from the idea of the programmability of a *CLI* versus the *static* nature of most GUIs in IT. Furthermore, a utility such as screen will make our life in the CLI a lot easier, as we're going to be able to deal with multiple virtual screens at the same time, thus enhancing our productivity.

So, in short, we're going to deal with the following recipes:

- Accessing the shell
- Setting up the user shell
- Setting up the Bash shell

- Using the most common shell commands
- Using `screen`

Technical requirements

For these recipes, we're going to use two Linux machines – in our case, it's going to be two VMware virtual machines with *Ubuntu* (*20.04 Focal Fossa*). Let's call them `cli1` and `gui1`, and as the book progresses, we're going to add some more, as our topics grow in complexity. So, all in all, we need the following:

- VMware Player, Workstation, Fusion, or ESXi
- Ubuntu 20.04 Focal Fossa installation ISO file
- A bit of time to install these two virtual machines without the GUI (the `cli1` machine) and with the GUI (`gui1`)

After the installation process is finished, we're going to start with shell basics – our very next topic.

Accessing the shell

First, let's briefly discuss various **shell** access methods. It can be as simple as just installing a virtual machine with Linux that's running text mode only, but it could also be a virtual machine with the GUI. That would require us doing something to have access to text mode; so, let's learn about these different ways of getting access to the shell.

It's also important to understand *why* accessing the shell is so important. The reasoning behind this is simple, which is that we can do a lot more in the shell than in the GUI. In the second part of this book, we're going to dig deeper into the concept of shell scripting, and then it's going to become obvious how that applies to our statement about being able to do a lot more in the shell.

Getting ready

For starters, we need to deploy our two virtual machines. We can actually install both of these machines as text-mode machines with the OpenSSH server (the installation process asks about OpenSSH at one point). Then, we could add the GUI to the `gui1` machine so that we can work with that, too. We do that by typing a couple of commands into the `gui1` machine after logging in as `student` (which is the username we came up with for this example):

```
sudo apt-get -y install tasksel
tasksel
```

`sudo` is going to ask us for the `student` user password (it can be any user that you created during the installation process; `student` is just something that we used in our example). When the **tasksel TUI** interface starts, we will select the **Ubuntu desktop** package set, as shown in the following screenshot:

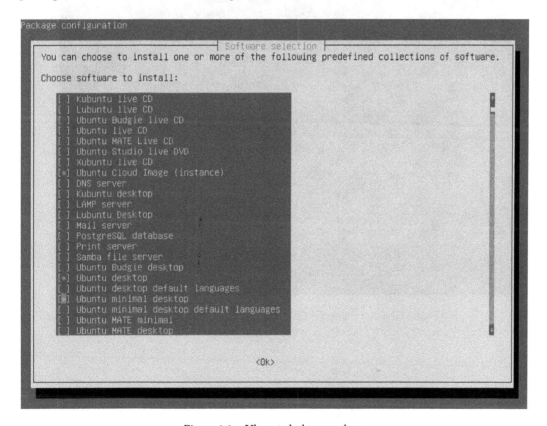

Figure 1.1 – Ubuntu desktop package

As you can see from *Figure 1.1*, you need to select **Ubuntu desktop** by using the arrow keys and spacebar to select from the menu, then use *Tab* or the arrow keys to select **Ok** and press *Enter*.

Let's now discuss how to access the shell.

How to do it...

If we deployed our Ubuntu machine with default options, we're going to be faced with text mode by default. To access the shell and have the capability to do something with our Linux virtual machine, we need to type in our *username* and *password*. It needs to be the username or password that we typed in during the installation process. In our virtual machine, the user named student was created, with the preassigned student password. When we successfully log in, we're faced with the regular text mode and associated shell, as in the following screenshot:

```
Ubuntu 20.04.3 LTS cli1 tty3

cli1 login: student
Password:
Welcome to Ubuntu 20.04.3 LTS (GNU/Linux 5.4.0-97-generic x86_64)

 * Documentation:  https://help.ubuntu.com
 * Management:     https://landscape.canonical.com
 * Support:        https://ubuntu.com/advantage

  System information as of Thu 03 Feb 2022 01:21:31 PM UTC

  System load:  0.07              Processes:             355
  Usage of /:   49.4% of 18.57GB  Users logged in:       1
  Memory usage: 16%               IPv4 address for ens33: 192.168.175.154
  Swap usage:   0%


2 updates can be applied immediately.
To see these additional updates run: apt list --upgradable


Last login: Thu Feb  3 13:20:57 UTC 2022 on tty3
student@cli1:~$ _
```

Figure 1.2 – Accessing the CLI from text mode after logging in

If, however, we did a GUI installation, there are three different ways of accessing the shell:

1. We could start a text Terminal in **GNOME (GNOME Terminal)** and use the shell from there. On the plus side, it gives us a GUI-like look and feel that might be a bit more user-friendly to a lot of people. On the downside, we're rarely going to find a GUI on Linux servers in production environments, so it might be a case of *learning bad habits*. To start GNOME Terminal, we can either use the built-in GNOME search function (by pressing the *WIN* key) or just right-click on the desktop and open **Terminal**. The result will look like this:

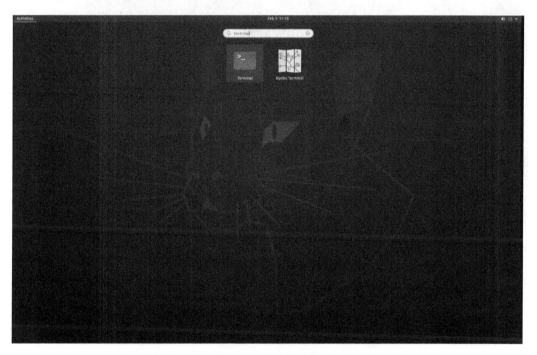

Figure 1.3 – Finding Terminal in the GNOME GUI by using
the WIN key to search for the Terminal keyword

2. We could just switch to a text-based text console, as Linux doesn't stop us from using text consoles when we deploy the GUI. For that, we need to press a dedicated key combination that is going to get us to one of those text consoles. For example, we can press the *Ctrl + Alt + F3* key combination. That is going to take us to text mode, specifically, to the text console number.

There, we can log in and start typing our commands. The result will look like this:

```
Ubuntu 20.04.3 LTS cli1 tty3

cli1 login: student
Password:
Welcome to Ubuntu 20.04.3 LTS (GNU/Linux 5.4.0-97-generic x86_64)

 * Documentation:  https://help.ubuntu.com
 * Management:     https://landscape.canonical.com
 * Support:        https://ubuntu.com/advantage

  System information as of Thu 03 Feb 2022 01:21:31 PM UTC

  System load:  0.07             Processes:             355
  Usage of /:   49.4% of 18.57GB  Users logged in:       1
  Memory usage: 16%              IPv4 address for ens33: 192.168.175.154
  Swap usage:   0%


2 updates can be applied immediately.
To see these additional updates run: apt list --upgradable


Last login: Thu Feb  3 13:20:57 UTC 2022 on tty3
student@cli1:~$ _
```

Figure 1.4 – Switching to the text Terminal from the GUI

We're in the shell again. We can now start using any commands we want to.

3. We could use the `systemctl` command to switch to text mode as a default mode for the current session (until the next reboot). We could even use it to make the text mode permanent, regardless of the fact that we have full GUI installation. To achieve that, in our GUI, we need to log in and then type the following sequence of commands into GNOME Terminal:

```
systemctl set-default multi-user.target
systemctl isolate multi-user.target
```

We can use the first command if we want to set our Linux virtual machine to boot to text mode *by default*. We can use the second command to re-configure our Linux machine to switch to text mode *immediately*.

How it works...

There are a couple of sets of commands that we used in the previous recipe, so let's explain what these commands do so that we can have a clear picture of what we were doing.

The first set of commands is as follows:

```
sudo -i
apt-get -y install tasksel
tasksel
```

These three commands are going to do the following:

- `sudo -i` is going to ask us for the current user's password. If that user has been added to the `sudoers` system (`/etc/sudoers`), that means that we can use the current user's password to log in as root and use administrative privileges.

- `apt-get -y install tasksel` will install the `tasksel` application. The main purpose of this application is to simplify package deployment. Specifically, we are going to use it in the next step to deploy an *Ubuntu desktop* set of packages (multiple hundreds of packages). Imagine typing all the `apt-get` commands for that deployment procedure manually!

- The `tasksel` command is going to start the tasksel application, which will be used to deploy the necessary packages.

The second set of commands do the following things:

- `systemctl set-default multi-user.target` will set text mode as the default boot target. The translation of this is that our Linux machine will boot in text mode by default only when it becomes valid after the next reboot.

- `systemctl isolate multi-user.target` will switch us to text mode immediately. It's completely different from the `set-default` procedure, as it has nothing to do with the state of our Linux machine post-reboot.

See also

If you need more information about `apt-get`, `tasksel`, or `systemctl`, we suggest that you visit these links:

- `apt-get`: `https://help.ubuntu.com/community/AptGet/Howto`

- `tasksel`: `https://help.ubuntu.com/community/Tasksel`

- `systemctl`: `https://www.liquidweb.com/kb/what-is-systemctl-an-in-depth-overview/`

Setting up the user shell

Now that we have learned about how to access the shell, let's configure it for our comfortable use. We're going to see a couple of examples so that we can understand how customizable the Linux shell is. Specifically, we're going to customize the look and feel of our prompt.

Getting ready

We just need to keep our virtual machines up and running.

How to do it...

We're going to edit a file called `/home/student/.bashrc`. Before we do that, let's create a backup copy of the `.bashrc` file, just in case we make some mistakes:

```
cp /home/student/.bashrc /home/student/.bashrc.tmp
```

Before we edit this file, make sure that you take note of how the prompt looks at this point. If you're logged in as `student` to the `cli1` machine, your prompt should look like this:

```
student@cli1:~$
```

Let's edit the `.bashrc` file by using `nano`. Type in the following command:

```
nano /home/student/.bashrc
```

When we type in this command, we're going to open `.bashrc` in the nano editor. Let's scroll all the way to the end of this file, which should look like this:

```
  GNU nano 4.8                                    /home/student/.bashrc
# ~/.bashrc: executed by bash(1) for non-login shells.
# see /usr/share/doc/bash/examples/startup-files (in the package bash-doc)
# for examples

# If not running interactively, don't do anything
case $- in
    *i*) ;;
      *) return;;
esac

# don't put duplicate lines or lines starting with space in the history.
# See bash(1) for more options
HISTCONTROL=ignoreboth

# append to the history file, don't overwrite it
shopt -s histappend

# for setting history length see HISTSIZE and HISTFILESIZE in bash(1)
HISTSIZE=1000
HISTFILESIZE=2000

# check the window size after each command and, if necessary,
# update the values of LINES and COLUMNS.
shopt -s checkwinsize

# If set, the pattern "**" used in a pathname expansion context will
# match all files and zero or more directories and subdirectories.
#shopt -s globstar
```

Figure 1.5 – .bashrc default content

Let's go all the way below the last `fi` and add the following statement:

```
PS1="MyCustomPrompt> "
```

Next, use *Ctrl + X* to save the file. Then, when we're back in the shell, let's type in the following command:

```
source .bashrc
```

If we did everything correctly, our prompt should look like this now:

```
MyCustomPrompt>
```

This can be further customized by using PS1 parameters. Let's locate the following:

```
PS1="MyCustomPrompt> "
```

We'll change it to the following:

```
PS1="\u@\H> \A "
```

The \u@\H part represents the username@host part of the prompt. The \A part is for 24-hour time. So, when we do the following:

```
source .bashrc
```

Again, we should get the following state of our prompt:

```
[student@cli1> 19:30]
```

19:30 represents time. We could also customize things such as the type of font (underlined, normal, dim, bold) and color (black, red, green, and so on). Let's do that now. For example, let's edit .bashrc again and set PS1 like this:

```
PS1="\e[0;31m[\u@\H \A] \e[0m"
```

Our prompt should now look like this:

```
[student@cli1 19:39]
```

In this particular example, \e[tells the PS1 variable that we want to change the color of our prompt. 0;31m means red (30 is black, 34 is blue, and so on). The [] enclosed part is our regular prompt, as previously discussed. The last bit, \e[0m, tells the PS1 variable that we're done with color modification for our PS1 output.

As we can see, touching just one shell variable (PS1) can drastically change the way we consume text mode in our Linux virtual machine.

How it works...

As a shell variable, PS1 can be used to customize the look and feel of our shell. Think of it in the way that most users customize their GUI using different wallpapers, text sizes, colors, and so on, and it's a natural thing that we do because we like what we like. PS1 is often called a **primary prompt display variable**, as described in the previous section.

The `source` command that we used *executes* `.bashrc`, in the sense that it will apply the settings from `.bashrc`. That's why we didn't need to log off and log in again, as it would be a waste of time, and the `source` command can help us with that.

Let's now add some more settings to our `.bashrc` file, as there are many more things that we can customize.

Setting up the Bash shell

We played with the `PS1` variable and configured it so that it's more to our liking. Let's now use more `.bashrc` settings to configure our Bash shell even further.

Getting ready

We need to leave our virtual machines running. If they are not powered on, we need to power them back on.

How to do it...

Let's discuss how to change the following shell parameters:

1. Add some custom aliases.

    If we open the `.bashrc` file again, we can do some additional magic with it. First, let's add a couple of aliases. Close to the end of the `.bashrc` file, there's a section with a couple of aliases (`ll`, `la`, and `l`). Let's add the following lines to that part of the `.bashrc` file:

    ```
    alias proc="ps auwwx"
    alias pfilter="ps auwwx | grep "
    alias start="systemctl start "
    alias stop="systemctl stop "
    alias ena="systemctl enable "
    ```

    This code will introduce five new aliases:

*   To see a list of processes
*   To filter processes according to the keyword to be typed in after `pfilter`
*   To start the service; service name to be typed after `start`
*   To stop the service; service name to be typed after `stop`
*   To enable the service; service name to be typed after `ena`

As we can see, using aliases can make our typing shorter and the administration process simpler.

2. Adjust the size of the Bash history.

    At the top of the `.bashrc` file, there's a section similar to this:

    ```
    HISTSIZE = 1000
    HISTFILESIZE = 1000
    ```

    If we want the Bash shell to remember more than the last 1,000 typed commands in the current session (`HISTSIZE`), and to save more than 1,000 commands in the history file (called `.bash_history`), we can change the values of these variables, let's say to `2000` and `2000`.

3. Adjust the `PATH` variable.

    Let's say that we want to add a custom path to our existing `PATH` variable. For example, we installed our custom application in the `/opt/bin` directory, and we don't want to call that application by using the full path every single time.

    We need to edit the `.profile` file for this, as the `PATH` variable for our current user is set there. So, open the `.profile` file in the editor and add the following line to the end of this file:

    ```
    PATH=$PATH:/opt/bin
    ```

4. Set our default editor.

    Let's add the following two lines to the `.bashrc` file, at the end of the file:

    ```
    export VISUAL=nano
    export EDITOR=nano
    ```

    This would set nano as our preferred default editor.

How it works...

The Bash shell has a set of reserved variables that we can use solely for Bash purposes. Some of these reserved variables include the following:

- PS1, PS2, PS3, and PS4
- HISTFILESIZE and HISTSIZE
- VISUAL and EDITOR
- OLDPWD
- PWD

These are reserved for specific Bash functions, so we shouldn't create custom variables with these names. You can learn more about these reserved variables from the link provided in the *There's more...* section.

As far as PS variables are concerned, we can consider them to be our entry into customization of the Bash shell. That especially goes for the PS1 variable, as it's the variable that's most commonly used. We can use all of these variables to set Bash to suit our own needs as we don't have to use predefined global configuration only. As time goes by, more and more Linux system administrators create their own, customized configurations for Bash as it increases the convenience of using the Bash shell and their own productivity.

There's more...

If we need to learn more about Bash reserved variables and PS variables, we can check the following links:

- `https://tldp.org/LDP/Bash-Beginners-Guide/html/sect_03_02.html`
- `https://access.redhat.com/solutions/505983`

Using the most common shell commands

Let's now switch our attention to learning a basic set of **Linux shell commands**. We're going to discuss commands to manipulate files and folders, processes, archives, and links. We'll do that via a scenario that is going to involve many steps.

Getting ready

We still need the same virtual machines as with our previous recipes.

How to do it...

In order for us to be able to use shell commands, we have to start the shell. If we're using the CLI, we just need to log in and we're going to enter the shell session. If we're using a GUI approach, we have to find a GUI Terminal in our application menu. After that, we can start typing commands:

1. For starters, let's use a basic set of commands to work with files and directories.

    Let's list the content of the current directory:

    ```
    ls -al
    ```

The output will look similar to this:

```
[student@cli1 13:27] ls -al
total 80
drwxr-xr-x 15 student student 4096 Feb  3 13:27 .
drwxr-xr-x  3 root    root    4096 Feb  3 10:40 ..
-rw-------  1 student student   84 Feb  3 13:27 .bash_history
-rw-r--r--  1 student student  220 Feb 25  2020 .bash_logout
-rw-r--r--  1 student student 3675 Feb  3 13:27 .bashrc
drwx------ 10 student student 4096 Feb  3 11:16 .cache
drwx------ 10 student student 4096 Feb  3 11:17 .config
drwxr-xr-x  2 student student 4096 Feb  3 11:15 Desktop
drwxr-xr-x  2 student student 4096 Feb  3 11:15 Documents
drwxr-xr-x  2 student student 4096 Feb  3 11:15 Downloads
drwx------  3 student student 4096 Feb  3 11:16 .gnupg
drwxr-xr-x  3 student student 4096 Feb  3 11:15 .local
drwxr-xr-x  2 student student 4096 Feb  3 11:15 Music
drwxr-xr-x  2 student student 4096 Feb  3 11:15 Pictures
-rw-r--r--  1 student student  807 Feb 25  2020 .profile
drwxr-xr-x  2 student student 4096 Feb  3 11:15 Public
drwx------  2 student student 4096 Feb  3 11:16 .ssh
-rw-r--r--  1 student student    0 Feb  3 10:41 .sudo_as_admin_successful
drwxr-xr-x  2 student student 4096 Feb  3 11:15 Templates
drwxr-xr-x  2 student student 4096 Feb  3 11:15 Videos
-rw-------  1 student student 1482 Feb  3 13:27 .viminfo
[student@cli1 13:28] _
```

Figure 1.6 – Standard output for the ls -al command with all the pertinent information

2. Now, let's create a directory called directory1 and a stack of five files called test1 to test5. That's what the touch command does – it creates empty files. Then, let's copy these files to that directory:

```
mkdir directory1
touch test1
touch test2
touch test3
touch test4
touch test5
cp test* directory1
```

3. After that, let's create a directory called directory2 and move files 1-5 to directory2:

```
mkdir directory2
mv test* directory2
```

4. Let's check the amount of used disk space in `directory1` and `directory2`:

```
[student@clil 21:47] du directory1
4          directory1
[student@clil 21:47] du -hs directory2
4.0K     directory2
[student@clil 21:48]
```

5. Let's check the capacity of our current disk (the `-h` switch provides us with a nice, human-readable output):

```
df -h .
Filesystem                        Size  Used Avail Use%
Mounted on
/dev/mapper/ubuntu--vg-ubuntu--lv  19G  4.5G   14G  26%
/
```

The next stack of commands is related to links known as **hard links** and **soft links**.

6. For creating hard and soft links, let's log in to our `clil` virtual machine and log in as `root`. The overall concept of hard and soft links will be explained a bit later in this chapter. So, create a temporary directory and use some files. We'll just use an existing file as it's more than enough for this scenario (the `.bashrc` file):

```
mkdir links
cd links
cp /root/.bashrc content.cfg
ln content.cfg hardlink.cfg
ln -s content.cfg softlink.cfg
ls -al
cd /root
ln -s links links2
ln links links3
ln: links: hard link not allowed for directory
cp .bashrc /tmp
cd /tmp
ln .bashrc /root/notworking
ln: failed to create hard link '/root/notworking' =>
'.bashrc': invalid cross-device link
ln -s .bashrc /root/working.cfg
ls /root/working.cfg
/root/working.cfg
```

7. Let's now check the beginning and end of one of these files. For example, let's use / tmp/.bashrc:

```
head /tmp/.bashrc
[student@cli1 22:28] head /tmp/.bashrc
# ~/.bashrc: executed by bash(1) for non-login shells.
# see /usr/share/doc/bash/examples/startup-files (in the
package bash-doc)
# for examples

# If not running interactively, don't do anything
case $- in
    *i*) ;;
      *) return;;
esac
```

Let's now check the tail end of the same file:

```
[student@cli1 22:29] tail /tmp/.bashrc
  if [ -f /usr/share/bash-completion/bash_completion ];
then
      . /usr/share/bash-completion/bash_completion
    elif [ -f /etc/bash_completion ]; then
      . /etc/bash_completion
    fi
fi
PS1="\e[0;31m[\u@\H \A] \e[0m"
export VISUAL=nano
export EDITOR=nano
```

8. The next step is going to involve checking running processes and the system state.

Let's now use commands to check the load our system currently has, find some processes, and kill some of them for fun.

First, let's check the load (uptime command) and find the top 20 most time-consuming processes (the ps command):

```
uptime
22:35:48 up  3:16,  2 users,  load average: 0.00, 0.00,
0.00
ps   auwwx | head -20
```

```
[student@cli1 22:35] ps auwwx | head -20
```

USER	PID	%CPU	%MEM	VSZ	RSS	TTY	STAT
START	TIME	COMMAND					
root	1	0.0	0.5	103252	11844	?	Ss
19:18	0:02	/sbin/init					
root	2	0.0	0.0	0	0	?	S
19:18	0:00	[kthreadd]					
root	3	0.0	0.0	0	0	?	I<
19:18	0:00	[rcu_gp]					
root	4	0.0	0.0	0	0	?	I<
19:18	0:00	[rcu_par_gp]					
root	6	0.0	0.0	0	0	?	I<
19:18	0:00	[kworker/0:0H-kblockd]					
root	9	0.0	0.0	0	0	?	I<
19:18	0:00	[mm_percpu_wq]					
root	10	0.0	0.0	0	0	?	S
19:18	0:00	[ksoftirqd/0]					
root	11	0.0	0.0	0	0	?	I
19:18	0:04	[rcu_sched]					
root	12	0.0	0.0	0	0	?	S
19:18	0:00	[migration/0]					

Next, let's find a specific process by name and kill it:

```
student@gui1:~$ ps auwwx | grep -i firefox
student    47198 22.1 21.3 2825436 426736 ?        R1
22:38   0:12 /usr/lib/firefox/firefox -new-window
student    47253  1.5  6.8 2427560 137988 ?        S1
22:38   0:00 /usr/lib/firefox/firefox -contentproc
-childID 1 -isForBrowser -prefsLen 1 -prefMapSize 223938
-parentBuildID 20210222142601 -appdir /usr/lib/firefox/
browser 47198 true tab
student    47266  1.0  5.5 2402216 110116 ?        S1
22:38   0:00 /usr/lib/firefox/firefox -contentproc
-childID 2 -isForBrowser -prefsLen 85 -prefMapSize 223938
-parentBuildID 20210222142601 -appdir /usr/lib/firefox/
browser 47198 true tab
student    47304  1.3  6.6 2468136 133340 ?        S1
22:38   0:00 /usr/lib/firefox/firefox -contentproc
-childID 3 -isForBrowser -prefsLen 1246 -prefMapSize
223938 -parentBuildID 20210222142601 -appdir /usr/lib/
firefox/browser 47198 true tab
```

```
student    47363  0.6  4.1 2386184 83588 ?          S1
22:38   0:00 /usr/lib/firefox/firefox -contentproc
-childID 4 -isForBrowser -prefsLen 10270 -prefMapSize
223938 -parentBuildID 20210222142601 -appdir /usr/lib/
firefox/browser 47198 true tab
student    48047  0.0  0.0    5168    880 pts/1    S+
22:39   0:00 grep --color=auto -i firefox
student@guil:~$ killall firefox
student@guil:~$ ps auwwx | grep -i firefox
student    48323  0.0  0.0    5168    884 pts/1    S+
22:40   0:00 grep --color=auto -i firefox
```

We're done with this part of the recipe. Let's now discuss the next part of this recipe, which is about administering users and groups.

9. Using shell commands for user and group administration, let's first go through the list of commands that we're going to use in this recipe:

- useradd: Command that's used to *create* a local user account

- usermod: Command that's used to *modify* a local user account

- userdel: Command that's used to *delete* a local user account

- groupadd: Command that's used to *create* a local group

- groupmod: Command that's used to *modify* a local group

- groupdel: Command that's used to *delete* a local group

- passwd: Command that's most often used to assign passwords to user accounts, but it can be used for some other scenarios (for example, locking user accounts)

- chage: Command that's used to manage user password expiry.

    So, let's create our first users and groups by using the useradd and groupadd commands with a scenario. Let's say that our task is as follows:

- Create four users called jack, joe, jill, and sarah.

- Create two user groups called profs and pupils.

- Re-configure the jack and jill user accounts to be members of the profs group.

- Re-configure the joe and sarah user accounts to be members of the pupils group.

- Assign a standard password to all the accounts (we're going to use `P@ckT2021` for this purpose).

- Configure user accounts so that they have to change their password on the next login.

- Set a specific expiry data for the `profs` users group – the minimum number of days before password change set to `15`, the maximum number of days before forced password change set to `30`, the warning for password change needs to start a week before it expires, and set the expiry date for accounts to 2022/01/01 (January 1st, 2022).

- Set a specific expiry data for the `pupils` users group – the minimum number of days before password change set to `7`, the maximum number of days before forced password change set to `30`, the warning for password change needs to start 10 days before it expires, and set the expiry date for accounts to 2021/09/01 (September 1st, 2021).

- Modify the `profs` group to be called `professors` and the `pupils` group to be called `students`.

10. The first task is to create the user accounts:

```
useradd jack
useradd joe
useradd jill
useradd sarah
```

This will create entries for these four users in the `/etc/passwd` file (where most of the users' information is stored – username, user ID, group ID, default home directory, and default shell) and the `/etc/shadow` file (where users' passwords and aging information are stored).

11. Then, we need to create the groups:

```
groupadd profs
groupadd pupils
```

This will create entries for these groups in the `/etc/group` file, where the system keeps all the system groups.

12. The next step is to manage users' group membership, for both the `professors` and `students` user groups.

    Before we do that, we need to be aware of one fact. There are two distinctive local group types, a **primary group** and a **supplementary group**. A primary group is important in terms of being the key parameter used when creating new files and directories, as the users primary group will be used by default for that (there are exceptions, as we'll mention in *the Setting up the Bash shell recipe* in this chapter, about umask, permissions, and ACLs). A supplementary group is important when dealing with sharing files and folders and related scenarios and exceptions. This is what's usually used for some additional settings for more advanced scenarios. These scenarios are going to be explained partially in the aforementioned *Setting up the Bash shell recipe* in this chapter, as well as in recipes about NFS and Samba in *Chapter 9, An Introduction to Shell Scripting*.

    Primary and supplementary groups are stored in the `/etc/group` file.

13. Now that we've gotten that out of the way, let's modify our users' settings so that they belong to *supplementary* groups as assigned by the scenario:

    ```
    usermod -G profs jack
    usermod -G profs jill
    usermod -G pupils joe
    usermod -G pupils sarah
    ```

    Let's now check how that changes the `/etc/group` file:

    ```
    root@cli1:~# cat /etc/group | tail -6
    jack:x:1001:
    joe:x:1002:
    jill:x:1003:
    sarah:x:1004:
    professors:x:1005:jack,jill
    students:x:1006:joe,sarah
    root@cli1:~# _
    ```

    Figure 1.7 – Entries in the /etc/group file

    The first four entries in the `/etc/group` file were actually created when we used the `useradd` command to create these user accounts. The next two entries (except for the last part, after the `:` sign) were created by the `groupadd` commands. Entries after the `:` sign were created after the `usermod` commands.

14. Let's now set their initial password and set a forced password change on the next login. We can do this in a couple of different ways, but let's learn the more *programmatic* approach to doing this, by echoing a string and using it as the plain-text password for a user account:

```
echo "jack:P@ckT2021" | chpasswd
echo "joe:P@ckT2021" | chpasswd
echo "jill:P@ckT2021" | chpasswd
echo "sarah:P@ckT2021" | chpasswd
```

The echo part, without the rest of the command, would just mean typing P@ckT2021 in to a Terminal, like this:

```
echo "P@ckT2021"
P@ckT2021
```

In CentOS and similar distributions, we could use the passwd command with the --stdin parameter, which would mean that we want to add a password for the user account via standard input (keyboard, variables, and so on). In Ubuntu, this is not available. So, we can echo the username:P@ckT2021 string to the shell and pipe that to the chpasswd command, which achieves just that purpose; instead of outputting the string to our Terminal, the chpasswd command uses it as standard input into itself.

15. Let's set the expiry date for professors and students. For this purpose, we need to learn how to use the chage command and some of its parameters (-m, -M, -W, and -E). In short, they mean the following:

- If we use the -m parameter, that means that we want to assign the minimum number of days before the password change is *allowed*.

- If we use the -M parameter, that means that we want to assign the maximum number of days before the password change is *forced*.

- If we use the -W parameter, that means that we want to set the number of warning days before password expiration, which in turn means that the shell is going to start throwing us messages about needing to change our password before it expires.

- If we use the -E parameter, that means that we want to set account expiration to a certain date (YYYY-MM-DD format).

Let's now translate that into commands:

```
chage -m 15 -M 30 -W 7 -E 2022-01-01 jack
chage -m 15 -M 30 -W 7 -E 2022-01-01 jill
chage -m 7 -M 30 -W 10 -E 2021-09-01 joe
chage -m 7 -M 30 -W 10 -E 2021-09-01 sarah
```

16. Finally, let's modify the groups to their final settings:

```
groupmod -n professors profs
groupmod -n students pupils
```

These commands will only change group names, not their other data (such as group ID), which is going to be reflected in our users' information, as well:

```
root@cli1:~/links# id jack
uid=1001(jack) gid=1001(jack) groups=1001(jack),1005(professors)
root@cli1:~/links# id jill
uid=1003(jill) gid=1003(jill) groups=1003(jill),1005(professors)
root@cli1:~/links# id sarah
uid=1004(sarah) gid=1004(sarah) groups=1004(sarah),1006(students)
root@cli1:~/links# id joe
uid=1002(joe) gid=1002(joe) groups=1002(joe),1006(students)
root@cli1:~/links#
```

Figure 1.8 – Checking created users' settings

As we can see, jack and jill are members of a group that's now called professors, while joe and sarah are now members of a group called students.

We deliberately left the userdel and groupdel commands for last, as these come with some caveats and shouldn't be used lightly. Let's create a user called temp and a group called temporary, and then let's delete them:

```
useradd temp
groupadd temporary
userdel temp
groupdel temporary
```

This will work just fine. The thing is, because we used the `userdel` command without any parameters, it will leave the user's home directory intact. Since users' home directories are usually stored in the `/home` directory, by default that means that the `/home/temp` directory is still going to be there. When deleting users, we sometimes want to delete a user but not their files. If you specifically want to delete a user account and all the data from that user account, then use the `userdel -r username` command. But think twice before doing it!

How it works...

Let's now discuss the more complex part of the previous recipe, which is **symbolic links** and **hard links**.

Using the `ln` command without extra parameters tries to create a hard link. Using `ln` with the `-s` parameter tries to create a soft link. We can clearly see that there are some errors in that part of our recipe. Let's discuss them now by going back from the top.

When we've finished typing in the first six commands from the recipe (ending with `ls -al`), which we're using to list the folder contents, the end result should look similar to this:

```
root@cli1:~/links# ls -al
total 16
drwxr-xr-x 2 root root 4096 Mar 16 23:00 .
drwx------ 5 root root 4096 Mar 16 22:59 ..
-rw-r--r-- 2 root root 3106 Mar 16 23:00 content.cfg
-rw-r--r-- 2 root root 3106 Mar 16 23:00 hardlink.cfg
lrwxrwxrwx 1 root root   11 Mar 16 23:00 softlink.cfg -> content.cfg
root@cli1:~/links#
```

Figure 1.9 – Original file, hard link and soft link

There are some conclusions that can be reached just by interpreting this previous screenshot:

- The `content` file and the `hardlink` file have the same size (1,349 bytes, in our case).

- The `content` file and the `softlink` file don't have the same file size (1,349 bytes versus 11 bytes here).

- Soft links are marked differently by default (usually, a different color in the Terminal).

Now, for the purposes of building up this explanation, let's delete the original `content` file:

```
r m content.cfg
```

The end result will look like this:

```
root@cli1:~/links# rm content.cfg
root@cli1:~/links# ls -al
total 12
drwxr-xr-x 2 root root 4096 Mar 16 23:00 .
drwx------ 5 root root 4096 Mar 16 22:59 ..
-rw-r--r-- 1 root root 3106 Mar 16 23:00 hardlink.cfg
lrwxrwxrwx 1 root root   11 Mar 16 23:00 softlink.cfg -> content.cfg
root@cli1:~/links#
```

Figure 1.10 – Removing the original file leaves interesting consequences

We can see that the original file is gone, while the hard link is still here and has the same size. On the other hand, the soft link changed color (from green to red), indicating that there's some kind of problem. Interesting, isn't it?

If we open the `hardlink.cfg` file in the **vi editor**, the content is definitely there:

```
# ~/.bashrc: executed by bash(1) for non-login shells.
# see /usr/share/doc/bash/examples/startup-files (in the package bash-doc)
# for examples

# If not running interactively, don't do anything
case $- in
    *i*) ;;
      *) return;;
esac

# don't put duplicate lines or lines starting with space in the history.
# See bash(1) for more options
HISTCONTROL=ignoreboth

# append to the history file, don't overwrite it
shopt -s histappend

# for setting history length see HISTSIZE and HISTFILESIZE in bash(1)
HISTSIZE=1000
HISTFILESIZE=2000

# check the window size after each command and, if necessary,
# update the values of LINES and COLUMNS.
shopt -s checkwinsize

# If set, the pattern "**" used in a pathname expansion context will
# match all files and zero or more directories and subdirectories.
#shopt -s globstar

# make less more friendly for non-text input files, see lesspipe(1)
[ -x /usr/bin/lesspipe ] && eval "$(SHELL=/bin/sh lesspipe)"

# set variable identifying the chroot you work in (used in the prompt below)
if [ -z "${debian_chroot:-}" ] && [ -r /etc/debian_chroot ]; then
    debian_chroot=$(cat /etc/debian_chroot)
fi
"hardlink.cfg" 117L, 3675C                              1,1           Top
```

Figure 1.11 – The hardlink.cfg file still has the original content

The reason why this happens comes from the way in which filesystems work. When we delete a file, we don't delete the content of the file, we just delete an entry in the filesystem table (filename) that points to the content of the file. The reason for this is simple in that it's about speed and convenience. If the operating system actually removed the file content, it would need to free its blocks and write zeros to them. That would take a lot of time. Furthermore, it would complicate file recovery.

This is where hard links and soft links come into play. The main difference between them is something that we can easily deduce from the scenario. It's the fact that hard links point to the actual file content, while soft links point to the original filename. That also explains the size difference. Hard links must be the same size as the original file (as the original file and hard link point to the content of the same content, therefore the same size). The reason why `softlink.cfg` only consumes 11 bytes on the filesystem is simple; it's that the `content.cfg` string needs 11 bytes to be saved to the filesystem table.

This is also the reason why there are two other major differences between hard links and soft links:

- Hard links cannot point to a directory and they have to point to a file.
- Hard links cannot go across partitions. We can't reference/see data from the first mounted partition if we look from the perspective of the second partition. The second partition has its own filesystem table (which contains entries pointing to the actual content on that partition) that's completely independent of the filesystem table of the first partition.

The cool thing, going back to our recipe, is that we can easily recover the original file. If we go back to the `/root/links` directory, we can just copy the `hardlink.cfg` file to `content.cfg` and our original file and the corresponding symbolic link are back:

```
cd /root/links
cp hardlink.cfg content.cfg
```

The end result will be just like earlier, when we created the `content.cfg` file and the hard link and soft link pointing to it:

```
root@cli1:~/links# cp hardlink.cfg content.cfg
root@cli1:~/links# ls -al
total 16
drwxr-xr-x 2 root root 4096 Mar 16 23:04 .
drwx------ 5 root root 4096 Mar 16 23:04 ..
-rw-r--r-- 1 root root 3106 Mar 16 23:04 content.cfg
-rw-r--r-- 1 root root 3106 Mar 16 23:00 hardlink.cfg
lrwxrwxrwx 1 root root   11 Mar 16 23:00 softlink.cfg -> content.cfg
root@cli1:~/links#
```

Figure 1.12 – Our original file and soft link are back

We will use these commands throughout this book, so we need to make sure that we master using them before we move on to the next chapters. But for the time being, we'll add just one more command to the stack. It's the subject of our next recipe, known as **screen**.

Using screen

screen is one of those text utilities that was incredibly popular in the 1990s and 2000s, with its popularity shrinking after that. System administrators often have to open multiple consoles on the same machine or use any of those multiple consoles to connect to external machines. Let's see how screen fits into this scenario.

Getting ready

Before starting with this recipe, we need to make sure that we have screen on our Linux machine. So, we need to use the following command:

```
apt-get -y install screen
```

After that, we're ready to follow our recipe.

How to do it...

We need to start a regular text Terminal (this can be done in the GUI as well, but it can be considered as a bit of a less-effective way to use screen real estate). Then, we just need to type in the following command:

```
screen
```

When we start screen, it is going to throw us a long piece of text about licensing and other less-than-interesting subjects, with a couple of important pieces of information at the bottom of the screen. It will look similar to this:

```
GNU Screen version 4.08.00 (GNU) 05-Feb-20

Copyright (c) 2018-2020 Alexander Naumov, Amadeusz Slawinski
Copyright (c) 2015-2017 Juergen Weigert, Alexander Naumov, Amadeusz Slawinski
Copyright (c) 2010-2014 Juergen Weigert, Sadrul Habib Chowdhury
Copyright (c) 2008-2009 Juergen Weigert, Michael Schroeder, Micah Cowan, Sadrul Habib Chowdhury
Copyright (c) 1993-2007 Juergen Weigert, Michael Schroeder
Copyright (c) 1987 Oliver Laumann

This program is free software; you can redistribute it and/or modify it under the terms of the GNU
General Public License as published by the Free Software Foundation; either version 3, or (at your
option) any later version.

This program is distributed in the hope that it will be useful, but WITHOUT ANY WARRANTY; without
even the implied warranty of MERCHANTABILITY or FITNESS FOR A PARTICULAR PURPOSE. See the GNU
General Public License for more details.

You should have received a copy of the GNU General Public License along with this program (see the
file COPYING); if not, see https://www.gnu.org/licenses/, or contact Free Software Foundation,
Inc., 51 Franklin Street, Fifth Floor, Boston, MA  02111-1301  USA.

Send bugreports, fixes, enhancements, t-shirts, money, beer & pizza to screen-devel@gnu.org


Capabilities:
+copy +remote-detach +power-detach +multi-attach +multi-user +font +color-256 +utf8 +rxvt
+builtin-telnet
```

Figure 1.13 – Basic screen information

The only part of this output that's really interesting to us is **Capabilities**. It tells us that with screen, we can do some cool stuff, such as copy, detach, and work with fonts. But even without most of these advanced features, screen enables us to open multiple virtual text Terminals, within the limits of one text Terminal. Then, it enables us to detach (something like putting the screen process in the background), log off, come back later, log in, and re-attach our session to screen. That enables some cool things, such as leaving a permanent set of virtual text consoles open for the most common, mostly used use cases.

After we press the *Enter* key on the screen shown in the previous screenshot, we're going to be thrown into the text mode again. This is screen's first virtual text console. If we want to use additional virtual text consoles, we can create them by using the *Ctrl + A + C* key combination. Every one of these virtual text consoles is numbered from 0 onward. If we create five virtual text consoles in screen (numbered 0-4) and we're in screen 4 and want to jump to screen 0, we can easily do that in two ways. The first one involves *absolute* addressing, in other words, we can tell screen that we want to go specifically to screen 0 (by using *Ctrl + A + 0*). The second way to go from screen 4 to 0 is to use a *circular* approach. When we use the *Ctrl + A + spacebar* key combination, we're circling through screens in a subsequent fashion – 0, then 1, then 2, and so on. If we're on screen 4 and we want to go to 0, because we don't have a screen 5, we can just circle from 4 to 0 by using *Ctrl + A + spacebar*.

If we need to log off, we can detach our screen. The key combination for that is *Ctrl + A + D* (detach screen). If sometime later we want to go back to our screens, we need to type in the following command:

```
screen -R
```

We can also copy-paste in screen by using the *Ctrl + A +]* key combination, then scroll and find the bit of text that we want to start copying, use the spacebar to start copying and to end the copying process, and then the *Ctrl + A +]* combination if we want to paste text somewhere. It takes a bit of practice, but it's also very usable. *Just imagine doing stuff like that in 1996!*

> **Important Note**
> When working with screen, we suggest that you first press *Ctrl + A*, let those keys go, and then press whichever key you need to go wherever you want to go on the screen.

How it works...

screen works by creating multiple detachable virtual text consoles. These consoles remain active until there's a process that kills screen, or until the system reboots. Keeping in mind that most production environments based on Linux servers don't have a GUI, having the capability to connect to a server once and then open multiple screens comes in handy.

There's more...

screen requires a bit of trial and error and getting used to. We recommend that you check the following link to learn more:

```
https://www.howtogeek.com/662422/how-to-use-linuxs-screen-
command/
```

2
Using Text Editors

There's just no way around the topic of this chapter, as system administrators edit **text files** daily. Therefore, we decided to cover three commonly used editors – **vi**, **Vim**, and **nano**. If you're more into GUI tools, make sure that you check `gedit`, although we won't cover that editor here as it's practically the same as using Notepad on Microsoft Windows. There are various reasons why these **editors** were chosen, but most importantly, they are installed out of the box on almost all Linux distributions, so they're the most common pre-installed editors. There are situations where additional software installation is not an option, such as *air-gapped* environments.

We will cover the following recipes in this chapter:

- Learning the basics of the Vi(m) editor
- Learning the basics of the nano editor
- Going through the advanced Vi(m) settings

Technical requirements

For these recipes, we're going to use a Linux machine. We can use any virtual machine from our previous recipe. For example, let's say that we're going to use a `cli1` virtual machine as it's the most convenient to use, seeing as it's a command-line-interface-only machine. So, all in all, we need the following:

- A virtual machine with any distribution of Linux installed (in our case, it's going to be *Ubuntu 20.10*).

- A bit of time to digest the complexities of using the Vi(m) editor. nano is less complex; therefore, it's going to be easier to learn about that one.

So, start your virtual machine and let's get cracking!

Learning the basics of the Vi(m) Editor

Vi and Vim are *the text editors* of choice for many system administrators and engineers. In a nutshell, the difference between them is that **vim** (**vi improved**) has many more capabilities than the original vi (visual editor). You can find these editors everywhere – from all the Unixes and Linuxes to the commercial Linux- or Unix-based software of today. For example, VMware's *vSphere Hypervisor* has a version of the vi editor built in. The rationale for this is simple – you need to have some sort of *standardized editor* that can be used to edit various text files available on a filesystem. Over the years, you'll surely find some cutdown version of vi or Vim on various network devices such as switches and routers, and even more complex devices like firewalls. It's just the way it is. If something's Unix- or Linux-based, chances are it's using text configuration files, and text configuration files need a text editor. Pretty straightforward logic, isn't it?

Just as an example – the Vim editor has spinoffs that can be used in a variety of different ways, including **vim-athena** (created with *Athena GUI* support), **vim-gtk**, and **vim-gtk3** (created with *GTK/GTK3* support), **vim-tiny** (a slimmed-down version of Vim), and **vim-nox**. But still, most people that we know of prefer using the good old-fashioned vi or Vim in a CLI.

For this first part of our recipe, we're going to explain the way vi and Vim work and use them to do some most common things, such as the following:

- Three `vi(m)` modes – `insert`, `command`, and `ex` mode

- Moving around a text file that we want to edit by moving the cursor

- Deleting text (we could refer to it as *cutting* and *deleting* at the same time)

- Inserting additional content into a text file

- Saving and exiting in the vi(m) editors

- Finding content in a text file

- Copying and pasting text (what vi and Vim refer to as *yank* and *paste*)

That's going to be enough for this first recipe. We're going to go back to the **advanced vim capabilities** in the last recipe of this chapter, where we're going to dig much deeper into Vim and learn how to use much more advanced concepts, such as using regular expressions, line marking, buffers, and sorting.

Getting ready

We just need to check whether vi and Vim are installed on our system. The simplest way to do it is to just go brute-force and issue the following command:

```
sudo apt-get -y install Vim-tiny busybox Vim dictionaries-
common wamerican
```

Ubuntu doesn't have or use the vi editor by default, so we can just install the `Vim-tiny` package to kind of emulate the same thing. Another way to use the vi editor in Ubuntu would be to use the following command:

```
busybox vi
```

Seeing that `busybox` is a command-line tool that *embeds* multiple Linux command-line utilities into one, this command is something that we need to be aware of. But also, we need to remember that the intent of `busybox` is to have a way to embed multiple popular CLI tools into one, which in turn means that none of these tools are completely the same as their standalone versions.

After installation is done (if needed at all), we're going to start using Vim and learn how to use it via examples. Let's issue the following commands as `root`:

```
cp /etc/passwd /root
cp /usr/share/dict/words /root
```

Take note of the fact that between `cp`, `/etc/passwd`, and `/root` (the same thing applies to `cp`, `/usr/share/dict/words`, and `/root` in the second command), we need to hit *the spacebar* on our keyboard. We're effectively copying the `passwd` and `words` files to the `/root` directory to have some source files to play with.

When we have successfully finished copying these files, we'll start the Vim editor and start editing. First, we're going to use the `passwd` file. Type in the following:

```
Vim /root/passwd
```

Let's start learning!

How to do it...

Now that we have the `/root/passwd` file opened in our Vim editor, let's play with it a bit. Moving around in normal mode is straightforward. Let's just start by using the arrow keys on our keyboard to move up and down and left and right. After we're done with that, let's just jump to the top of our file by using the *gg* sequence (by pressing the *g* key twice).

First, we're going to delete the first line. Vi(m) starts in something called *normal* mode, and if we press the *d* key twice, we're going to delete the first line. Let's check the before state:

```
root:x:0:0:root:/root:/bin/bash
daemon:x:1:1:daemon:/usr/sbin:/usr/sbin/nologin
bin:x:2:2:bin:/bin:/usr/sbin/nologin
sys:x:3:3:sys:/dev:/usr/sbin/nologin
sync:x:4:65534:sync:/bin:/bin/sync
games:x:5:60:games:/usr/games:/usr/sbin/nologin
man:x:6:12:man:/var/cache/man:/usr/sbin/nologin
lp:x:7:7:lp:/var/spool/lpd:/usr/sbin/nologin
mail:x:8:8:mail:/var/mail:/usr/sbin/nologin
news:x:9:9:news:/var/spool/news:/usr/sbin/nologin
uucp:x:10:10:uucp:/var/spool/uucp:/usr/sbin/nologin
proxy:x:13:13:proxy:/bin:/usr/sbin/nologin
www-data:x:33:33:www-data:/var/www:/usr/sbin/nologin
backup:x:34:34:backup:/var/backups:/usr/sbin/nologin
list:x:38:38:Mailing List Manager:/var/list:/usr/sbin/nologin
irc:x:39:39:ircd:/var/run/ircd:/usr/sbin/nologin
```

Figure 2.1 – The top section of the /root/passwd file

And now, after we have pressed the *d* key twice, it should look like this (if we are still positioned at the first line, the root line):

```
daemon:x:1:1:daemon:/usr/sbin:/usr/sbin/nologin
bin:x:2:2:bin:/bin:/usr/sbin/nologin
sys:x:3:3:sys:/dev:/usr/sbin/nologin
sync:x:4:65534:sync:/bin:/bin/sync
games:x:5:60:games:/usr/games:/usr/sbin/nologin
man:x:6:12:man:/var/cache/man:/usr/sbin/nologin
lp:x:7:7:lp:/var/spool/lpd:/usr/sbin/nologin
mail:x:8:8:mail:/var/mail:/usr/sbin/nologin
news:x:9:9:news:/var/spool/news:/usr/sbin/nologin
uucp:x:10:10:uucp:/var/spool/uucp:/usr/sbin/nologin
proxy:x:13:13:proxy:/bin:/usr/sbin/nologin
www-data:x:33:33:www-data:/var/www:/usr/sbin/nologin
backup:x:34:34:backup:/var/backups:/usr/sbin/nologin
list:x:38:38:Mailing List Manager:/var/list:/usr/sbin/nologin
irc:x:39:39:ircd:/var/run/ircd:/usr/sbin/nologin
gnats:x:41:41:Gnats Bug-Reporting System (admin):/var/lib/gnats:/usr/sbin/nologin
nobody:x:65534:65534:nobody:/nonexistent:/usr/sbin/nologin
systemd-network:x:100:102:systemd Network Management,,,:/run/systemd:/usr/sbin/nologin
systemd-resolve:x:101:103:systemd Resolver,,,:/run/systemd:/usr/sbin/nologin
systemd-timesync:x:102:104:systemd Time Synchronization,,,:/run/systemd:/usr/sbin/nologin
messagebus:x:103:106::/nonexistent:/usr/sbin/nologin
syslog:x:104:110::/home/syslog:/usr/sbin/nologin
_apt:x:105:65534::/nonexistent:/usr/sbin/nologin
```

Figure 2.2 – After pressing the d key twice, the first line is gone

Let's now expand this use case further by pressing (just as an example) the *5dd* key sequence. The result should look something like this:

```
man:x:6:12:man:/var/cache/man:/usr/sbin/nologin
lp:x:7:7:lp:/var/spool/lpd:/usr/sbin/nologin
mail:x:8:8:mail:/var/mail:/usr/sbin/nologin
news:x:9:9:news:/var/spool/news:/usr/sbin/nologin
uucp:x:10:10:uucp:/var/spool/uucp:/usr/sbin/nologin
proxy:x:13:13:proxy:/bin:/usr/sbin/nologin
www-data:x:33:33:www-data:/var/www:/usr/sbin/nologin
backup:x:34:34:backup:/var/backups:/usr/sbin/nologin
list:x:38:38:Mailing List Manager:/var/list:/usr/sbin/nologin
irc:x:39:39:ircd:/var/run/ircd:/usr/sbin/nologin
gnats:x:41:41:Gnats Bug-Reporting System (admin):/var/lib/gnats:/usr/sbin/nologin
nobody:x:65534:65534:nobody:/nonexistent:/usr/sbin/nologin
systemd-network:x:100:102:systemd Network Management,,,:/run/systemd:/usr/sbin/nologin
systemd-resolve:x:101:103:systemd Resolver,,,:/run/systemd:/usr/sbin/nologin
systemd-timesync:x:102:104:systemd Time Synchronization,,,:/run/systemd:/usr/sbin/nologin
messagebus:x:103:106::/nonexistent:/usr/sbin/nologin
syslog:x:104:110::/home/syslog:/usr/sbin/nologin
_apt:x:105:65534::/nonexistent:/usr/sbin/nologin
```

Figure 2.3 – After the 5dd operation (deleting five lines), we deleted five lines after the cursor

As we can see, the first five lines after our cursor (the lines starting with daemon, bin, sys, sync, and games) are gone.

Let's now jump to the last line in our /root/passwd file, and copy and paste it behind the last line. First, we need to go to the end of our file, which can be achieved by using the *Shift + g* sequence (basically, the capital letter *G*). After that, if we want to copy the line after the cursor (in effect, complete the last line in the file), we need to first yank it (copy) and then paste it to a correct spot. Yanking can be achieved by using the *yy* sequence (pressing the *y* key twice). That puts the line after our cursor in a copy and paste buffer. If we want to paste it after our last line, we need to press the *p* key. Our copied line will automatically be pasted after the last line. The end result, if we used the same virtual machine as in *Chapter 1, Basics of Shell and Text Terminal*, should be something like this:

```
geoclue:x:123:128::/var/lib/geoclue:/usr/sbin/nologin
saned:x:124:130::/var/lib/saned:/usr/sbin/nologin
hplip:x:125:7:HPLIP system user,,,:/run/hplip:/bin/false
colord:x:126:131:colord colour management daemon,,,:/var/lib/colord:/usr/sbin/nologin
gnome-initial-setup:x:127:65534::/run/gnome-initial-setup/:/bin/false
gdm:x:128:132:Gnome Display Manager:/var/lib/gdm3:/bin/false
jack:x:1001:1001::/home/jack:/bin/sh
joe:x:1002:1002::/home/joe:/bin/sh
jill:x:1003:1003::/home/jill:/bin/sh
sarah:x:1004:1004::/home/sarah:/bin/sh
sarah:x:1004:1004::/home/sarah:/bin/sh
```

Figure 2.4 – The yank and paste of a single line

Now, let's select three lines beginning with sshd (so, the sshd, systemd-coredump, and student lines) and copy and paste them after the line beginning with joe. First, we're going to use cursor keys to position at the beginning of the sshd line. Then, we're going to type the *y3y* key sequence. This will start yanking (copying) from the cursor, copy the next three lines in the copy and paste buffer, and then end yanking. If we did that successfully, Vim is going to throw us a message at the bottom of the screen, saying 3 lines yanked.

After we have these lines in the copy and paste buffer, we need to paste them. Let's use the cursor keys to move to the line beginning with joe, and then press the *p* key. The result should look like this:

```
jack:x:1001:1001::/home/jack:/bin/sh
joe:x:1002:1002::/home/joe:/bin/sh
sshd:x:112:65534::/run/sshd:/usr/sbin/nologin
systemd-coredump:x:999:999:systemd Core Dumper:/:/usr/sbin/nologin
student:x:1000:1000:student:/home/student:/bin/bash
jill:x:1003:1003::/home/jill:/bin/sh
sarah:x:1004:1004::/home/sarah:/bin/sh
sarah:x:1004:1004::/home/sarah:/bin/sh
```

Figure 2.5 – Yank and paste, and multiple lines of text

Now that we have played with yank and paste and delete, it's time to add some content to this file. In order to do that, we need to enter **insert mode**. That can be achieved by typing the *i* key. So, let's add a bit of text after our cursor – press the *i* key and start typing. Let's add the following:

```
something:x:1400:1400::/home/something:/bin/bash
```

After we're done with inserting, press the *Esc* key (to go back to **normal mode**). The end result should look like this:

```
jack:x:1001:1001::/home/jack:/bin/sh
joe:x:1002:1002::/home/joe:/bin/sh
something:x:1400:1400::/home/something:/bin/bash
sshd:x:112:65534::/run/sshd:/usr/sbin/nologin
systemd-coredump:x:999:999:systemd Core Dumper:/:/usr/sbin/nologin
student:x:1000:1000:student:/home/student:/bin/bash
jill:x:1003:1003::/home/jill:/bin/sh
sarah:x:1004:1004::/home/sarah:/bin/sh
sarah:x:1004:1004::/home/sarah:/bin/sh
```

Figure 2.6 – Inserting additional text with insert mode

Now that we've done that successfully, the next logical step will be to save the file if we're happy with its contents. Let's say that we are and we're ready to save the file. In order for us to do that, we need to enter **ex mode** and tell Vim that we want to exit and save. There are several different key sequences that will make this happen for us. The first one is *:wq!* (write and quit – don't ask us for confirmation), and the second one is *:x*. There are other ways, such as using the *ZZ* key sequence, but let's stick to the more commonly used ones (*wq* and *x*). We need to make sure that we type these key sequences with a colon sign (*:*). As we will explain in a bit, using the colon sign means that we want to enter ex mode and do some final operations with our edited file. If we use this key sequence successfully, we should end up in shell, with our original file saved with all the changes that we made to it.

In truth, Vim has a spectacular number of key sequences that can be used for a variety of operations on text files. Feel free to translate this *spectacular* as either a very good or very bad thing, as it's all subjective – some of us like it a lot, some of us will hate it. Here are some commonly used key sequences:

- *dw* – delete a word
- *2dw* – delete two words
- *yw* – yank one word
- *u* – undo the last change
- *U* – undo changes made to the current line
- *a* – append text after the cursor
- *A* – append text to the end of the current line
- *Ctrl + f* – scroll the file forward by one screen
- *n Ctrl + f* – scroll the file forward by *n* screens
- *Shift + m* – move the cursor to the middle of the page
- *:50* – move the cursor to line 50 of the current file
- *$* – move the cursor to the end of the line
- *x* – delete the character at the cursor
- *X* – delete the character before the cursor
- *^* – go to the first character of the line
- *o* – insert a line after the current one
- *Ctrl + g* – print the file info

There are literally hundreds of other commands, and we deliberately selected only some of them that we feel are useful and commonly used. Let's now do some more complex things by using a built-in Vim teaching tool called **Vimtutor**. In the command line, start Vimtutor by typing the following:

```
Vimtutor
```

After that, Vimtutor is going to ask us about the intended output file for practice, and we can just press the *Enter* key here. We should have the following content on our screen:

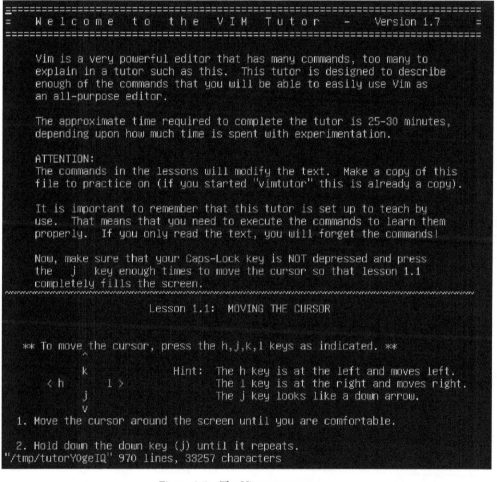

Figure 2.7 – The Vimtutor start page

Let's now use this file to practice a bit. The first thing that we're going to do is copy the first paragraph (starting with **Vim** and ending with **editor.**) before the paragraph starting with **The approximate time**.

Let's position our cursor at the beginning of the Vim line by using the arrow keys. After we have done that, we need to use the *y}* key sequence to instruct Vim to **yank** the paragraph starting at the cursor. Then, we need to move to the empty line between the first and second paragraphs by using the cursor keys and pressing the *p* key to paste the copied paragraph after the cursor. The result should look like this:

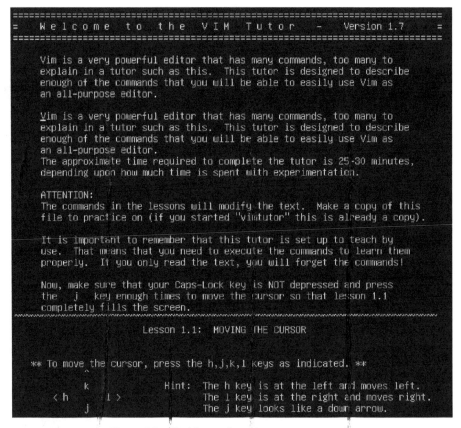

Figure 2.8 – Yanking and pasting a text paragraph

Let's say that we want to convert the complete file to lowercase characters. Of course, this operation involves several other operations:

- We need to move to the beginning of our file (*gg*).

- We need to turn on **visual mode** (more about this a bit later), achieved by pressing the *Shift* + *v* key sequence (uppercase *V*).

- We need to mark the text all the way to the end of our file, achieved by pressing the *Shift* + *g* key sequence (uppercase *G*).

- We need to make the text lowercase, achieved by pressing the *u* key.

So, the key sequence we're looking for is *ggVGu*. The result of our operation should look like this:

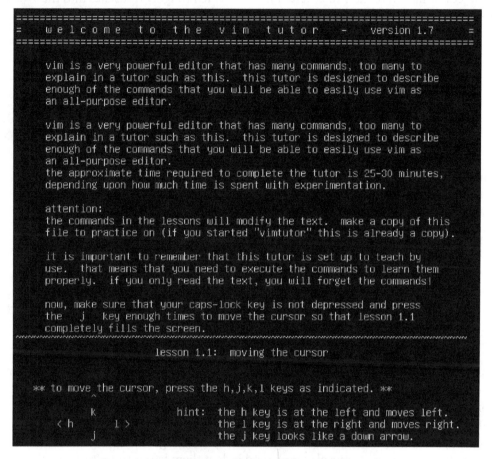

Figure 2.9 – Our Vimtutor file, with all lowercase characters

If we wanted to do the opposite (uppercase all the characters), we'd use the *ggVGU* key sequence (*U* is for uppercase and *u* is for lowercase characters).

We're going to take a short break from all of these key sequences by explaining how Vim works – specifically, we're going to focus on commonly used modes and briefly mention some of the lesser-used ones. Let's start with **normal mode** and work our way toward visual and **replace** modes.

How it works...

The Vim editor has more than 10 different modes, which roughly translates into the different ways in which it works. The most used modes are as follows:

- Normal mode
- Insert mode
- Ex mode
- Visual mode
- Replace mode

When we start Vim, we're in normal mode; we can sniff around in it by using the cursor, we can do a bit of yank and paste, and we can delete. So, it's used for general operations such as navigating through an edited file, a bit of rough-cut editing, and that's that.

If we want to add additional content to our text file, we usually switch to insert mode by using the *i* key. In insert mode, we can easily add a bit of text after our cursor and move around in our edited file. When we're ready to go back to normal mode, we can do that by pressing the *Esc* key. If, however, we're done with file editing and we just want to save the file and exit, we need to go to normal mode and then to ex mode. This is achieved by pressing the *Esc* key, followed by the colon (*:*). That puts us in ex mode, and then we can proceed to do *wq!*, *x*, or *ZZ*.

Visual and replace modes are quite a bit different. Visual mode has sub-modes (**character**, **line**, and **block**), and can be used to select (highlight) parts of the text that we want to work with and manipulate. For example, line and block modes can be useful for modifying YAML files when working with Ansible. Character mode can be used to highlight a part of code. YAML syntax is sensitive to indenting, so by using line mode, we can highlight portions of our playbooks and indent them left or right (by using the > and < keys) so that we don't have to do it manually. Block mode can be used efficiently to check indentation that was created by using line mode. These modes can be entered by using *Shift* + *V* (line mode) and *Ctrl* + *v* (block mode). Character mode can be entered by using the *v* key.

Replace mode allows us to type in our content in a text file over existing content. We can use the *R* key to enter replace mode (from normal mode).

See also

If you need more information about the basics of Vim, we suggest that you check out this content:

- Vim: `https://www.Vim.org/`

- *Mastering Vim*: `https://www.amazon.com/exec/obidos/` `ASIN/1789341094/stichtingiccfhol`

- An interactive Vim tutorial: `https://www.openVim.com/`

Learning the basics of the nano editor

If you feel that the Vim editor is too complicated for you, we can feel your pain. That's why choosing the editor you're going to work with is a subjective choice. We'd like to offer another much simpler editor to the table, called **nano**.

Getting ready

Keep the CLI1 virtual machine powered on and let's continue editing our files.

How to do it...

We're going to edit the `words` file that we copied in the previous recipe. Before that, let's just make sure that nano is installed by typing in the following command:

```
sudo apt-get -y install nano
```

Let's now open the file called `words` from the `root` directory by typing in the following command:

```
nano /root/words
```

Our file should be opened in the nano editor, as shown in the following screenshot:

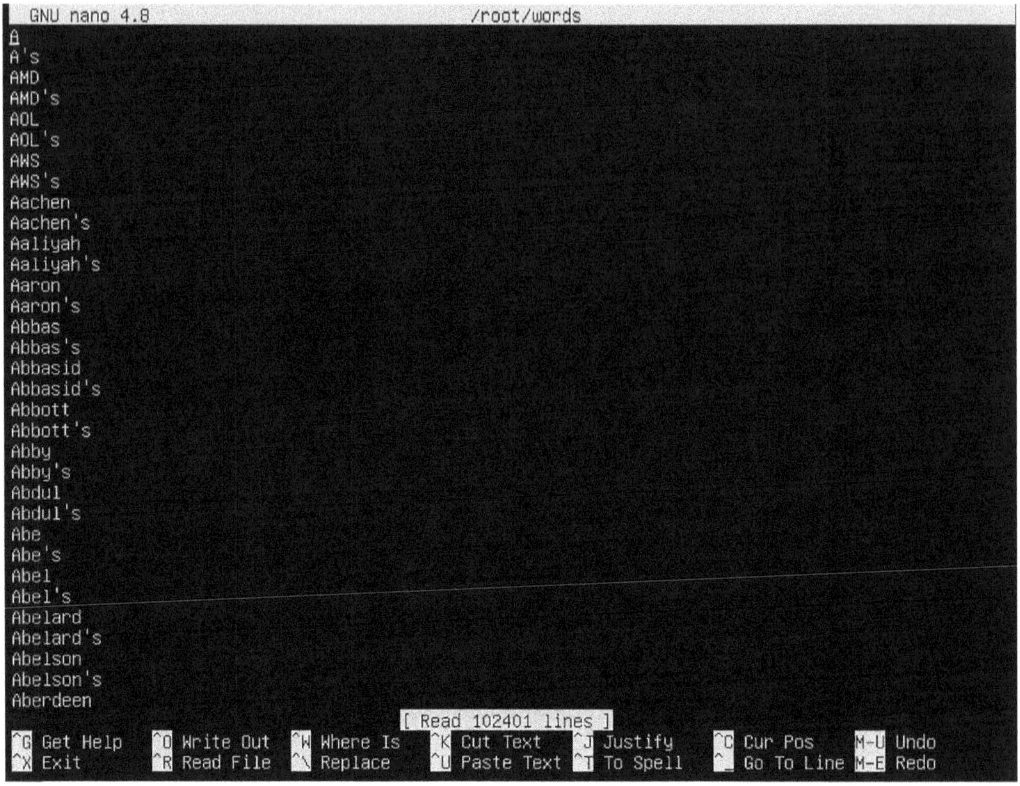

```
  GNU nano 4.8                              /root/words
A
A's
AMD
AMD's
AOL
AOL's
AWS
AWS's
Aachen
Aachen's
Aaliyah
Aaliyah's
Aaron
Aaron's
Abbas
Abbas's
Abbasid
Abbasid's
Abbott
Abbott's
Abby
Abby's
Abdul
Abdul's
Abe
Abe's
Abel
Abel's
Abelard
Abelard's
Abelson
Abelson's
Aberdeen
                              [ Read 102401 lines ]
^G Get Help   ^O Write Out   ^W Where Is   ^K Cut Text    ^J Justify    ^C Cur Pos     M-U Undo
^X Exit       ^R Read File   ^\ Replace    ^U Paste Text  ^T To Spell      Go To Line  M-E Redo
```

Figure 2.10 – Starting editing with the nano editor

For those of us who are more prone to using text editors such as Notepad or Wordpad, nano should be a bit more familiar territory. It doesn't have the scope of capabilities or advanced functionality that Vim has, but for the most part, that might not be so important, at least not for most text file editing operations. Or is it really that simple? Let's check it out.

Editing in nano works in the same fashion as with other *regular* editors – we just need to explain the lower part of the screenshot (the part where we can see Help, Exit, and so on). In nano, if we need help, we need to press *Ctrl + g*. This is the result that we'll get:

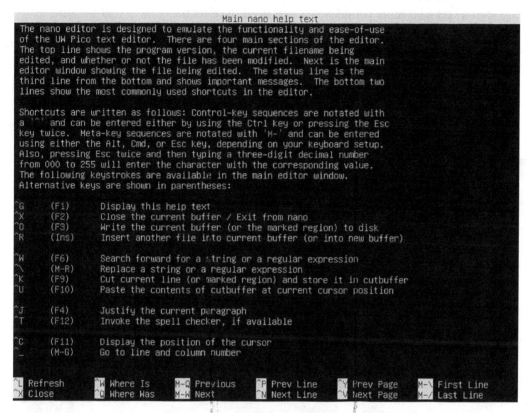

Figure 2.11 – nano help

We can spend our time scrolling through this help window if we want to. But, for starters, let's just say that this ^ character means *press the Ctrl key*.

So, on our first nano screenshot, ^G means *Ctrl + G*, ^X means *Ctrl + X*, and so on. It still isn't as easy as using some text editors that a lot of people use on Microsoft Windows, but it's a bit more user-friendly than Vim. If nothing else, some of the commonly used commands are right at the bottom of our screen so that we don't have to learn all of the key sequences or research them online before we consider using the editor.

If we want to close our help window from the second nano screenshot, we just need to press *Ctrl + X*. This will get us back to the state shown in the first nano screenshot.

If we want to delete a line, we need to use *Ctrl + K*. If we need to delete multiple lines, things start to get a bit more complex. We first need to select the content that we want to delete (*Ctrl + Shift + 6*), use the cursor to move to the place that we want to delete to, and then press *Ctrl + K*. Let's say that we want to delete five lines. So, selecting content that we want to delete looks like this:

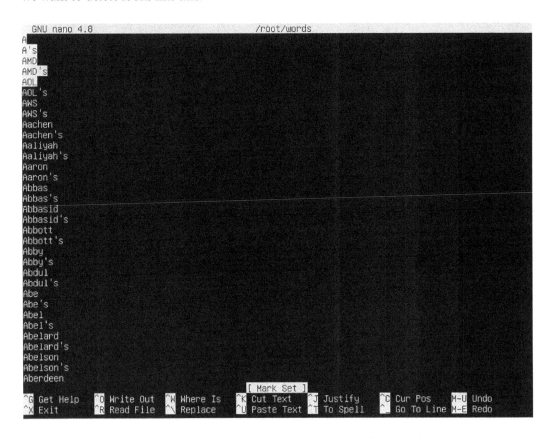

Figure 2.12 – After pressing Ctrl + Shift + 6 and using the cursor keys to
go down five lines, we're ready for Ctrl + K

After we have selected the correct text, we just need to press *Ctrl + K* to delete it. The result will look like this:

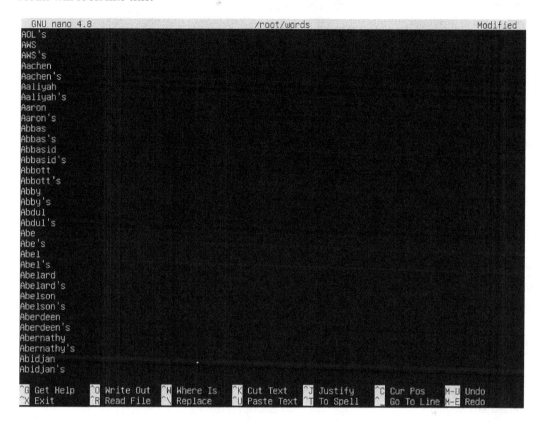

Figure 2.13 – Five lines successfully deleted – result!

The same idea applies to doing a copy and paste operation on a paragraph. We'd use *Ctrl + Shift + 6* and the cursor to mark the text, *Alt + 6* to put the copied text in the copy and paste buffer, and then use *Ctrl + U* to paste it wherever it needs to be pasted in nano. Saving the file is equivalent to using *Ctrl + X* to exit, and then confirming that we want the changes to be saved to a file.

There's more...

If you need to learn more about nano, check out the following links:

- The nano editor cheat sheet: `https://www.nano-editor.org/dist/latest/cheatsheet.html`

- How to use nano: `https://linuxize.com/post/how-to-use-nano-text-editor/`

Going through the advanced Vi(m) settings

In the first part of this chapter, we learned some basic Vim operations, which were moving around, copying and pasting, saving, and exiting. Let's take care of some more **advanced operations**, such as working with find and replace, regular expressions, and similar concepts.

Getting ready

We need to leave our CLI virtual machine running. If it's not powered on, we need to power it back on.

How to do it...

Finding content in Vim is a multi-step process, and it depends on a couple of things. First, it depends on the *direction* that we want to take, *forward* or *backward*, as there are different key sequences for these operations. Let's open the `/root/words` file again to find some text:

```
Vim /root/words
```

Let's start by finding the word `fast`. For that to work, we need to use the `/` character from normal mode, as it tells Vim that we're about to use the `search` function. So, `/fast` will search for the words `fast  forward` from our cursor. This is the expected result:

```
Behring's
Beiderbecke
Beiderbecke's
Beijing
Beijing's
Beirut
Beirut's
Bekesy
Bekesy's
Bela
Bela's
Belarus
Belarus's
Belau
Belau's
Belem
Belem's
Belfast
Belfast's
Belgian
Belgian's
Belgians
Belgium
Belgium's
Belgrade
Belgrade's
Belinda
Belinda's
Belize
Belize's
Bell
Bell's
Bella
Bella's
Bellamy
Bellamy's
/fast
```

Figure 2.14 – Finding a word in Vim

If we now press *Enter* and then the *n* key, we will search for the next appearance of the word fast. This is the expected result:

```
Behring's
Beiderbecke
Beiderbecke's
Beijing
Beijing's
Beirut
Beirut's
Bekesy
Bekesy's
Bela
Bela's
Belarus
Belarus's
Belau
Belau's
Belem
Belem's
Belfast
Belfast's
Belgian
Belgian's
Belgians
Belgium
Belgium's
Belgrade
Belgrade's
Belinda
Belinda's
Belize
Belize's
Bell
Bell's
Bella
Bella's
Bellamy
Bellamy's
/fast                                              1849,4            1%
```

Figure 2.15 – The next appearance of the word fast

However, if we want to find the 10th appearance of the word `fast`, we need to either press the correct key sequence or use a regular expression. Let's start with a key sequence, which is going to be (again from normal mode) `10/fast`. This is the expected result:

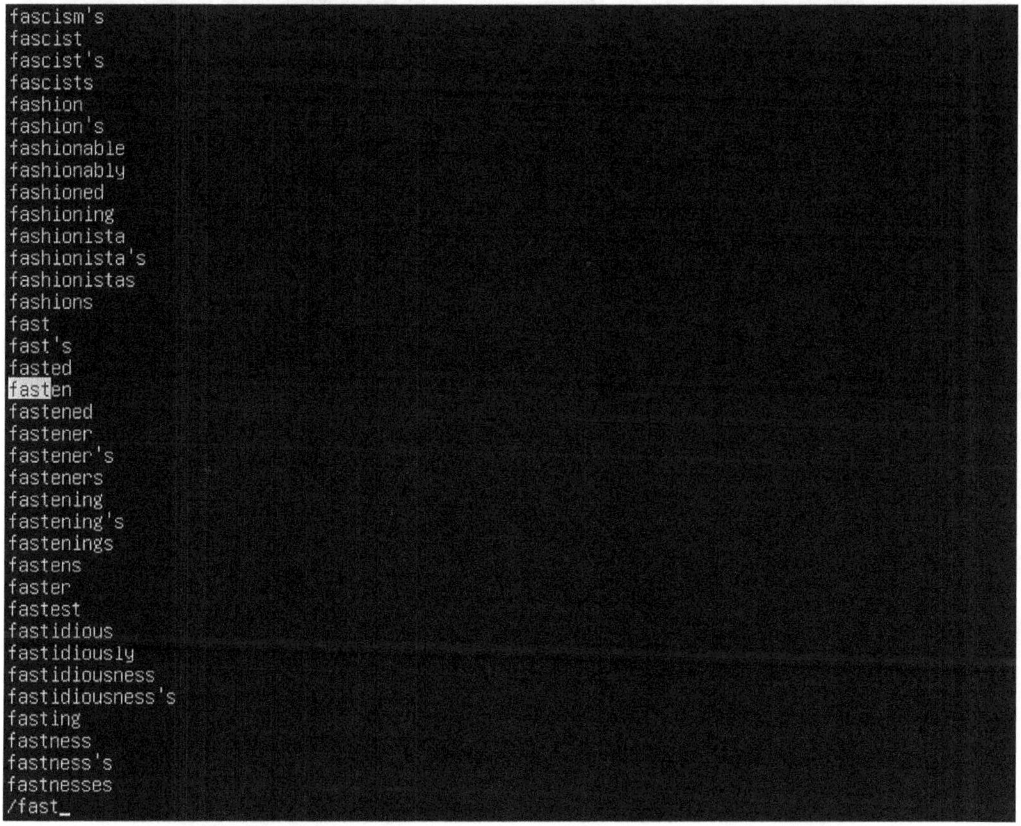

```
fascism's
fascist
fascist's
fascists
fashion
fashion's
fashionable
fashionably
fashioned
fashioning
fashionista
fashionista's
fashionistas
fashions
fast
fast's
fasted
fasten
fastened
fastener
fastener's
fasteners
fastening
fastening's
fastenings
fastens
faster
fastest
fastidious
fastidiously
fastidiousness
fastidiousness's
fasting
fastness
fastness's
fastnesses
/fast_
```

Figure 2.16 – Finding the n-th appearance of a word

If we want to find the previous appearance of our word (basically, search backward), we need to press the *N* key (capital *N*). This is the expected result:

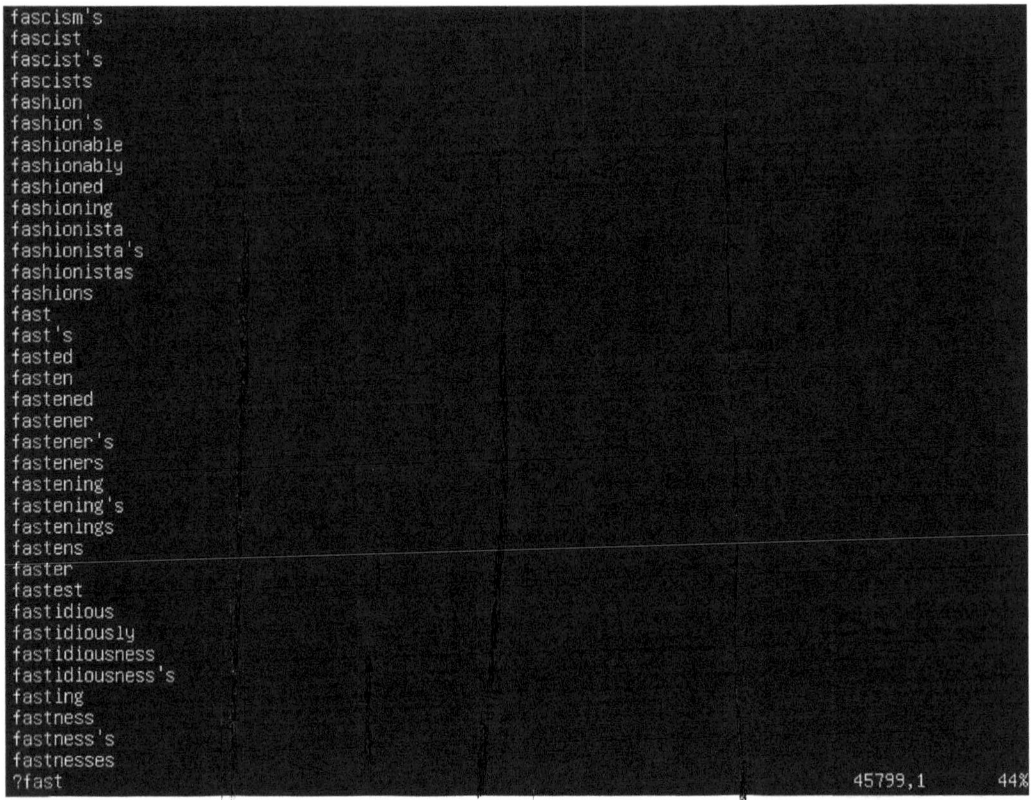

```
fascism's
fascist
fascist's
fascists
fashion
fashion's
fashionable
fashionably
fashioned
fashioning
fashionista
fashionista's
fashionistas
fashions
fast
fast's
fasted
fasten
fastened
fastener
fastener's
fasteners
fastening
fastening's
fastenings
fastens
faster
fastest
fastidious
fastidiously
fastidiousness
fastidiousness's
fasting
fastness
fastness's
fastnesses
?fast                                                         45799,1        44%
```

Figure 2.17 – Finding a word backward from the previous cursor

Let's now do a bit of search and replace. Let's say that we want to find all appearances of the word `airplane` and change them to `metro`, starting from the beginning of our file. The key sequence used for that would be *gg* (to go back to the file beginning) and then `:%s/airplane/metro/g`, followed by the *Enter* key. This is the expected result:

```
airlifts
airline
airline's
airliner
airliner's
airliners
airlines
airmail
airmail's
airmailed
airmailing
airmails
airman
airman's
airmen
metro
metro's
metros
airport
airport's
airports
airs
airship
airship's
airships
airsick
airsickness
airsickness's
airspace
airspace's
airstrip
airstrip's
airstrips
airtight
airwaves
airwaves's
3 substitutions on 3 lines                              20833,1        20%
```

Figure 2.18 – Replacing all appearances of a word with another word

This syntax presumes the automatic replacement of all occurrences of the word airplane with the word metro placed anywhere in the file. If we just wanted to replace the first appearance of a string in any line, we need to first find that word by using the / word key sequence. Then, we need to use the :s/word1/word2/ key sequence to only change the first appearance of word1 with word2. Let's use the word airship for that example and change that word to ship. If we type in /airship, followed by the *Enter* key, Vim will position us to the first next appearance of the word airship. If we then use the :s/airship/ship/ key sequence followed by the *Enter* key, we should get this result:

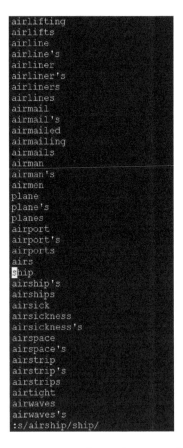

Figure 2.19 – Replacing one appearance of a word in a specific line with another word

It's a subtle difference, but an important one.

We could also use many more commands in vi – for example, using a dot sign (.). That can be used to repeat the last change made in normal mode, which you might also find to be very useful.

We're going to stop here, as we will cover more advanced text search patterns by using regular expressions in *Chapter 7, Network-Based File Synchronization*.

How it works...

String replacement in Vim works by using an external command called `sed`, a **stream editor**. This command is regularly used by system engineers all over the world to quickly replace simple or complex text patterns of any given file (or multiple files) to another complex text pattern. It uses regular expressions as a basis (explained in *Chapter 7, Network-Based File Synchronization*), which means that, by default, doing search and replace in Vim is quite powerful, albeit a bit complex, as we need to learn the ins and outs of sed and the way Vim treats it as a *plugin*.

That being said, most of us focus on the *quite powerful* part of the last paragraph, as using a Vim/sed combination to quickly replace complex text patterns yields fast and precise results – as long as we know what we're doing, of course.

There's more...

Using these concepts requires a bit of extra reading. So, we need to make sure that we check the following additional links:

- Vim Tips Wiki – search and replace: `https://Vim.fandom.com/wiki/Search_and_replace`

- *Vim tips: the basics of search and replace*: `https://www.linux.com/training-tutorials/Vim-tips-basics-search-and-replace/`

3
Using Commands and Services for Process Management

Managing processes is an important job of a Linux system administrator. That can be for a variety of reasons – maybe some processes got stuck and we need to finish them, or we want to set some process(es) to work in the background or even to be started periodically or at a later date. Whatever the scenario is, it's important to know how to administer processes and make them do the work that needs to be done efficiently and with regard to other processes running on the system.

In this chapter, we are going to learn about the following recipes:

- Process management tools
- Managing background jobs
- Managing process priorities
- Configuring `crond`

Technical requirements

For these recipes, we're going to use a Linux machine – we can use any virtual machine from our previous recipes. Again, we can just continue using the `cli1` machine that we used in the previous chapter. So, to sum this up, we need the following:

- A virtual machine with Linux installed, any distribution (in our case, it's going to be Ubuntu 20.10)

So, start your virtual machine and let's get cracking!

Process management tools

Managing processes means learning about the ways in which processes work and the specific text-mode tools that we can use to manage them. We are going to start by introducing some simple concepts – explain what processes are and which states they can be in – and then we're going to move on to commands and how to use them to manage processes from an administrative standpoint. That means that we are going to learn 10+ new commands/concepts that are necessary to understand how all of this works.

Getting ready

The vast majority of commands and utilities that we are going to use in this recipe come pre-installed with our Linux distribution. That being said, there are a couple of cool additional tools we can use to further drive the point of managing processes and system resources home. So, let's install one more utility as it's capable of being used as a tool to monitor system resources along with low-level stuff, such as working with processes. It's called `glances`; let's install it by typing the following:

```
apt-get -y install glances
```

That should cover everything that we need in this recipe, so let's get cracking!

How to do it...

The first two commands that we must cover are `ps` and `top`. These are commands that Linux system administrators use dozens of times on a daily basis if they're managing a Linux server. Both of these commands are very valuable, as we can get a lot of information about our system if we know how to use them properly, especially `ps`.

So, let's first use `ps` as a command without any additional options (of which there are many):

```
root@cli1:~# ps
    PID TTY          TIME CMD
    857 tty1     00:00:00 login
   1191 tty1     00:00:00 sudo
   1192 tty1     00:00:00 su
   1193 tty1     00:00:00 bash
  81165 tty1     00:00:00 ps
```

Figure 3.1 – Default ps command output

By default, `ps` gives us a report about currently running processes. By starting it in a shell without any additional options, we can get a list of processes running in our current shell. We can already see some interesting information in this output. For starters, we can see five processes and their IDs (the `PID` field on the left side). Then, we can see where they're running, which is what the `TTY` field is all about. The `TIME` field tells us how much accumulated CPU time the process has used so far. Furthest to the right, we can see the `CMD` field, which tells us the name of an actual process that was started.

To fully appreciate the power of the `ps` command, we really need to look to its man page. There's a really nice `EXAMPLES` section in it. Here's an excerpt from that section:

```
EXAMPLES
       To see every process on the system using standard syntax:
              ps -e
              ps -ef
              ps -eF
              ps -ely

       To see every process on the system using BSD syntax:
              ps ax
              ps axu

       To print a process tree:
              ps -ejH
              ps axjf

       To get info about threads:
              ps -eLf
              ps axms

       To get security info:
              ps -eo euser,ruser,suser,fuser,f,comm,label
              ps axZ
              ps -eM
```

Figure 3.2 – Example for using the ps command

Let's use an extreme derivative of one of these examples. Let's type the following command:

```
ps auwwx | less
```

We used the | less part of this command to output just the first page of the ps command output. The output should look something like this:

```
USER         PID %CPU %MEM    VSZ   RSS TTY      STAT START   TIME COMMAND
root           1  1.1  0.5 102936 11668 ?        Ss   09:23   0:03 /sbin/init
root           2  0.0  0.0      0     0 ?        S    09:23   0:00 [kthreadd]
root           3  0.0  0.0      0     0 ?        I<   09:23   0:00 [rcu_gp]
root           4  0.0  0.0      0     0 ?        I<   09:23   0:00 [rcu_par_gp]
root           5  0.0  0.0      0     0 ?        I    09:23   0:00 [kworker/0:0-mm_percpu_wq]
root           6  0.0  0.0      0     0 ?        I<   09:23   0:00 [kworker/0:0H-kblockd]
root           7  0.0  0.0      0     0 ?        I    09:23   0:00 [kworker/0:1-events]
root           9  0.0  0.0      0     0 ?        I<   09:23   0:00 [mm_percpu_wq]
root          10  0.0  0.0      0     0 ?        S    09:23   0:00 [ksoftirqd/0]
root          11  0.0  0.0      0     0 ?        I    09:23   0:00 [rcu_sched]
root          12  0.0  0.0      0     0 ?        S    09:23   0:00 [migration/0]
root          13  0.0  0.0      0     0 ?        S    09:23   0:00 [idle_inject/0]
root          14  0.0  0.0      0     0 ?        S    09:23   0:00 [cpuhp/0]
root          15  0.0  0.0      0     0 ?        S    09:23   0:00 [cpuhp/1]
root          16  0.0  0.0      0     0 ?        S    09:23   0:00 [idle_inject/1]
root          17  0.0  0.0      0     0 ?        S    09:23   0:00 [migration/1]
root          18  0.0  0.0      0     0 ?        S    09:23   0:00 [ksoftirqd/1]
root          19  0.0  0.0      0     0 ?        I    09:23   0:00 [kworker/1:0-events]
root          20  0.0  0.0      0     0 ?        I<   09:23   0:00 [kworker/1:0H-kblockd]
root          21  0.0  0.0      0     0 ?        S    09:23   0:00 [kdevtmpfs]
root          22  0.0  0.0      0     0 ?        I<   09:23   0:00 [netns]
root          23  0.0  0.0      0     0 ?        S    09:23   0:00 [rcu_tasks_kthre]
root          24  0.0  0.0      0     0 ?        S    09:23   0:00 [rcu_tasks_rude_]
root          25  0.0  0.0      0     0 ?        S    09:23   0:00 [rcu_tasks_trace]
```

Figure 3.3 – ps auwwx command output (much more verbose)

As we can clearly see, there's a lot more detail in this output sorted by PID. Some of the newly added fields include the following:

- USER: This field tells us the name of the user who started the process.
- %CPU: This field tells us how much CPU time the process uses.
- %MEM: This field tells us how much memory the process uses.
- VSZ: This field tells us how much virtual memory the process uses.
- RSS: Resident Set Size, the amount of non-swapped memory used by the process.
- STAT: Process status code.
- START: Time when the process was started.

Just as an example, a lot of system administrators use the %CPU and %MEM fields to find processes that are using too much CPU or memory.

Let's say that we need to find a process by name. There are multiple ways of doing this, the most common two being using either the `ps` command or the `pgrep` command. Let's see how that would work:

```
root@cli1:~# pgrep sshd
894
root@cli1:~# ps auwwx | grep sshd
root         894  0.0  0.3  13084  6940 ?        Ss   09:23
0:00 sshd: /usr/sbin/sshd -D [listener] 0 of 10-100 startups
root        1893  0.0  0.0   5172   880 tty2     S+   09:40
0:00 grep --color=auto sshd
```

Figure 3.4 – Using pgrep or ps to find a process by name

As a command, we tend to use `grep` to create a filter that will find a text sample by going through text output. We can see that both commands gave us the result that we needed – it's just formatted differently and with a different level of detail. We can also use the `pidof` command to find a PID for any given process, similar to `pgrep`:

```
root@cli1:~# pidof sshd
894
root@cli1:~#
```

Figure 3.5 – Using the pidof command

Let's now explain the idea of the `top` command. After we start the `top` command, we should get something similar to this:

```
top - 11:40:07 up 1 day,  2:16,  2 users,  load average: 0.00, 0.00, 0.00
Tasks: 214 total,   1 running, 213 sleeping,   0 stopped,   0 zombie
%Cpu(s):  0.2 us,  0.2 sy,  0.0 ni, 99.7 id,  0.0 wa,  0.0 hi,  0.0 si,  0.0 st
MiB Mem :   1955.0 total,     97.2 free,    326.7 used,   1531.1 buff/cache
MiB Swap:   2048.0 total,   2045.2 free,      2.8 used.   1442.6 avail Mem

    PID USER      PR  NI    VIRT    RES    SHR S  %CPU  %MEM     TIME+ COMMAND
    753 root      20   0  236964   8164   6736 S   1.7   0.4   2:22.64 vmtoolsd
    840 root      20   0  235324   7492   6564 S   0.3   0.4   0:04.25 accounts-daemon
      1 root      20   0  104256  13068   8672 S   0.0   0.7   0:04.68 systemd
      2 root      20   0       0      0      0 S   0.0   0.0   0:00.02 kthreadd
      3 root       0 -20       0      0      0 I   0.0   0.0   0:00.00 rcu_gp
      4 root       0 -20       0      0      0 I   0.0   0.0   0:00.00 rcu_par_gp
      6 root       0 -20       0      0      0 I   0.0   0.0   0:00.00 kworker/0:0H-kblockd
      9 root       0 -20       0      0      0 I   0.0   0.0   0:00.00 mm_percpu_wq
     10 root      20   0       0      0      0 S   0.0   0.0   0:00.13 ksoftirqd/0
     11 root      20   0       0      0      0 I   0.0   0.0   0:49.67 rcu_sched
     12 root      rt   0       0      0      0 S   0.0   0.0   0:00.39 migration/0
     13 root     -51   0       0      0      0 S   0.0   0.0   0:00.00 idle_inject/0
     14 root      20   0       0      0      0 S   0.0   0.0   0:00.00 cpuhp/0
     15 root      20   0       0      0      0 S   0.0   0.0   0:00.00 cpuhp/1
     16 root     -51   0       0      0      0 S   0.0   0.0   0:00.00 idle_inject/1
     17 root      rt   0       0      0      0 S   0.0   0.0   0:00.56 migration/1
     18 root      20   0       0      0      0 S   0.0   0.0   0:00.77 ksoftirqd/1
     20 root       0 -20       0      0      0 I   0.0   0.0   0:00.00 kworker/1:0H-kblockd
     21 root      20   0       0      0      0 S   0.0   0.0   0:00.00 kdevtmpfs
     22 root       0 -20       0      0      0 I   0.0   0.0   0:00.00 netns
```

Figure 3.6 – Using the top command

There are multiple things happening in this interactive output at the same time:

1. The top line is actually output from the uptime command. If we add the next four lines (beginning with Tasks, %Cpu(s), Mib Mem, and MiB Swap), that is what we call the top *summary area*.

2. After that, we can clearly see that top acts as a frontend to the ps command but is implemented in an interactive sense.

The interactive part of the top command stems from the fact that it actually refreshes regularly – by default, every 3 seconds. We can change that default refresh interval by pressing the *S* key, which will make top ask us to change the delay from 3.0 to any number. If we want to change the refresh interval to 1 second, we just press 1 and *Enter*.

We can ask top to show us processes by a single user (by pressing *U* and typing in the user's login name) and to kill processes (by pressing *K* and typing in the PID and the signal we want to send to that PID). We can also manipulate process priority, which we will cover in our *third* recipe of this chapter. All in all, top is a very useful and often-used command to do process management. It acts as a frontend to many different commands, such as nice, renice, and kill.

The next set of commands that we need to learn about is kill and killall. We shouldn't use the literal translation of these utilities to try to instinctively understand what they do as we'll be surprised that that translation doesn't apply. Specifically, the kill command is used when we want to kill a process by its corresponding PID. killall, in contrast, is used to kill processes by name. There are – of course – viable use cases for both. To show an example for both of these commands, we are going to use the following top output:

```
top - 11:56:24 up 1 day,  2:32,  5 users,  load average: 0.13, 0.05, 0.01
Tasks: 224 total,   1 running, 223 sleeping,   0 stopped,   0 zombie
%Cpu(s):   0.0 us,   0.0 sy,   0.0 ni,100.0 id,   0.0 wa,   0.0 hi,   0.0 si,   0.0 st
MiB Mem :   1955.0 total,     75.3 free,    341.9 used,   1537.8 buff/cache
MiB Swap:   2048.0 total,   2045.2 free,      2.8 used.   1426.9 avail Mem

  PID USER      PR  NI    VIRT    RES    SHR S  %CPU  %MEM     TIME+ COMMAND
41246 student   20   0    9024   4124   3328 S   0.3   0.2   0:00.02 top
 1173 student   20   0   19432   9728   8176 S   0.0   0.5   0:00.04 systemd
 1174 student   20   0  105936   3396      0 S   0.0   0.2   0:00.00 (sd-pam)
 1180 student   20   0    7092   3796   2260 S   0.0   0.2   0:00.02 bash
 1840 student   20   0    7092   3912   2328 S   0.0   0.2   0:00.01 bash
41223 student   20   0    7092   5136   3416 S   0.0   0.3   0:00.01 bash
41235 student   20   0    5608   2968   2724 S   0.0   0.1   0:00.00 screen
41236 student   20   0    6048   2824   2000 S   0.0   0.1   0:00.00 screen
41237 student   20   0    7016   5044   3448 S   0.0   0.3   0:00.00 bash
41247 student   20   0    7016   5008   3416 S   0.0   0.3   0:00.01 bash
41254 student   20   0    9028   4008   3208 S   0.0   0.2   0:00.02 top
```

Figure 3.7 – top output – notice the top command being started twice by the student user

Let's kill both of these `top` processes in a separate shell. If we want to kill the first one by using the `kill` command, we need to type the following:

```
kill 41246
```

If we want to kill all of the started `top` commands by name, we can type the following:

```
killall top
```

When using the `kill` command, we're killing a single PID. When using the `killall` command, we are killing all the started `top` processes. Of course, in order for us to be able to kill a process by using either of these commands, we have to log in as either root or `student`. Only the user that started the process and root can kill a user process. We need to remember that the default signal of both of these commands is the `SIGTERM` signal (signal number `15`). If we want to kill a process by using a custom signal, we can achieve that by adding that number to any of these two commands preceded by a minus sign. Here's an example:

```
kill -9 41246
```

This will send the `SIGKILL` signal to the process. Both of these signals are explained in the *How it works...* section of this recipe.

It's also good to note the fact that sometimes we need to find the PID of a currently running shell or a parent PID of a shell. We can do that by using the following two commands:

```
root@cli1:~# echo $$
1851
root@cli1:~# echo $PPID
1850
```

Figure 3.8 – PID of our current shell process, the parent process

Let's now check how `glances` can help us check what's happening with our system. If we just start the command, we're going to get the following output:

```
cli1 (Ubuntu 20.10 64bit / Linux 5.8.0-63-generic) - IP 192.168.0.5/24          Uptime: 0:00:59

CPU  [  1.0%]    CPU \      1.0%    MEM -    25.5%    SWAP -    0.0%    LOAD      2-core
MEM  [ 25.5%]    user:      0.7%    total:   1.91G    total:   2.00G    1 min:    0.50
SWAP [  0.0%]    system:    0.3%    used:     498M    used:        0    5 min:    0.17
                 idle:     99.0%    free:    1.42G    free:    2.00G    15 min:   0.06

NETWORK          Rx/s    Tx/s    TASKS 253 (289 thr), 1 run, 135 slp, 117 oth sorted automatically
ens33            5Kb     0b
lo               4Kb     4Kb     CPU%    MEM%      PID USER       THR  NI S Command
                                 1.7     2.8      1427 root         1   0 R /usr/bin/python3 /usr/bin/glan
DefaultGateway           8ms     0.3     0.9       766 root         7   0 S /sbin/multipathd -d -s
                                 0.0     2.6       920 root         1   0 S /usr/bin/python3 /usr/bin/glan
DISK I/O         R/s     W/s     0.0     1.7       941 root        10   0 S /usr/lib/snapd/snapd
dm-0             0       0       0.0     1.6       495 root         1  -1 S /lib/systemd/systemd-journald
sda              0       0       0.0     1.0      1022 root         2   0 S /usr/bin/python3 /usr/share/un
sda1             0       0       0.0     0.9      1033 root         1   0 S /usr/sbin/apache2 -k start
sda2             0       0       0.0     0.9       936 root         1   0 S /usr/bin/python3 /usr/bin/netw
sda3             0       0       0.0     0.6         1 root         1   0 S /sbin/init
sr0              0       0       0.0     0.5       823 root         1   0 S /usr/bin/VGAuthService
                                 0.0     0.5      1394 student      1   0 S /lib/systemd/systemd --user
FILE SYS         Used    Total   0.0     0.4       896 systemd-n    1   0 S /lib/systemd/systemd-networkd
/                7.65G   18.6G   0.0     0.4       945 root         1   0 S /lib/systemd/systemd-logind
                                 0.0     0.4       824 root         3   0 S /usr/bin/vmtoolsd
                                 0.0     0.4       909 root         3   0 S /usr/lib/accountsservice/accou
                                 0.0     0.3      1017 root         1   0 S sshd: /usr/sbin/sshd -D [liste
                                 0.0     0.3       814 systemd-t    2   0 S /lib/systemd/systemd-timesyncd
                                 0.0     0.3      1034 root         3   0 S /usr/libexec/polkitd --no-debu
                                 0.0     0.3      1043 www-data     1   0 S /usr/sbin/apache2 -k start
                                 0.0     0.3      1046 www-data     1   0 S /usr/sbin/apache2 -k start
                                 0.0     0.3      1048 www-data     1   0 S /usr/sbin/apache2 -k start
                                 0.0     0.3      1049 www-data     1   0 S /usr/sbin/apache2 -k start
                                 0.0     0.3      1050 www-data     1   0 S /usr/sbin/apache2 -k start
                                 0.0     0.3       538 root         1   0 S /lib/systemd/systemd-udevd
                                 0.0     0.3       939 syslog       4   0 S /usr/sbin/rsyslogd -n -iNONE

2021-12-13 21:23:02 UTC
```

Figure 3.9 – glances default output

It's easy to see the level of detail, as well as the different formats, that `glances` uses. Furthermore, we really appreciate the fact that it uses color output as default, which makes the information a bit easier to read. We can go into different methods of displaying data. For example, we can type `1` to switch between per-CPU core versus aggregated statistics. We can also use it in server mode (by starting it with the `-s` switch) so that we can monitor remote hosts. So, from the server perspective, we would start it with the following:

```
glances -s
```

From the client perspective, we would start `glances` with the following:

```
glances -c @servermachine
```

`glances` is cross-platform (it's Python-based), as it supports Linux, OS X, Windows, and FreeBSD. It also has a built-in web UI that can be used via a web browser if we're more into using the GUI than the CLI. But one of the most convenient features that it has is the ability to export data in various different formats – CSV, Elasticsearch, RabbitMQ, Cassandra, and others.

How it works...

A process is a unit of command execution that an operating system initializes so that it can be managed both from the operating system standpoint and from our standpoint as system administrators. That means that a process acts like an instance of any given program, and it has some common properties (state, PID, and many others that we will describe in this chapter), as well as some tasks that it needs to do. For example, after we start a command (`process`), that command can open and read from the file, user input, or other programs, do something with that input, and then terminate after the work is done.

It is important to note that processes aren't *paused* if we reboot a machine – they get stopped and then start as our Linux machine boots or we start it manually after the reboot. So, *there's no process persistence across reboots*. Most of the time (except for most of the processes that are a part of the operating system startup procedure), they don't even keep the same PID across a reboot.

In terms of process types, we have five different types:

- **Parent and child processes**: In simple terms, a parent process is a process that creates additional processes that we call child processes. Child processes exit when the parent process exits. A parent process doesn't exit if a child process exits.

- **Zombie and orphan processes**: There are situations where the parent process gets killed before the child process exits. The remaining child process is called an orphan process. On the other hand, a zombie process is a situation where a process is killed but still exists in the process table.

- **Daemons**: Daemons are usually related to some system tasks that usually involve working with other processes and servicing them. They also don't use a terminal as they run in the background.

In terms of states, we have these:

- **Running/runnable**: Running state refers to a state where a process is being executed by a CPU. Runnable state, on the other hand, means that a process is ready to be executed but is currently not consuming CPU or queued to be executed by a CPU.

- **Interruptible/uninterruptible sleep**: In an interruptible sleep state, a process can be awakened and it can accept **signals** aimed at it. In an uninterruptible sleep state, that doesn't happen, and the process remains asleep. This scenario often includes a **system call** – a process can't do a system call and can't be paused or killed until it finishes its job.

- **Stopped**: A process is often stopped when it receives a signal and when we're debugging a process.

- **Zombie**: A dead process that's been halted but still exists in the process table is in a zombie state.

From the operating system perspective, processes are units of execution – for a program or service. Processes get scheduled by the operating system and that means assigning them resources so that they can run from a programmatic perspective (context), and some basic attributes so that they can be managed from the system administrative perspective. That includes creating an entry in a process table with a PID (number of the process) and other types of attribute data. We are going to explain these attributes and how to notice process states a bit later in this chapter, when we start discussing practical aspects of working with commands such as `top` and `ps`.

We mentioned the concept of a signal. When we're dealing with different ways of establishing communication between a kernel and userspace program, there are two ways of achieving that – via either a system call or a signal. Usually, we use commands such as `kill` or `killall` if we want to send a signal to a process by assigning a signal number or name along with the command. Let's take a look at an excerpt from the signal list:

Signal	Standard	Action	Comment
SIGABRT	P1990	Core	Abort signal from abort(3)
SIGALRM	P1990	Term	Timer signal from alarm(2)
SIGBUS	P2001	Core	Bus error (bad memory access)
SIGCHLD	P1990	Ign	Child stopped or terminated
SIGCLD	-	Ign	A synonym for SIGCHLD
SIGCONT	P1990	Cont	Continue if stopped
SIGEMT	-	Term	Emulator trap
SIGFPE	P1990	Core	Floating-point exception
SIGHUP	P1990	Term	Hangup detected on controlling terminal or death of controlling process
SIGILL	P1990	Core	Illegal Instruction
SIGINFO	-		A synonym for SIGPWR
SIGINT	P1990	Term	Interrupt from keyboard
SIGIO	-	Term	I/O now possible (4.2BSD)
SIGIOT	-	Core	IOT trap. A synonym for SIGABRT
SIGKILL	P1990	Term	Kill signal
SIGLOST	-	Term	File lock lost (unused)
SIGPIPE	P1990	Term	Broken pipe: write to pipe with no readers; see pipe(7)
SIGPOLL	P2001	Term	Pollable event (Sys V); synonym for SIGIO
SIGPROF	P2001	Term	Profiling timer expired
SIGPWR	-	Term	Power failure (System V)
SIGQUIT	P1990	Core	Quit from keyboard
SIGSEGV	P1990	Core	Invalid memory reference
SIGSTKFLT	-	Term	Stack fault on coprocessor (unused)
SIGSTOP	P1990	Stop	Stop process
SIGTSTP	P1990	Stop	Stop typed at terminal
SIGSYS	P2001	Core	Bad system call (SVr4); see also seccomp(2)
SIGTERM	P1990	Term	Termination signal
SIGTRAP	P2001	Core	Trace/breakpoint trap

Figure 3.10 – Excerpt from the signal man page

As we can see, there are many of them, and approximately 30 of these signals have been implemented by the Linux kernel. Also, there are two types of signals from the process perspective:

- **Signals that can be handled by processes**: For example, the SIGHUP signal (number 1)

- **Signals that cannot be handled by processes, but directly by the kernel**: For example, the SIGKILL signal (number 9)

The word *handled* is used in a programmatic sense here – handling something means using some kind of *handler* to write a piece of code that's going to intercept the signal message and redirect it to something, such as a function or subroutine.

There are major differences between these two types. Let's use an example of a daemon process such as an *Apache web server*. If a daemon process receives the SIGHUP signal and it supports it (it has a routine in its source code handling the SIGHUP signal, like Apache does), the most common thing that it will do after receiving SIGHUP is to refresh its state by re-reading its configuration. To quote the Apache manual:

Sending the HUP or restart signal to the parent causes it to kill off its children like in TERM, but the parent doesn't exit. It re-reads its configuration files, and re-opens any log files. Then it spawns a new set of children and continues serving hits.

Unlike this scenario, when you send the SIGKILL signal to Apache, it will be terminated without giving any regard to refreshing its configuration, content, or anything of the sort. We can't write a handle to redirect this signal to anything other than the process being killed. We can think of it as a *kernel sucking the life out of a process* type of scenario, as the process can't get access to resources to run and is effectively eradicated by the system (kernel).

The third commonly used signal is SIGTERM (number 15). It's also used to terminate the process (such as SIGKILL), but it does it in a graceful way. We can think of it as a *Hello, Mr. Process, would you please be so kind as to terminate yourself gracefully? Thank you very much!* message from the kernel. Then the process does what it needs to do and shuts itself down.

Now that we've had a brief primer on how processes and signals work, let's continue our quest for knowledge about processes by learning about the management of background processes. As we already explained the basics of background processes, that shouldn't be a difficult task.

See also

If you need more information about processes, signals, and similar concepts, make sure that you check out the following:

- **Basics of Linux processes**: http://www.science.unitn.it/~fiorella/guidelinux/tlk/node45.html
- **Linux command basics** – Seven commands for process management: https://www.redhat.com/sysadmin/linux-command-basics-7-commands-process-management

- **Signal man page**: https://man7.org/linux/man-pages/man7/signal.7.html
- **Basics of glances**: https://www.tecmint.com/glances-an-advanced-real-time-system-monitoring-tool-for-linux/
- **TLDP Chapter 4: Processes**: https://tldp.org/LDP/tlk/kernel/processes.html

Managing background jobs

There are various types of situations where we would like to start a process and run it in the background. For example, let's say that we want to start a process, log off, and then come back tomorrow and check the result of that process. Let's learn how this works by using an example.

Getting ready

Keep the cli1 virtual machine powered on and let's use the shell to explain how the idea of a background process works, as opposed to a foreground process. We will make sure that we also explain the concept in the *How it works...* section.

How to do it...

Let's imagine a scenario in which we want to download a large file by using shell tools. The *usual suspect* that we'd use for this kind of task in Linux is a program called wget. We want to start a wget session (wget is a shell command that enables us to download files from the http and ftp URIs) to download a large ISO file, but we want to log off (or do something else) while the download is taking place. This is achieved by putting the wget process in the background. This is just one common example of using a background process to our advantage.

First, we need to install wget. Let's do that by using the following command:

```
apt-get -y install wget
```

wget is a common utility, and it's mostly installed by default. But either way, by using this command, we'll make sure that it's installed.

Let's use the Ubuntu 20.04 ISO file as the file that we want to download by using two examples. The first one is going to be running wget as a *foreground* process, and the second one is going to be running wget as a *background* process. The second example can actually be done in two different ways as wget has a built-in option that can be used to put it in the background. Of course, as we're trying to explain the *system-wide* concept, not a specific utility, let's make sure that we do both.

At the time of writing, the `Ubuntu 20.04 ISO` file can be found here:

`https://releases.ubuntu.com/20.04/ubuntu-20.04.3-live-server-amd64.iso`

Let's use `wget` to download it as a foreground process, by typing in the following command:

```
wget https://releases.ubuntu.com/20.04/ubuntu-20.04.3-live-
server-amd64.iso
```

The result should look something like this:

```
root@cli1:~# wget https://releases.ubuntu.com/20.04/ubuntu-20.04.2-live-server-amd64.iso
--2021-07-28 13:30:57--  https://releases.ubuntu.com/20.04/ubuntu-20.04.2-live-server-amd64.iso
Resolving releases.ubuntu.com (releases.ubuntu.com)... 91.189.91.123, 91.189.91.124, 2001:67c:1562::
28, ...
Connecting to releases.ubuntu.com (releases.ubuntu.com)|91.189.91.123|:443... connected.
HTTP request sent, awaiting response... 200 OK
Length: 1215168512 (1.1G) [application/x-iso9660-image]
Saving to: 'ubuntu-20.04.2-live-server-amd64.iso'

ubuntu-20.04.2-live-serv    1%[                            ]  22.79M  1.52MB/s    eta 12m 41s^
```

Figure 3.11 – Foreground process – exclusively locks the shell access

As we can clearly see, the download is working, but the problem is the fact that for the next 12+ minutes, we can't do anything in this shell, as the underlying shell session is being exclusively used by `wget`. We can't write commands, get command results – nothing. The only thing that we could do to prevent that would be to use a *Ctrl + C* sequence to quit the download and be thrown into the shell. But that's not what we want to do. What we want to do is the following:

1. Start the download.

2. Be thrown back into the shell with the download still working.

This is a situation in which running a process as a background task can be very helpful. So, let's add one additional parameter to the previous command:

```
wget https://releases.ubuntu.com/20.04/ubuntu-20.04.2-live-
server-amd64.iso&
```

The `&` sign at the end of this command tells the kernel to put this process in the background. Let's see what the end result is:

```
root@cli1:~# wget https://releases.ubuntu.com/20.04/ubuntu-20.04.2-live-server-amd64.iso &
[1] 43787
root@cli1:~#
Redirecting output to 'wget-log'.
root@cli1:~#
```

Figure 3.12 – Starting the process in the background

We can clearly see that we've been thrown back into the shell (`root@cli1` prompt) and that we can keep writing additional commands. We can also see that a `wget` process was started with PID `43787`, which we could use to issue a `kill` command if we so choose.

Obviously, we can issue multiple commands with & at the end, and then we'd have multiple processes running in the background. This is where the `[1]` part of the previous output comes in handy. This number represents an index number assigned to the background process. In other words, the `wget` that we started with PID `43787` is the first background process. If we were to start multiple background processes, each new background process would get the next number – 2, 3, and so on.

Obviously, we need to learn how to manage multiple background jobs. This is what the `jobs` command is all about. Let's see how that works. First, we are going to start multiple background jobs:

```
root@cli1:~# wget https://releases.ubuntu.com/20.04/ubuntu-20.04.2-live-server-amd64.iso &
[1] 43919
root@cli1:~#
Redirecting output to 'wget-log.1'.

root@cli1:~# wget https://releases.ubuntu.com/20.04/ubuntu-20.04.2-live-server-amd64.iso &
[2] 43922
root@cli1:~#
Redirecting output to 'wget-log.2'.

root@cli1:~# wget https://releases.ubuntu.com/20.04/ubuntu-20.04.2-live-server-amd64.iso &
[3] 43923
root@cli1:~#
Redirecting output to 'wget-log.3'.
```

Figure 3.13 – Starting multiple background processes

Then, let's use the `jobs` and `kill` commands to work out which background jobs we have and kill them by index (not by using their PID). This is the way to do it:

```
root@cli1:~# jobs
[1]   Running                 wget https://releases.ubuntu.com/20.04/ubuntu-20.04.2-live-server-amd6
4.iso &
[2]-  Running                 wget https://releases.ubuntu.com/20.04/ubuntu-20.04.2-live-server-amd6
4.iso &
[3]+  Running                 wget https://releases.ubuntu.com/20.04/ubuntu-20.04.2-live-server-amd6
4.iso &
root@cli1:~# kill %3
root@cli1:~# kill %2
[3]+  Terminated              wget https://releases.ubuntu.com/20.04/ubuntu-20.04.2-live-server-amd6
4.iso
root@cli1:~# kill %1
[2]+  Terminated              wget https://releases.ubuntu.com/20.04/ubuntu-20.04.2-live-server-amd6
4.iso
root@cli1:~#
[1]+  Terminated              wget https://releases.ubuntu.com/20.04/ubuntu-20.04.2-live-server-amd6
4.iso
root@cli1:~#
```

Figure 3.14 – Checking and killing multiple background processes

By using the `kill %index_number` syntax, we were able to kill background jobs by their index number, instead of their PIDs. This syntax is shorter and shouldn't be discounted in everyday life as it makes a lot of things easier – as long as we don't log off. If we log off, the whole idea changes a bit as we can't access these processes by using their index numbers, but we can definitely manage them by using PIDs. So, let's imagine for a second that we started two `wget` sessions as background processes, and then logged off and logged back on. Let's try to list these processes as background processes, then as just regular, general processes, and kill them by PID. This is what happens after that:

```
root@cli1:~#
root@cli1:~#
root@cli1:~# jobs
root@cli1:~# ps auwwx | grep -i wget
root        44056  3.7  0.3  11452  6548 tty2      S     13:48   0:00 wget https://releases.ubuntu.com/
20.04/ubuntu-20.04.2-live-server-amd64.iso
root        44058  3.6  0.3  11452  6416 tty2      S     13:48   0:00 wget https://releases.ubuntu.com/
20.04/ubuntu-20.04.2-live-server-amd64.iso
root        44078  0.0  0.0   5168   884 tty2      S+    13:49   0:00 grep --color=auto -i wget
root@cli1:~# kill -9 44056 44058
```

Figure 3.15 – jobs provides no output, but the ps command does

We can clearly see that the `jobs` command provides no output (can't find index numbers of background jobs), but our processes are still running. Why? Well, background processes that we started were created in the shell that's no longer active. After we logged off, we started a new shell, and, because of the way in which the `jobs` command works, we can't see those background jobs anymore. But we can definitely see them as processes running on the system, and, if we want to do so, we can kill them successfully by using their PIDs, as we did with the `kill` command. We used the `ps` command here and filtered its output by using `grep` – a command that is able to search specific pieces of text from a text-based output (in our case, we were searching through the whole table of processes by using `ps auwwx`, created a serial pipeline by using the pipe sign (|), and then threw the output from the `ps` command in the `grep` command.

We mentioned that the `wget` command has the capability to start itself in the background by using a command-line option (`-b`). This is not all that common, but it's definitely useful. So, say we were to use the following command:

```
wget -b https://releases.ubuntu.com/20.04/ubuntu-20.04.2-live-
server-amd64.iso
```

This should be the end result:

```
root@cli1:~# wget -b https://releases.ubuntu.com/20.04/ubuntu-20.04.2-live-server-amd64.iso
Continuing in background, pid 44316.
Output will be written to 'wget-log.6'.
root@cli1:~# jobs
root@cli1:~# ps auwwx | grep wget
root       44316  4.7  0.2  11456  5308 ?        Ss   13:58   0:00 wget -b https://releases.ubuntu.c
om/20.04/ubuntu-20.04.2-live-server-amd64.iso
root       44324  0.0  0.0   5172   884 tty2     S+   13:59   0:00 grep --color=auto wget
root@cli1:~# kill -9 44316
root@cli1:~# ps auwwx | grep wget
root       44344  0.0  0.0   5172   816 tty2     S+   13:59   0:00 grep --color=auto wget
```

Figure 3.16 – wget can be started in the background by using the -b switch

What's really interesting about this procedure is the following:

- wget clearly states that it's starting itself in the background, but it doesn't give us an index number.

- If we use the jobs command, we can't see it as a background process.

- We can kill it by using regular means, a kill command.

This is a bit of a different concept, as wget effectively achieves this jobs *command invisibility* by creating a wget child process and terminating the parent process. Since the parent process is no longer there, it's no longer associated with a specific shell, and therefore not indexed. The result is that it's not visible in the jobs table for the current shell. We can achieve something similar by using the disown command. Let's start a process in the current shell, and then do the thing that wget basically does:

```
root@cli1:~# wget https://release..ubuntu.com/20.44/ubuntu-20.04.--live-server-am664.iso &
[1] 44854
root@cli1:~#
Redirecting output to 'wget-log.1'.

root@cli1:~# jobs
[1]+  Running                 wget https://releases.ubuntu.com/20.04/ubuntu-20.04.2-live-server-amd6
4.iso &
root@cli1:~# disown %1
root@cli1:~# jobs
root@cli1:~# ps auwwx | grep -i wget
root       44854  6.6  0.3  11452  6436 tty2     S    14:19   0:00 wget https://releases.ubuntu.com/
20.04/ubuntu-20.04.2-live-server-amd64.iso
root       44860  0.0  0.0   5168   884 tty2     S+   14:19   0:00 grep --color=auto -i wget
root@cli1:~# kill -9 44854
```

Figure 3.17 – Disowning a background process

There are other ways of making sure that a process goes to the background. The most common scenario is we want to start a process in the background, we forget to put the & sign at the end of our command, and we're stuck with the foreground process. What to do then?

The answer is simple – we press *Ctrl + Z* (to put the process in the suspended state), and then type in the bg command. It's going to put the process in the background, as if we started it with the & sign from the start. Combining all of that with jobs, disown and kill would look like this:

```
root@cli1:~# wget https://releases.ubuntu.com/20.04/ubuntu-20.04.2-live-server-amd64.iso
--2021-07-28 14:25:13--  https://releases.ubuntu.com/20.04/ubuntu-20.04.2-live-server-amd64.iso
Resolving releases.ubuntu.com (releases.ubuntu.com)... 91.189.91.123, 91.189.91.124, 2001:67c:1562::
28, ...
Connecting to releases.ubuntu.com (releases.ubuntu.com)|91.189.91.123|:443... connected.
HTTP request sent, awaiting response... 200 OK
Length: 1215168512 (1.1G) [application/x-iso9660-image]
Saving to: 'ubuntu-20.04.2-live-server-amd64.iso'

untu-20.04.2-live-server   0%[                                    ]   4.13M  1.42MB/s              ^
Z
[1]+  Stopped                 wget https://releases.ubuntu.com/20.04/ubuntu-20.04.2-live-server-amd6
4.iso
root@cli1:~# bg
[1]+ wget https://releases.ubuntu.com/20.04/ubuntu-20.04.2-live-server-amd64.iso &
root@cli1:~#
Redirecting output to 'wget-log'.
jobs
[1]+  Running                 wget https://releases.ubuntu.com/20.04/ubuntu-20.04.2-live-server-amd6
4.iso &
root@cli1:~# disown %1
root@cli1:~# ps auwwx | grep wget
root        45006  3.6  0.3  11456   6280 tty2     S    14:25   0:00 wget https://releases.ubuntu.com/
20.04/ubuntu-20.04.2-live-server-amd64.iso
root        45014  0.0  0.0   5172    880 tty2     S+   14:25   0:00 grep --color=auto wget
root@cli1:~# kill 45006
```

Figure 3.18 – Using Ctrl + Z and bg to put a process in the background

We started wget in the foreground and put it in a suspended state by typing *Ctrl + Z*. Then, we moved that process to the background by using the bg command. Since it's job number 1 in our shell, we disowned it, used ps to find its PID, and killed it.

If, for some reason, we wanted to go from the background to the foreground with a process (providing that it has an index number and was started in the current shell), we can do that by using the fg command. So, if we use the previous procedure as an example, it would look like this:

```
root@cli1:~# wget https://releases.ubuntu.com/20.04/ubuntu-20.04.2-live-server-amd64.iso
--2021-07-28 14:29:27--  https://releases.ubuntu.com/20.04/ubuntu-20.04.2-live-server-amd64.iso
Resolving releases.ubuntu.com (releases.ubuntu.com)... 91.189.91.123, 91.189.91.124, 2001:67c:1562::
25, ...
Connecting to releases.ubuntu.com (releases.ubuntu.com)|91.189.91.123|:443... connected.
HTTP request sent, awaiting response... 200 OK
Length: 1215168512 (1.1G) [application/x-iso9660-image]
Saving to: 'ubuntu-20.04.2-live-server-amd64.iso'

     ubuntu-20.04.2-1   0%[                                   ] 952.00K  1.13MB/s
Z
[1]+  Stopped                 wget https://releases.ubuntu.com/20.04/ubuntu-20.04.2-live-server-amd6
4.iso
root@cli1:~# bg
[1]+ wget https://releases.ubuntu.com/20.04/ubuntu-20.04.2-live-server-amd64.iso &
root@cli1:~#
Redirecting output to 'wget-log'.

root@cli1:~#
root@cli1:~# fg
wget https://releases.ubuntu.com/20.04/ubuntu-20.04.2-live-server-amd64.iso
--2021-07-28 14:29:27--  https://releases.ubuntu.com/20.04/ubuntu-20.04.2-live-server-amd64.iso
Resolving releases.ubuntu.com (releases.ubuntu.com)... 91.189.91.123, 91.189.91.124, 2001:67c:1562::
25, ...
Connecting to releases.ubuntu.com (releases.ubuntu.com)|91.189.91.123|:443... connected.
HTTP request sent, awaiting response... 200 OK
Length: 1215168512 (1.1G) [application/x-iso9660-image]
Saving to: 'ubuntu-20.04.2-live-server-amd64.iso'

server-amd64.iso         0%[                                   ]  10.05M  1.50MB/s    eta 14m 25s
C
```

Figure 3.19 – Using Ctrl + Z, bg, and fg to move a process to the background and back to the foreground

We can clearly see that the wget process went to the background (*Ctrl + Z* and bg commands), then went to the foreground (the fg command), and was terminated at the end by using *Ctrl + C*. If we have multiple background processes in our current shell, we can also use indexing with the fg command (fg index_number).

How it works...

Processes can run in two different ways:

- **Foreground**: If we start a process from the shell, that process is going to occupy our current shell and will not allow us to type in additional commands. A kind of exception to that rule is a scenario in which the started process requires additional user input, but that input needs to be baked into the core of the process that we're executing (a part of the programming code). In this scenario, the shell is exclusively used by the started process until either the process finishes, we put it in the background, or it gets killed by other external factors (such as other processes or the kernel, or if it crashes for some reason).

- **Background**: If we start a process in the background, it runs and frees up our shell so that we can continue using it to type other commands.

When a process goes to the background in the current shell, it gets an index number so that we have the capability to manage it by using its index number. We can use fg, kill, and similar commands by using this index number (for example, kill %1 would kill the first job in the job index table).

As we saw in our practical demonstration, there are multiple ways of making sure that processes are started in the background – either when they are started or after they are started. What makes this concept plausible is the fact that we can easily put processes in the background, to be handled by the operating system, while we're away from it, which sometimes means freeing our precious time.

There's more...

If we need to learn more about foreground and background processes, we can check the following links:

- **Linux commands** – jobs, bg, and fg: https://www.redhat.com/sysadmin/jobs-bg-fg

- **Linux command basics** – seven commands for process management: https://www.redhat.com/sysadmin/linux-command-basics-7-commands-process-management

Managing process priorities

When we were explaining how to work with the top command, we intentionally omitted some details to give them their own time and place to discuss them later on in this chapter. We'll discuss one of these details here: the difference between the PR and NI fields in the top output. Let's do that now.

Getting ready

Keep the cli1 virtual machine powered on and let's continue using our shell.

How to do it...

We are going to learn how to use the top, nice, and renice commands to manage process scheduling in accordance with our wishes. First, let's use the top command. Let's renice a running process to a more negative value and a more positive value. Let's use the following top output for that:

```
 PID USER       PR  NI    VIRT    RES    SHR S  %CPU  %MEM     TIME+ COMMAND
1173 student    20   0   19432   9720   8168 S   0.0   0.5   0:00.04 systemd
1174 student    20   0  105936   3308      0 S   0.0   0.2   0:00.00 (sd-pam)
1180 student    20   0    7092   3636   2260 S   0.0   0.2   0:00.02 bash
1840 student    20   0    7092   4564   3020 S   0.0   0.2   0:00.01 bash
41223 student   20   0    7092   5080   3416 S   0.0   0.3   0:00.01 bash
41235 student   20   0    5608   2968   2724 S   0.0   0.1   0:00.03 screen
41236 student   20   0    6048   2824   2000 S   0.0   0.1   0:00.04 screen
41237 student   20   0    7016   5108   3448 S   0.0   0.3   0:00.00 bash
41247 student   20   0    7016   5072   3416 S   0.0   0.3   0:00.01 bash
47160 student   20   0    9024   4004   3208 S   0.0   0.2   0:00.00 top
```

Figure 3.20 – Starting point – processes started by the student user

Let's now change the priority of the process with PID 47160 (top). Press the *R* key and the top output will change this output to something such as the following:

```
PID to renice [default pid = 1173]
```

Figure 3.21 – Let's renice a PID

Then, again, type in the number 47160 and press *Enter*, followed by the niceness level – let's say, -10. We should get something like this as a result:

```
 PID USER       PR  NI    VIRT    RES    SHR S  %CPU  %MEM     TIME+ COMMAND
1174 student    20   0  105936   3308      0 S   0.0   0.2   0:00.00 (sd-pam)
1180 student    20   0    7092   3636   2260 S   0.0   0.2   0:00.02 bash
1840 student    20   0    7092   4564   3020 S   0.0   0.2   0:00.01 bash
41223 student   20   0    7092   5080   3416 S   0.0   0.3   0:00.01 bash
41235 student   20   0    5608   2968   2724 S   0.0   0.1   0:00.03 screen
41236 student   20   0    6048   2824   2000 S   0.0   0.1   0:00.06 screen
41237 student   20   0    7016   5108   3448 S   0.0   0.3   0:00.00 bash
41247 student   20   0    7016   5072   3416 S   0.0   0.3   0:00.01 bash
47160 student   10 -10    9024   4004   3208 S   0.0   0.2   0:00.21 top
```

Figure 3.22 – End result – more negative niceness level, higher priority

We can clearly see that the NI field of our PID changed from 0 to -10. Since we gave it a more negative niceness level, that means higher priority.

This example explains how to renice an already running process. But clearly, we can't use that to set the niceness level before we start the process. That's why we have the nice command. Here's an example of using the nice command:

```
nice -n -10 top
```

If we start this command as root and check the output, after we give it a bit of time, we should see something like this:

```
 PID USER       PR  NI    VIRT    RES    SHR S  %CPU  %MEM      TIME+ COMMAND
47529 root        5 -15    9024   4064   3272 R   0.3   0.2   0:00.02 top
    1 root       20   0  104256  13032   8672 S   0.0   0.7   0:04.79 systemd
```

Figure 3.23 – Using nice to pre-assign priority to a process at process start

Obviously, there's a caveat here – if we tried to start this command as a regular user, that wouldn't work. Regular users don't have a right to use the `nice` command – imagine how many different possibilities of abusing the system and/or crashing it would be given to any user if they did have that right. There are ways around this if we want to give the right to use the `nice` command to some users – we can do it via PAM modules or the `sudo` system. But for the time being, let's agree that this is not something that needs to be *urbi et orbi*, just as an exception.

Let's now explain how these concepts work.

How it works...

Let's start by executing the `top` command and checking the important part of its output:

```
  PID USER       PR   NI
  698 root       rt    0
  753 root       20    0
46372 root       20    0
    1 root       20    0
    2 root       20    0
    3 root        0  -20
    4 root        0  -20
    6 root        0  -20
    9 root        0  -20
   10 root       20    0
   11 root       20    0
   12 root       rt    0
   13 root      -51    0
   14 root       20    0
   15 root       20    0
   16 root      -51    0
```

Figure 3.24 – top output related to process priority

Let's briefly explain the difference between these two fields:

- PR (priority field): Real, kernel-scheduled priority at the moment of looking, assigned by the kernel. The `rt` mark means real time; it ranges between 0 and 139, although it can have negative static values for real-time processes.

- NI (niceness field): The process priority that it should have, assigned in user space (not kernel space) by default or by additional commands (`nice` and `renice`). The lower the number, the higher the priority, on a scale from -20 to +19.

Obviously, there's a big difference between these two numbers, seeing that one of them is *the real deal* (PR) and the other one is like *advice* (NI). Explaining process priorities is relatively easy in theory, but it becomes a bit more challenging to put into practice because of some architectural reasons. So, we'll try to explain this with an example that used to be possible but is not anymore as the speed of modern CPUs and memory is many times over what it used to be 10 years ago. So, let's first discuss the theoretical concept and then use an example.

Theoretically speaking, when we use the `nice` and `renice` commands, what we're doing is assigning a specific amount of CPU context to a process – a running one (`renice`) or a soon-to-be-running one (`nice`). We use the word *context* here in a programmatic sense from a CPU perspective. Translation – if we want to run a process, the kernel needs to assign it some CPU. If we have a running process and we `renice` it to a more negative value, that is going to tell the kernel and its scheduler to pay more attention to that specific process, therefore giving it more access to the CPU. If we `renice` it to a more positive value, that's going to tell the kernel to pay less attention to giving CPU resources to that process. By assigning more CPU to a process, there's a chance that that process is going to work faster and do its job faster as a result.

Obviously, this is a bit of a simplification as there are other factors at play here. For example, every process needs some memory to work, and the busier the memory is with other processes, the less speed the process has to access memory content, therefore lowering the memory bandwidth and increasing the latency of memory access. So, assigning a more negative value to niceness doesn't always directly translate to more performance out of a process. Also, what if the process doesn't need more CPU as it's currently idling and not doing anything of significance? There can be many more factors here – **Non-Unified Memory Access (NUMA)** operating system/application compatibility, effective usage of multiple threads/cores, locking mechanisms, and others. So, this is more of a general, academic discussion that can have exceptions that can happen for a variety of reasons, the state of the system being one of the most common ones.

Now that we've taken care of the theoretical background, let's use an example that used to be very easy to demonstrate as a lot of people had these sorts of experiences in the past, when CPU and memory were much, much less capable than they are today.

10 years ago, if we were to use an average computer of the day to display a high-resolution Flash video from YouTube, we had a problem. CPUs were kind of strong enough to do that, but only just. So, in order for us to be able to watch them in Linux by using a web browser (for example, Firefox), we had to tune the system to do it. So, we started the web browser, found a video that we wanted to watch, and pressed the *play* button, and it would work for a couple of seconds and then stutter. Then, it would work again, then stutter again. This was a frustrating experience. In those days, we didn't have GPU acceleration for Flash, so the CPU was the only device that could be of any assistance in these scenarios.

But if we knew how to set process priority, we could've solved that problem in most cases, depending on the CPU speed. We could go to the shell, find the PID of the Firefox process, and `renice` it to a much more negative value. All of a sudden, the kernel would instruct its scheduler to pay more attention to Firefox as a process, and lo and behold – our video would stop stuttering. Why? Because the kernel – by virtue of us using the `renice` command – realized that we wanted to give that process more priority and therefore ordered the CPU scheduler to make it happen.

There are many more aspects of tuning CPU performance. Modern Linux distributions have many options for this as the kernel gets more and more programmatic in terms of its approach to CPU scheduling. That's why various Linux distributions introduced concepts such as **tuned**, a profile-driven system that's able to tune our system performance based on pre-assigned or manually created profiles, and **tuna**, a utility that enables deep application-specific tuning. We always need to have the capability to go deep with tuning so that our system can have optimized performance for any specific use case.

There's more...

If we need to learn more about these concepts, we can check the following links:

- **A guide to the Linux top command**: `https://www.booleanworld.com/guide-linux-top-command/`

- **CPU scheduling**: `https://access.redhat.com/documentation/en-us/red_hat_enterprise_linux/6/html/performance_tuning_guide/s-cpu-scheduler`

- **top man page**: `https://man7.org/linux/man-pages/man1/top.1.html`

- **nice man page**: https://man7.org/linux/man-pages/man1/ nice.1.html

- **renice man page**: https://man7.org/linux/man-pages/man1/ renice.1.html

- **Getting started with tuned**: https://access.redhat.com/ documentation/en-us/red_hat_enterprise_linux/8/html/ monitoring_and_managing_system_status_and_performance/ getting-started-with-tuned_monitoring-and-managing-system- status-and-performance

- **Reviewing a system using the tuna interface**: https://access.redhat. com/documentation/en-us/red_hat_enterprise_linux/8/html/ monitoring_and_managing_system_status_and_performance/ reviewing-a-system-using-tuna-interface_monitoring-and- managing-system-status-and-performance

Configuring crond

Having the capability to run jobs on a schedule is very important for everyday system administration. We schedule backups, run cleanup procedures, send reports, do antivirus checks, and do other tasks that business procedures need. Scheduling them means a certain level of automation and getting rid of the manual approach to things, which in turn again gives us more time to focus on more important tasks. Generally speaking, we use either commands or scripts as a way to do these scheduled tasks, and to execute them, we use cron daemon (crond). Let's learn how to use crond to schedule jobs in accordance with our needs.

Getting ready

Keep the cli1 virtual machine powered on and let's create some scheduled jobs via crond.

How to do it...

Let's start by using root to create a cron job. We are going to achieve that by typing in the following command as root:

```
crontab -e
```

In Ubuntu, we are going to be asked to select which editor we want to use. For continuity reasons, let's say that we choose the vi editor (`vi.basic`). Let's add the following entry to the end of the file that we're editing in vi:

```
* * * * * ls -al /root > /tmp/root.txt
```

If we save the file as is, we just created the first root cron job – the one that's going to execute every minute. The * signs are actually *frequency fields* in these crontab files. Look at the following:

```
man 5 crontab
```

Now scroll a bit lower on the man page; we're going to find the following examples:

```
field           allowed values
-----           --------------
minute          0-59
hour            0-23
day of month    1-31
month           1-12 (or names, see below)
day of week     0-7 (0 or 7 is Sun, or use names)

A field may be an asterisk (*), which always stands for ``first-last''.

Ranges of numbers are allowed.  Ranges are two numbers separated with a hyphen.  The spec-
ified range is inclusive.  For example, 8-11 for an ``hours'' entry specifies execution at
hours 8, 9, 10 and 11.

Lists are allowed.  A list is a set of numbers (or ranges) separated by commas.  Examples:
``1,2,5,9'', ``0-4,8-12''.

Step values can be used in conjunction with ranges.  Following a range with ``/<number>''
specifies  skips  of the number's value through the range.  For example, ``0-23/2'' can be
used in the hours field to specify command execution every other hour (the alternative  in
the V7 standard is ``0,2,4,6,8,10,12,14,16,18,20,22'').  Steps are also permitted after an
asterisk, so if you want to say ``every two hours'', just use ``*/2''.

Names can also be used for the ``month'' and ``day of week'' fields. Use the first  three
letters  of  the  particular day or month (case doesn't matter).  Ranges or lists of names
are not allowed.

The ``sixth'' field (the rest of the line) specifies the command to be  run.   The  entire
command  portion  of the line, up to a newline or % character, will be executed by /bin/sh
or by the shell specified in the SHELL variable of the crontab file.  Percent-signs (%) in
the  command,  unless escaped with backslash (\), will be changed into newline characters,
and all data after the first % will be sent to the command as standard input.  There is no
way to split a single command line onto multiple lines, like the shell's trailing "\".
```

Figure 3.25 – Excerpt from man 5 crontab page

The first part explains the fields, as there are six of them. The first five are related to the frequency of executing the six fields. We can use space or *Tab* as delimiters between these fields.

The first five fields are as follows:

- The `minute` field: At which minute the job should be executed
- The `hour` field: At which hour the job should be executed
- The `day of month` field: On which day of the month the job should be executed
- The `month` field: On which month the job should be executed
- The `day of week` field: On which day of the week the job should be executed

All of these fields support various types of syntax:

- **Number**: We can type any number that makes sense for a given field. For example, if we type the number `75` in the `minute` field, that doesn't make any sense as single numbers in the `minute` field go from `0` to `59`.
- **Range**: We can type any range that makes sense for a given field. For example, if we type a range of `3-47` in the `hour` field, that won't work as the number range for hours is `0-23`.
- **Step values**: We can type things such as `0-10/2` in the first field, which `cron` translates to *from the 0th to 10th minute, every second minute.*
- **Lists**: If we use the previous example, we could've configured that so that we type `0,2,4,6,8,10`.
- **Combination thereof**: Any combination of previously mentioned syntaxes.

All of these syntaxes give us many different options to configure which jobs we want to execute and when, down to the minute. We also need to remember that `cron` doesn't allow frequency smaller than a minute. If we need to do something like that, we need to work *around* that fact (using the `sleep` function, the `at` command, and others).

If we are logged in as root, that gives us the capability to manage a user's cron jobs as well. For example, say we type in the following command:

```
crontab -e -u student
```

We are going to be editing `crontab` for the `student` user.

Furthermore, if we want to remove `student`'s cron jobs, we can do it like this:

```
crontab -r -u student
```

If we just need to list cron jobs from the `student` user, we can use the following command:

```
crontab -l -u student
```

Let's go back to system cron jobs, the second most common type of cron jobs. These are configured in system folders under the `/etc` directory, such as `/etc/cron.daily` and `/etc/cron.hourly`. The keyword that follows `cron.` in the directory name tells us the frequency for all of the jobs configured in that folder. For example, let's take a look at the `cron.daily` folder:

```
root@cli1:/etc/cron.daily# ls -al
total 40
drwxr-xr-x   2 root root 4096 Jul 28 19:07 .
drwxr-xr-x 102 root root 4096 Jul 28 19:29 ..
-rw-r--r--   1 root root  102 Feb 13  2020 .placeholder
-rwxr-xr-x   1 root root  311 Jul 16  2019 0anacron
-rwxr-xr-x   1 root root  376 Jun  1  2020 apport
-rwxr-xr-x   1 root root 1478 Aug 11  2020 apt-compat
-rwxr-xr-x   1 root root 1100 Jun 27  2020 dpkg
-rwxr-xr-x   1 root root  377 Sep 10  2020 logrotate
-rwxr-xr-x   1 root root 1123 Jul  5  2020 man-db
-rwxr-xr-x   1 root root  543 Jul 16  2019 mlocate
```

Figure 3.26 – Daily system cron jobs

Some of these filenames probably sound familiar. `logrotate` does log rotation, the `mlocate` job updates the file/folder database used by the `updatedb` command, and so on. At any given moment, there can be dozens of these system-wide cron jobs in those directories, depending on which packages we installed on our Linux server and how many additional system jobs we created ourselves. Let's use the `logrotate` file as an example:

```
root@cli1:/etc/cron.daily# cat logrotate
#!/bin/sh

# skip in favour of systemd timer
if [ -d /run/systemd/system ]; then
    exit 0
fi

# this cronjob persists removals (but not purges)
if [ ! -x /usr/sbin/logrotate ]; then
    exit 0
fi

/usr/sbin/logrotate /etc/logrotate.conf
EXITVALUE=$?
if [ $EXITVALUE != 0 ]; then
    /usr/bin/logger -t logrotate "ALERT exited abnormally with [$EXITVALUE]"
fi
exit $EXITVALUE
```

Figure 3.27 – logrotate cron.daily job, which actually executes a simple shell script

As we can see, this cron job actually executes a shell script, and its configuration is the polar opposite of what we saw in user cron jobs where `crontab` files have a very strict syntax. Here, we have much more freedom and the fact that we can just write outright shell script code in these files makes our job easier. We have a lot of additional content about shell scripting coming up later in this book, from *Chapter 9, An Introduction to Shell Scripting*, all the way to the end of the book. So, let's table that discussion for a later time when we introduce all the necessary concepts – variables, loops, functions, arrays, and so on. It is going to be a lot of fun indeed.

How it works...

`crond` is a spooling type of service – it creates queues of tasks that it needs to do and then executes them in accordance with specified criteria. In larger enterprises, we might consider `crond` criteria to be a part of a bigger picture that we usually call policy. Businesses rely on IT-related policies to implement standards and levels of service, and in that sense, IT policies are nothing more than objects describing a certain need. A need to do scheduled, daily backups, a need to run regular security checks by using an **intrusion detection system (IDS)** or antivirus, you name it. It doesn't really matter if we're talking about some general policies or security-related policies – policies are a standardized way of bringing IT chaos to order in a streamlined, compliance-oriented fashion. When we take a look at `crond`, it's one of those essential tools to deliver those policies.

`crond` takes care of different types of scheduled jobs:

- **System-related scheduled jobs**: Daily, hourly, weekly, and other jobs that are executed in order for the system to work properly.

- **Delayed or deferred jobs**: If a server has been turned off at the time when scheduled jobs should've been started, then it's up to `crond` to make sure to execute them later if we install the `anacron` package, which takes care of scenarios such as constant server shutdowns that would lead to periodic tasks not being executed regularly.

- **User-based crond jobs**: These are per-user jobs that regular system users can create so that they can execute a job at a given moment.

First and foremost, we need to learn how user-based cron jobs work, as these are the most common jobs on multi-user servers.

When we use the `crontab -e` command for any given user, a `crontab` file gets created in one of the `crond` directories. These files are nothing special in terms of complexity – there's obviously a bit of syntax involved, but they're really well documented even when starting from scratch so that most users won't have problems with figuring this out. In Ubuntu, `crontab` files are created in the `/var/spool/cron/crontabs` directory, where all user cron jobs get saved as a text file per user. If a file in that directory is named as root, that means that the root user has a cron job. If there's a file named `student`, that means that the user called `student` has a cron job scheduled. This makes it easier to debug if it ever comes to that. Also, we need to take note of the fact that there are people who prefer to edit those files rather than using the `crontab` command. At the end of the day, whichever way we solve the IT problem that we need to solve is good, as long as it works properly. Let's take a look at one of those files – in this case, it's going to be an excerpt from the root user's `crontab` file:

```
# For example, you can run a backup of all your user accounts
# at 5 a.m every week with:
# 0 5 * * 1 tar -zcf /var/backups/home.tgz /home/
#
# For more information see the manual pages of crontab(5) and cron(8)
#
# m h  dom mon dow   command
* * * * * ls -al /root > /tmp/root.txt
```

Figure 3.28 – Root's crontab file

We can clearly see that root scheduled a job to be executed every minute here. That job lists the content of the `/root` directory and saves it to a file in the `/tmp` directory called `root.txt`. It's simple enough, but it clearly shows the way in which `crontab` configuration files are created.

`crond` regularly checks these files and executes configuration stored in them at the scheduled time. This is the reason why it's very important to be careful what we put in these files. We really shouldn't put plain-text passwords, login information, or anything similar to these concepts in `crontab` files. By using the first five fields in the user's `crontab` file, `crond` determines the frequency for any given scheduled job. It parses these files line by line, which means that we can easily schedule multiple cron jobs as users, without any problems.

If we run into problems with users scheduling too many cron jobs eating away at the performance of our server, we can always ban them from using `crontab`. For example, if we want to deny the `student` user the capability of creating user cron jobs, we just need to edit the `/etc/cron.deny` file and add a user name per line, like this:

```
student
```

If we do that, and the user called `student` tries to create a cron job by using `crontab -e`, this is the expected result:

```
student@cli1:~$ crontab -e
You (student) are not allowed to use this program (crontab)
See crontab(1) for more information
student@cli1:~$
```

Figure 3.29 – Using cron.deny to disable a user's right to use crontab

That's a wrap for this chapter. The next chapter is going to be all about using the shell to configure network settings, which includes both network interfaces and firewalls. Stay tuned!

There's more...

If we need to learn more about these concepts, we can check the following links:

- **Automating system tasks**: https://access.redhat.com/ documentation/en-us/red_hat_enterprise_linux/7/html/ system_administrators_guide/ch-automating_system_tasks

- **Crontab man page, chapter 5**: https://man7.org/linux/man-pages/ man5/crontab.5.html

- **How I use cron in Linux**: https://opensource.com/article/17/11/ how-use-cron-linux

- **Use anacron for a better crontab**: https://opensource.com/ article/21/2/linux-automation

4
Using Shell to Configure and Troubleshoot a Network

Managing processes is an important job of a Linux system administrator. That can be for a variety of reasons – maybe some processes got stuck and we need to finish them, or we want to put some process(es) to work in the background, or even to be started periodically or at a later date. Whatever the scenario is, it's important to know how to administer processes and make them do the work that needs to be done efficiently and with regards to other processes running on the system.

In this chapter, we are going to learn about the following recipes:

- Using `nmcli` and `netplan`
- Using `firewall-cmd` and `ufw`
- Working with open ports and connections

- Configuring /etc/hosts and DNS resolving
- Using network diagnostic tools

Technical requirements

For these recipes, we're going to use two Linux machines. We can use the client1 virtual machine from our previous recipes. We'll also use another two virtual machines running **CentOS 8 2105** to cover CentOS-based scenarios (nmcli and firewall-cmd). Let's call them server1 and client2. All in all, we need the following:

- A virtual machine running Ubuntu 20.10
- Two virtual machines with CentOS 8 2105

So, let's start our virtual machines and let's get cracking!

Using nmcli and netplan

Network configuration has changed significantly in the past couple of releases – for all Linux distributions. It doesn't really matter whether we are discussing Red Hat and its clones or Debian and its clones – these changes happened across all of them. For example, Red Hat and its clones went from a network service to a mixture of network and NetworkManager services to a fully NetworkManager-based configuration. Ubuntu was using a networking service until recently when it switched to netplan. Let's explain all of these concepts so that we can have a full overview of these configuration methods and cover any situations you might end up in. We will also cover a scenario in which someone might want to turn off netplan and go back to using the networking service on Ubuntu.

Getting ready

We just need one Ubuntu and one CentOS machine for this recipe. Let's say we are going to use server1 and client1 to master nmcli and netplan. Furthermore, on CentOS, we need to deploy the net-tools package to get access to some of the commands used in this recipe (for example, the ifconfig command). Let's do that by using the following command:

```
dnf install net-tools
```

After that, we're ready to go.

How to do it

Let's first work with the two most common CentOS scenarios – implementing network configuration via nmcli for both **static IP address** and **Dynamic Host Configuration Protocol (DHCP)** scenarios. Let's just say that we are going to use network interface ens39 to create a network connection called static (for the static IP address, for example, 192.168.2.2/24 with gateway 192.168.2.254 and DNS servers 8.8.8.8 and 8.8.4.4) and, later on, a network connection called dynamic (for DHCP configuration). We just need to run a few commands as root per scenario for that:

```
nmcli connection add con-name static ifname ens39 type
ethernet ipv4.address 192.168.2.2/24 ipv4.gateway
192.168.2.254 ipv4.dns 8.8.8.8,8.8.4.4

nmcli connection reload

systemctl restart NetworkManager
```

The expected result should look like this:

```
[root@client2 ~]# nmcli connection add con-name static ifname ens39 type ethernet ipv4.addresses 192
.168.2.2/24 ipv4.gateway 192.168.2.254 ipv4.dns 8.8.8.8,8.8.4.4
Connection 'static' (84b7f9b1-61f4-4613-ac81-24628a4ee753) successfully added.
[root@client2 ~]# nmcli con show
NAME     UUID                                    TYPE      DEVICE
static   84b7f9b1-61f4-4613-ac81-24628a4ee753   ethernet  ens39
[root@client2 ~]# ifconfig ens39
ens39: flags=4163<UP,BROADCAST,RUNNING,MULTICAST>  mtu 1500
        inet 192.168.2.2  netmask 255.255.255.0  broadcast 192.168.2.255
        inet6 fe80::3e27:5f75:7f84:105f  prefixlen 64  scopeid 0x20<link>
        ether 00:0c:29:36:93:f2  txqueuelen 1000  (Ethernet)
        RX packets 3008  bytes 385179 (376.1 KiB)
        RX errors 0  dropped 0  overruns 0  frame 0
        TX packets 918  bytes 91053 (88.9 KiB)
        TX errors 0  dropped 0 overruns 0  carrier 0  collisions 0
```

Figure 4.1 – Adding a static IP configuration via nmcli

Let's now remove that connection and define a DHCP-based configuration. For that, we need to have a DHCP server available on our network, so that it can assign the necessary network configuration information to client2 (IP address, netmask, gateway, DNS server addresses, and so on). We need to type in the following commands:

```
nmcli connection delete static

nmcli connection add con-name dynamic ifname ens39 type
ethernet

nmcli connection reload

systemctl restart NetworkManager
```

This is our expected result:

```
[root@client2 ~]# nmcli connection add con-name dynamic ifname ens39 type ethernet
Connection 'dynamic' (0809bb3a-c94c-4848-96a7-1fb1b9c2603d) successfully added.
[root@client2 ~]# nmcli connection reload
[root@client2 ~]# systemctl restart NetworkManager
[root@client2 ~]# ifconfig ens39
ens39: flags=4163<UP,BROADCAST,RUNNING,MULTICAST>  mtu 1500
        inet 192.168.159.141  netmask 255.255.255.0  broadcast 192.168.159.255
        inet6 fe80::618f:53a3:7055:f0e3  prefixlen 64  scopeid 0x20<link>
        ether 00:0c:29:36:93:f2  txqueuelen 1000  (Ethernet)
        RX packets 77  bytes 8694 (8.4 KiB)
        RX errors 0  dropped 0  overruns 0  frame 0
        TX packets 157  bytes 18778 (18.3 KiB)
        TX errors 0  dropped 0 overruns 0  carrier 0  collisions 0

[root@client2 ~]# ping 8.8.8.8
PING 8.8.8.8 (8.8.8.8) 56(84) bytes of data.
64 bytes from 8.8.8.8: icmp_seq=1 ttl=128 time=58.0 ms
^C
--- 8.8.8.8 ping statistics ---
1 packets transmitted, 1 received, 0% packet loss, time 0ms
rtt min/avg/max/mdev = 58.048/58.048/58.048/0.000 ms
```

Figure 4.2 – Adding a DHCP configuration via nmcli

If everything is configured correctly on our network, we should've gotten an IP address and other networking information and have internet access.

In terms of netplan on Ubuntu, this configuration method is more in line with the currently popular *infrastructure as code* paradigm, so it's all about configuration files. So, we will again implement two of the most common scenarios – a static IP address and DHCP, but we will also cover a scenario with multiple network interfaces so that we can see what the syntax looks like.

First, let's start with the netplan static networking configuration. Let's say that we need to assign IP address 192.168.1.1/24 to network interface ens33, with the default gateway being 192.168.1.254 and DNS servers 8.8.8.8, and 8.8.4.4. We can just change the existing YAML configuration file that's already there, called 00-installer-config.yaml:

```
root@client1:/etc/netplan# cat 00-installer-config.yaml
# This is the network config written by 'subiquity'
network:
  ethernets:
    ens33:
      dhcp4: no
      dhcp6: no
      addresses: [192.168.1.1/24]
      gateway4: 192.168.1.254
      nameservers:
        addresses: [8.8.8.8, 8.8.4.4]
  version: 2
```

Figure 4.3 – Adding a static configuration via netplan

That covers our static IP address scenario. It's relatively obvious what we need to do in terms of a `netplan` DHCP scenario, so the configuration file needs to look like this for that specific scenario:

```
root@client1:/etc/netplan# cat 00-installer-config.yaml
# This is the network config written by 'subiquity'
network:
  ethernets:
    ens33:
      dhcp4: yes
  version: 2
```

Figure 4.4 – Adding a dynamic configuration via netplan

The last part of our recipe, as we mentioned, is about having multiple network interfaces and configuring them properly. Let's say that we have a network interface called ens33 that needs to be DHCP-configured, and an interface called ens38 that needs to be assigned an IP address, 192.168.1.1/24, with the same config data for gateway and DNS servers as before. The configuration file would look like this:

```
root@client1:/etc/netplan# cat 00-installer-config.yaml
# This is the network config written by 'subiquity'
network:
  ethernets:
    ens33:
      dhcp4: yes
    ens38:
      dhcp4: no
      dhcp6: no
      addresses: [192.168.1.1/24]
      gateway4: 192.168.1.254
      nameservers:
        addresses: [8.8.8.8, 8.8.4.4]
  version: 2
```

Figure 4.5 – Configuring multiple network interfaces via netplan

For some of the latest versions of Ubuntu, this yes/no configuration will be changed to true/false, so if you get an error here, you just need to make that change. Basically, it looks like a merge of the previous two files, with a couple of lines stripped so that we don't have unnecessary repetitions.

Let's now see how these two concepts work. It's simple enough, but still, it requires a bit of background knowledge, so let's dive in.

How it works

Now that we've done a brief primer on how processes and signals work, let's continue our quest for knowledge about processes by learning about the management of background processes. As we've already explained the basics of background processes, that shouldn't be a difficult task.

In terms of NetworkManager and its command-line configuration interface (nmcli), NetworkManager does its configuration via configuration files in the /etc/sysconfig/network-scripts directory. Let's show an example from our previous CentOS session – where we created an interface called dynamic. In that directory, there's a file called ifcfg-dynamic, with the following content:

```
[root@client2 network-scripts]# cat ifcfg-dynamic
TYPE=Ethernet
PROXY_METHOD=none
BROWSER_ONLY=no
BOOTPROTO=dhcp
DEFROUTE=yes
IPV4_FAILURE_FATAL=no
IPV6INIT=yes
IPV6_AUTOCONF=yes
IPV6_DEFROUTE=yes
IPV6_FAILURE_FATAL=no
IPV6_ADDR_GEN_MODE=stable-privacy
NAME=dynamic
UUID=0809bb3a-c94c-4848-96a7-1fb1b9c2603d
DEVICE=ens39
ONBOOT=yes
```

Figure 4.6 – Regular NetworkManager configuration file

That's quite a big configuration file for a simple configuration. Actually, if we were to polish this file a bit, we could make it at least two thirds shorter, and it would still work, for example, like this:

```
[root@client2 ~]# cat /etc/sysconfig/network-scripts/ifcfg-dynamic
BOOTPROTO=dhcp
NAME=dynamic
DEVICE=ens39
ONBOOT=yes
```

Figure 4.7 – Shortened configuration file

These configuration options aren't all that difficult to understand, as well as some other configuration options that are needed for static IP configuration (IPADDR, PREFIX or NETMASK, GATEWAY – these are all pretty self-explanatory). But the fact remains that – at least in part – NetworkManager still uses this bulky syntax as a history leftover, as we've been using this /etc/sysconfig/network-scripts directory and files in that directory to configure network interfaces for years and years now.

When comparing that to netplan, we can clearly see that netplan puts much more importance on declarative syntax with all of the structured code and indentation that it needs to have, which is what YAML is known for. It will frustrate you at the beginning, at least until you learn how to use the vim editor for editing YAML files, as it then becomes much easier. Check out the link in the *There's more* section to learn how to set up vim to help you with YAML syntax.

Both of these services – when the system gets booted up – read the aforementioned configuration files and set the network interfaces in accordance to the settings in them. A pretty straightforward process, as long as we understand the syntax. But we'd still recommend using nmcli for NetworkManager configuration as its syntax gets under your fingers quickly.

The next stop is firewalling, by using firewalld and ufw.

There's more

If you need more information about networking in CentOS and Ubuntu, make sure that you check the following:

- Configuring and managing networking: https://access.redhat.com/documentation/en-us/red_hat_enterprise_linux/8/html-single/configuring_and_managing_networking/index

- Netplan reference: https://netplan.io/reference/

- nmcli: https://developer-old.gnome.org/NetworkManager/stable/nmcli.html

- Setting up vim for YAML editing: https://www.arthurkoziel.com/setting-up-vim-for-yaml/

Using firewall-cmd and ufw

Using built-in firewalls has been a de facto standard in Linux for more than two decades now. Ever since the *invention* of ipfwadm (kernel v2.0), Linux kernel developers have been piling up functionality and a firewall has been one of those things. ipfwadm was followed by ipchains (kernel v2.2), iptables (kernel v2.4), and today it's all about firewalld (CentOS) and ufw (Ubuntu). Let's go through both of these concepts so that we can use them when we need them regardless of the Linux distribution we're working on.

Getting ready

As a part of this recipe, we are going to go through a list of dozens of different scenarios covering `firewalld` and `ufw`. In other words, we are going to introduce the necessary commands to do configuration changes for some of the most commonly used scenarios. First, let's install the necessary packages for CentOS (on our client2 machine) and Ubuntu (client1 machine). So, for CentOS, we need to type the following command:

```
yum -y install firewalld
```

Also, use this command for Ubuntu:

```
apt-get -y install ufw
```

In terms of services, for CentOS, we have these – just in case we used `iptables` previously, as `iptables` firewall is supported in CentOS 8:

```
systemctl stop iptables
systemctl disable iptables
systemctl mask iptables
systemctl enable firewalld
systemctl start firewalld
```

For Ubuntu, it's the same idea:

```
systemctl stop iptables
systemctl disable iptables
systemctl mask iptables
systemctl enable ufw
systemctl start ufw
ufw enable
```

Now that the services are configured, let's start!

How to do it

For `firewalld`, we are going to use its default command, which is `firewall-cmd`. For `ufw`, the command has the same name – `ufw`. First, let's take care of some basic commands. Let's first list all the rules:

```
firewall-cmd --list-all
```

Depending on how many rules we added previously, we should get an output similar to this:

```
[root@client2 ~]# firewall-cmd --list-all
public (active)
  target: default
  icmp-block-inversion: no
  interfaces: ens39
  sources:
  services: cockpit dhcpv6-client ssh
  ports:
  protocols:
  masquerade: yes
  forward-ports:
  source-ports:
  icmp-blocks:
  rich rules:
```

Figure 4.8 – firewall-cmd --list-all output

Let's now add and delete a couple of rules. This is what we are going to do:

- Add a rule that allows `192.168.2.254/24` to access everything on our `client2` machine.

- Add a rule that allows network subnet `192.168.1.0/24` to access the `SSH` service on our `client2` machine.

- Block the `192.168.3.0/24` network from accessing HTTP/HTTPS services.

- Allow the same subnet to access the DNS service.

- Forward port `900` to port `9090`.

- Configure `masquerade` so that `client1` can be used as a router/gateway.

As the last step, we are going to delete every single one of these rules, one by one.

Now that we're clear on what we are trying to do, let's punch in all of the necessary commands to make this happen. First, let's start by using the default configuration and default zone, which means that we need to check which zone it is. We can see in the previous screenshot that the public zone is active, so – for the time being – we are going to add all of the rules to that zone and explain zones and rich rules a bit later:

```
firewall-cmd --permanent --zone=public --add-
source=192.168.2.254/32
firewall-cmd --permanent --zone=public --add-rich-
rule='rule family="ipv4" source address="192.168.1.0/24" port
protocol="tcp" port="22" accept'
```

```
firewall-cmd --permanent --zone=public --add-rich-
rule='rule family="ipv4" source address="192.168.3.0/24" port
protocol="tcp" port="80" reject'

firewall-cmd --permanent --zone=public --add-rich-
rule='rule family="ipv4" source address="192.168.3.0/24" port
protocol="tcp" port="443" reject'

firewall-cmd --permanent --zone=public --add-rich-
rule='rule family="ipv4" source address="192.168.3.0/24" port
protocol="udp" port="53" accept'

firewall-cmd --permanent --zone=public --add-forward-port=port=
900:proto=tcp:toport=9090

firewall-cmd --permanent --zone=public --add-masquerade

echo "1" > /proc/sys/net/ipv4/ip_forward

echo "net.ipv4.ip_forward = 1" > /etc/sysctl.d/50-forward.conf

firewall-cmd --reload
```

Echo commands are used to enable IP forwarding right now (first echo), and to enable it permanently (second echo), by using a `sysctl` configuration file that's going to be loaded as our system boots up. The last command is to apply these settings to the current running state of `firewalld`. When we now type in the `firewall-cmd --list-all` command, we should get output like this:

```
[root@client2 ~]# firewall-cmd --list-all
public (active)
  target: default
  icmp-block-inversion: no
  interfaces: ens39
  sources: 192.168.2.254
  services: cockpit dhcpv6-client ssh
  ports:
  protocols:
  masquerade: no
  forward-ports:
        port=900:proto=tcp:toport=9090:toaddr=
  source-ports:
  icmp-blocks:
  rich rules:
        rule family="ipv4" source address="192.168.1.0/24" port port="22" protocol="tcp" accept
        rule family="ipv4" source address="192.168.3.0/24" port port="53" protocol="udp" accept
        rule family="ipv4" source address="192.168.3.0/24" port port="80" protocol="tcp" reject
        rule family="ipv4" source address="192.168.3.0/24" port port="443" protocol="tcp" reject
```

Figure 4.9 – End result of our configuration in firewalld

It's very important to learn how to remove these rules, as well. So, let's now remove these rules, one by one, going in the opposite direction:

```
firewall-cmd --permanent --zone=public --remove-masquerade
firewall-cmd --permanent --zone=public --remove-forward-port=po
```

```
rt=900:proto=tcp:toport=9090
```

```
firewall-cmd --permanent --zone=public --remove-rich-
rule='rule family="ipv4" source address="192.168.1.0/24" port
protocol="tcp" port="22" accept'
```

```
firewall-cmd --permanent --zone=public --remove-rich-
rule='rule family="ipv4" source address="192.168.3.0/24" port
protocol="udp" port="53" accept'
```

```
firewall-cmd --permanent --zone=public --remove-rich-
rule='rule family="ipv4" source address="192.168.3.0/24" port
protocol="tcp" port="443" reject'
```

```
firewall-cmd --permanent --zone=public --add-rich-
rule='rule family="ipv4" source address="192.168.3.0/24" port
protocol="tcp" port="80" reject'
```

```
firewall-cmd --permanent --zone=public --remove-
source=192.168.2.254/24
```

```
firewall-cmd --reload
```

That should remove all of the rules and apply the starting state as the current state. Let's check:

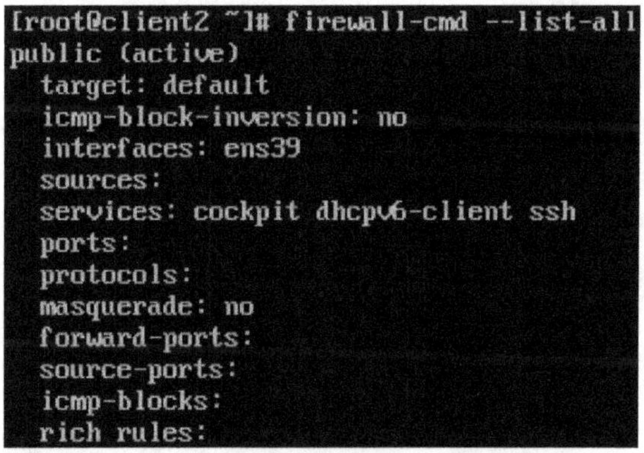

Figure 4.10 – The firewalld rule set after rule removal

Everything is back to its original state, so we can call this a success. Let's now apply the same ruleset to the Ubuntu virtual machine called client1 by using ufw. Let's first check the status by using the ufw status verbose command. We should get a result like this:

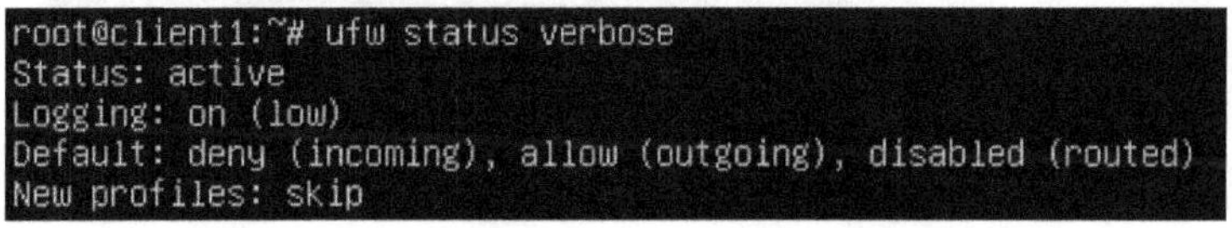

Figure 4.11 – ufw configuration starting point

Let's now add the same set of rules that we added in `firewalld`, to see how it's done via `ufw` and to be able to check the syntax differences:

```
ufw allow from 192.168.2.254/32
ufw allow from 192.168.1.0/24 proto tcp to any port 22
ufw deny from 192.168.3.0/24 proto tcp to any port 80
ufw deny from 192.168.3.0/24 proto tcp to any port 443
ufw allow from 192.168.3.0/24 proto udp to any port 53
```

For port forwarding, we need to edit the `/etc/ufw/before.rules` file and add the following configuration options before the `*filter` section:

```
*nat
:PREROUTING ACCEPT [0:0]
-A PREROUTING -p tcp --dport 900 -j REDIRECT --to-port 9090
COMMIT
```

We can now check the status of our work by using the `ufw` and `iptables` commands:

```
root@client1:~# ufw status verbose
Status: active
Logging: on (low)
Default: deny (incoming), allow (outgoing), disabled (routed)
New profiles: skip

To                         Action      From
--                         ------      ----
Anywhere                   ALLOW IN    192.168.2.254
22/tcp                     ALLOW IN    192.168.1.0/24
80/tcp                     DENY IN     192.168.3.0/24
443/tcp                    DENY IN     192.168.3.0/24
53/udp                     ALLOW IN    192.168.3.0/24
900/tcp                    ALLOW IN    Anywhere
9090/tcp                   ALLOW IN    Anywhere
900/tcp (v6)               ALLOW IN    Anywhere (v6)
9090/tcp (v6)              ALLOW IN    Anywhere (v6)

root@client1:~# iptables -t nat -L
Chain PREROUTING (policy ACCEPT)
target     prot opt source               destination
REDIRECT   tcp  --  anywhere             anywhere             tcp dpt:900 redir ports 9090

Chain INPUT (policy ACCEPT)
target     prot opt source               destination

Chain POSTROUTING (policy ACCEPT)
target     prot opt source               destination

Chain OUTPUT (policy ACCEPT)
target     prot opt source               destination
```

Figure 4.12 – ufw status verbose after all the added rules

We need to add two configuration options to `ufw` configuration for masquerading to work. First, we need to change the default policy for forwarding in the `/etc/default/ufw` file. It's set to `DROP` by default. We just need to change it to `ACCEPT`. It's at the beginning of this file, so the end result should look like this:

```
DEFAULT_FORWARD_POLICY="ACCEPT"
```

This is rather than looking like this:

```
DEFAULT_FORWARD_POLICY="DROP"
```

The next configuration option is actually in the same file that we previously edited, `/etc/ufw/before.rules`. We need to add one additional part to the `*nat` section called the `POSTROUTING` subsection, to the same place that we used before. So, similar to the previous example, we need to add the following configuration options, again before the `*filter` section:

```
*nat
:POSTROUTING ACCEPT [0:0]
-A POSTROUTING -o ens33 -j MASQUERADE
COMMIT
```

The last part of the `ufw` configuration is making sure that the kernel knows that it needs to turn on `ip` forwarding. For that to happen, we need to edit a file called `/etc/ufw/sysctl.conf` and uncomment the following configuration option by removing the comment mark (#) before it:

```
net/ipv4/ip_forward=1
```

This changes a value in the `/proc` filesystem, in a file called `/proc/sys/net/ipv4/ip_forward`. If we want to make sure that it works even without rebooting the machine, we need to issue the following command, as well:

```
echo "1" > /proc/sys/net/ipv4/ip_forward
```

That will enable IP forwarding on the kernel level and enable us to use masquerading in ufw. Let's now check the end result:

```
root@client1:~# ufw status verbose
Status: active
Logging: on (low)
Default: deny (incoming), allow (outgoing), allow (routed)
New profiles: skip

To                         Action       From
--                         ------       ----
Anywhere                   ALLOW IN     192.168.2.254
22/tcp                     ALLOW IN     192.168.1.0/24
80/tcp                     DENY IN      192.168.3.0/24
443/tcp                    DENY IN      192.168.3.0/24
53/udp                     ALLOW IN     192.168.3.0/24
900/tcp                    ALLOW IN     Anywhere
9090/tcp                   ALLOW IN     Anywhere
900/tcp (v6)               ALLOW IN     Anywhere (v6)
9090/tcp (v6)              ALLOW IN     Anywhere (v6)

root@client1:~# iptables -t nat -L
Chain PREROUTING (policy ACCEPT)
target     prot opt source                destination
REDIRECT   tcp  --  anywhere              anywhere             tcp dpt:900 redir ports 9090

Chain INPUT (policy ACCEPT)
target     prot opt source                destination

Chain POSTROUTING (policy ACCEPT)
target     prot opt source                destination
MASQUERADE  all  --  anywhere              anywhere

Chain OUTPUT (policy ACCEPT)
target     prot opt source                destination
```

Figure 4.13 – ufw ruleset after configuring masquerading

If we need to remove any of these configuration options, it's going to be done in two parts:

- Configuration options that we added to files directly via the editor will need to be removed via the editor.

- Configuration options that we added by using the ufw command can be easily reversed by using ufw's rule indexing.

Let's type in the command ufw status numbered. This is the expected result:

```
root@client1:~# ufw status numbered
Status: active

          To                     Action      From
          --                     ------      ----
[ 1]  Anywhere                   ALLOW IN    192.168.2.254
[ 2]  22/tcp                     ALLOW IN    192.168.1.0/24
[ 3]  80/tcp                     DENY IN     192.168.3.0/24
[ 4]  443/tcp                    DENY IN     192.168.3.0/24
[ 5]  53/udp                     ALLOW IN    192.168.3.0/24
[ 6]  900/tcp                    ALLOW IN    Anywhere
[ 7]  9090/tcp                   ALLOW IN    Anywhere
[ 8]  900/tcp (v6)               ALLOW IN    Anywhere (v6)
[ 9]  9090/tcp (v6)              ALLOW IN    Anywhere (v6)
```

Figure 4.14 – ufw ruleset indexed by number

We can see that every rule that we punched in has a number attached to it. Rules 8 and 9 are rules for ports `900` and `9090` that were automatically added for the IPv4 and IPv6 stack. We can remove all of these rules easily by using the numbers attached to them. The thing is, `ufw` doesn't have a mechanism for deleting multiple rules one by one, so we would need to delete them one by one, something like this:

```
ufw delete 9
<confirm by pressing y>
ufw delete 8
<confirm by pressing y>
....
ufw delete 1
<confirm by pressing y>
```

Yes, we could've simplified this with a `for` loop, something like this:

```
for i in {9..1};do yes|ufw delete $i;done
```

But shell scripting and using loops is yet to come in this book so... let's treat this as an *example in advance*.

Let's now explain how all of this works – firewalld zones, rich rules, `ufw` syntax – so that we can get an understanding of the background services and capabilities making it happen.

> **Note**
>
> `Firewalld` is available on Ubuntu, as well. It just needs to be installed, enabled, started, and configured, and the opposite needs to be done to `ufw`. If you're more prone to using `firewalld`, we suggest that you try doing that as it's easy and won't take much of your time.

How it works

There's a fundamental difference between `ufw` and `firewalld`, with `ufw` basically being just a frontend to `iptables`, while `firewalld` is much more dynamic, having the capability to work with zones to which we can assign various trust levels. The syntax is also different, `ufw` being a bit more namespace-based while `firewalld` requires a bit more effort in terms of rule typing. But, at the same time, `firewalld` has a D-Bus interface, which makes configuration easier in terms of applications, services, and users making configuration changes, on top of the fact that we don't have to restart the firewall for every change to take effect. It also interacts well with NetworkManager and `nmcli`, `libvirt`, Docker, and Podman, and other utilities such as fail2ban (although fail2ban works with `iptables` just as well). Sometimes it's a matter of preference; sometimes it's habits. Generally speaking, if you're more of an Ubuntu/Debian user, you're probably going to be more inclined to use `ufw`. By the same token, if you're more of a CentOS/Red Hat/Fedora/*SuSe user, you'll definitely be more inclined to use `firewalld`.

Firewalld's concept of using zones, to which we can assign network interfaces or IP addresses, is certainly useful, as it gives us much more freedom in terms of configuration. If we type in the `firewall-cmd --get-zones` command, we'll see the list of available zones at that point in time:

```
[root@client2 ~]# firewall-cmd --get-zones
block dmz drop external home internal libvirt nm-shared public trusted work
```

Figure 4.15 – Default firewalld zones

By default, `firewalld` is configured in the *deny everything, add exceptions* manner, again, for ease of use. We can use it to allow or deny connections based on ports, IP addresses, subnets, and services, which enables us to do everything we need to do, in terms of host-based firewall functionality. It also has a concept called rich rules (as shown in our examples in this recipe), that enables us to create complex rules with intricate levels of granularity. These rules can be based on source address, destination address, ports, protocols, services, port forwarding, and masquerade per subnet. They can be used for rate limiting, which allows us to set the number of accepted SSH connections to 5 or 10 per minute to make it much harder to do SSH brute force attacks on our Linux servers. All in all, it's a very well-thought-out and feature-rich firewall that's free for us to use. We just need to configure it.

`ufw`, being just a frontend of `iptables` – or, as people usually describe it, a command-line interface for `iptables` – is less feature-packed, but is definitely easier to configure, at least for the most commonly used scenarios. Its command-line interface is more human-readable (less complex) and easier to learn. Seeing that it's just a frontend for `iptables`, it's basically a user-space utility that manages Linux kernel filtering rules provided by the netfilter module stack.

Now that we have discussed some key concepts in Linux firewalling, it's time to move on to our next recipe, which is about checking open ports and connections. Let's see what that's all about.

There's more

If you need to learn more about `firewalld` and `ufw`, we recommend the following links:

- `firewall-cmd`: https://firewalld.org/documentation/man-pages/firewall-cmd.html

- A beginner's guide to `firewalld` in Linux: https://www.redhat.com/sysadmin/beginners-guide-firewalld

- `firewalld` rich language: https://firewalld.org/documentation/man-pages/firewalld.richlanguage.html

- `ufw`: https://help.ubuntu.com/community/UFW

Working with open ports and connections

Checking open ports on our local and/or remote machine is often part of security and configuration auditing processes. It's something that we use to check if we can connect to some remote ports to verify that a service works, whether a firewall is configured properly, or whether routing works – just regular, everyday tasks. Of course, it can also be a part of some hacking processes, which often start by using nmap and similar utilities to check for open ports and OS fingerprints. But, let's check how we can use utilities such as netstat, lsof, ss, and nmap to do good for our network and security.

Getting ready

Keep the client1 virtual machine powered on and let's continue using our shell. Generally speaking, if we're doing this on Ubuntu, we need to install some packages such as traceroute and nmap using apt-get:

```
apt-get -y install traceroute nmap
```

If, however, we are using CentOS, we need to use yum or dnf:

```
yum -y install traceroute nmap
```

After that, we are ready for our recipe.

How to do it

Let's first learn the usual ways of checking which ports and sockets are opened on our local Linux machine, starting with the netstat command. Yes, it's a common thing to check the routing table with netstat (the netstat -rn command), but we can also learn many more interesting details about our local Linux machine by using it in a different way. First, let's check all opened connections and ports by using the netstat -a command:

```
Proto Recv-Q Send-Q Local Address           Foreign Address         State
tcp       0      0 localhost:domain        0.0.0.0:*               LISTEN
tcp       0      0 0.0.0.0:ssh             0.0.0.0:*               LISTEN
tcp6      0      0 [::]:ssh                [::]:*                  LISTEN
udp       0      0 localhost:domain        0.0.0.0:*
raw6      0      0 [::]:ipv6-icmp          [::]:*                  7
Active UNIX domain sockets (servers and established)
Proto RefCnt Flags       Type       State         I-Node   Path
unix  2      [ ACC ]     SEQPACKET  LISTENING     29393    /run/udev/control
unix  2      [ ]         DGRAM                    40729    /run/user/1000/systemd/notify
unix  2      [ ACC ]     STREAM     LISTENING     40732    /run/user/1000/systemd/private
unix  2      [ ACC ]     STREAM     LISTENING     40740    /run/user/1000/bus
unix  2      [ ACC ]     STREAM     LISTENING     29378    @/org/kernel/linux/storage/multipathd
unix  2      [ ACC ]     STREAM     LISTENING     40741    /run/user/1000/gnupg/S.dirmngr
unix  2      [ ACC ]     STREAM     LISTENING     40742    /run/user/1000/gnupg/S.gpg-agent.browser
unix  2      [ ACC ]     STREAM     LISTENING     40743    /run/user/1000/gnupg/S.gpg-agent.extra
unix  2      [ ACC ]     STREAM     LISTENING     36658    /run/irqbalance/irqbalance833.sock
unix  2      [ ACC ]     STREAM     LISTENING     40744    /run/user/1000/gnupg/S.gpg-agent.ssh
unix  2      [ ACC ]     STREAM     LISTENING     40745    /run/user/1000/gnupg/S.gpg-agent
unix  2      [ ACC ]     STREAM     LISTENING     40746    /run/user/1000/pk-debconf-socket
unix  2      [ ACC ]     STREAM     LISTENING     40747    /run/user/1000/snapd-session-agent.socket
unix  2      [ ACC ]     STREAM     LISTENING     34767    /var/snap/lxd/common/lxd/unix.socket
unix  2      [ ACC ]     STREAM     LISTENING     39740    /var/run/vmware/guestServicePipe
unix  3      [ ]         DGRAM                    28582    /run/systemd/notify
```

Figure 4.16 – A part of the netstat -a output – the result is much longer so we stripped it a bit for formatting reasons

A lot of details are here. Let's see if we can format that a bit better. First, let's show all opened TCP ports by using the netstat -atp command:

```
root@client1:~# netstat -atp
Active Internet connections (servers and established)
Proto Recv-Q Send-Q Local Address           Foreign Address         State       PID/Program name
tcp       0      0 localhost:domain        0.0.0.0:*               LISTEN      2838/systemd-resolv
tcp       0      0 0.0.0.0:ssh             0.0.0.0:*               LISTEN      1116/sshd: /usr/sbi
tcp6      0      0 [::]:ssh                [::]:*                  LISTEN      1116/sshd: /usr/sbi
```

Figure 4.17 – netstat with the TCP port list

Then, let's show the same thing, but for opened UDP ports, by using the netstat -aup command:

```
root@client1:~# netstat -aup
Active Internet connections (servers and established)
Proto Recv-Q Send-Q Local Address           Foreign Address         State       PID/Program name
udp       0      0 localhost:domain        0.0.0.0:*                           2838/systemd-resolv
```

Figure 4.18 – netstat with the UDP port list

We can also show a subset of the information above in terms of listening ports (a port that an application or process is listening on). That's what the netstat -l command is all about:

```
root@client1:~# netstat -l
Active Internet connections (only servers)
Proto Recv-Q Send-Q Local Address           Foreign Address         State
tcp        0      0 localhost:domain        0.0.0.0:*               LISTEN
tcp        0      0 0.0.0.0:ssh             0.0.0.0:*               LISTEN
tcp6       0      0 [::]:ssh                [::]:*                  LISTEN
udp        0      0 localhost:domain        0.0.0.0:*
raw6       0      0 [::]:ipv6-icmp          [::]:*                  7
Active UNIX domain sockets (only servers)
Proto RefCnt Flags       Type       State         I-Node   Path
unix  2      [ ACC ]     SEQPACKET  LISTENING     29393    /run/udev/control
unix  2      [ ACC ]     STREAM     LISTENING     40732    /run/user/1000/systemd/private
unix  2      [ ACC ]     STREAM     LISTENING     40740    /run/user/1000/bus
unix  2      [ ACC ]     STREAM     LISTENING     29378    @/org/kernel/linux/storage/multipathd
unix  2      [ ACC ]     STREAM     LISTENING     40741    /run/user/1000/gnupg/S.dirmngr
unix  2      [ ACC ]     STREAM     LISTENING     40742    /run/user/1000/gnupg/S.gpg-agent.browser
unix  2      [ ACC ]     STREAM     LISTENING     40743    /run/user/1000/gnupg/S.gpg-agent.extra
unix  2      [ ACC ]     STREAM     LISTENING     36658    /run/irqbalance/irqbalance833.sock
unix  2      [ ACC ]     STREAM     LISTENING     40744    /run/user/1000/gnupg/S.gpg-agent.ssh
unix  2      [ ACC ]     STREAM     LISTENING     40745    /run/user/1000/gnupg/S.gpg-agent
unix  2      [ ACC ]     STREAM     LISTENING     40746    /run/user/1000/pk-debconf-socket
unix  2      [ ACC ]     STREAM     LISTENING     40747    /run/user/1000/snapd-session-agent.socket
unix  2      [ ACC ]     STREAM     LISTENING     34767    /var/snap/lxd/common/lxd/unix.socket
unix  2      [ ACC ]     STREAM     LISTENING     39740    /var/run/vmware/guestServicePipe
unix  2      [ ACC ]     STREAM     LISTENING     28585    /run/systemd/private
unix  2      [ ACC ]     STREAM     LISTENING     28587    /run/systemd/userdb/io.systemd.DynamicUse
r
unix  2      [ ACC ]     STREAM     LISTENING     29376    /run/lvm/lvmpolld.socket
unix  2      [ ACC ]     STREAM     LISTENING     29388    /run/systemd/journal/stdout
unix  2      [ ACC ]     STREAM     LISTENING     30015    /run/systemd/journal/io.systemd.journal
unix  2      [ ACC ]     STREAM     LISTENING     34766    @ISCSIADM_ABSTRACT_NAMESPACE
unix  2      [ ACC ]     STREAM     LISTENING     34764    /run/dbus/system_bus_socket
unix  2      [ ACC ]     STREAM     LISTENING     34769    /run/snapd.socket
unix  2      [ ACC ]     STREAM     LISTENING     34771    /run/snapd-snap.socket
unix  2      [ ACC ]     STREAM     LISTENING     34773    /run/uuidd/request
```

Figure 4.19 – Checking listening ports via netstat

We can do similar things with ss and lsof. Let's first use ss:

```
root@client1:~# ss -tulnp
Netid     State      Recv-Q     Send-Q         Local Address:Port                Peer Address:Port
Process
udp       UNCONN     0          0              127.0.0.53%lo:53                      0.0.0.0:*
  users:(("systemd-resolve",pid=2838,fd=12))
tcp       LISTEN     0          4096           127.0.0.53%lo:53                      0.0.0.0:*
  users:(("systemd-resolve",pid=2838,fd=13))
tcp       LISTEN     0          128                  0.0.0.0:22                      0.0.0.0:*
  users:(("sshd",pid=1116,fd=3))
tcp       LISTEN     0          128                     [::]:22                         [::]:*
  users:(("sshd",pid=1116,fd=4))
```

Figure 4.20 – Checking active connections via ss

Next on our list is `lsof`, a command that can be used to determine which files are being opened by their corresponding processes:

```
root@client1:~# lsof -nP -iTCP -sTCP:LISTEN
COMMAND      PID           USER    FD    TYPE DEVICE SIZE/OFF NODE NAME
sshd        1116           root    3u    IPv4  39729       0t0  TCP *:22 (LISTEN)
sshd        1116           root    4u    IPv6  39731       0t0  TCP *:22 (LISTEN)
systemd-r   2838 systemd-resolve  13u    IPv4  46764       0t0  TCP 127.0.0.53:53 (LISTEN)
```

Figure 4.21 – The same idea as ss, but with more details about actual commands/services using ports

The options that we used are as follows:

- `-n` for using port numbers, not port names

- `-P` for using numerical addresses, without DNS resolving

- `-iTCP -sTCP:LISTEN` to show only files that have an opened port in the TCP state `LISTEN`

Then, if we wanted to narrow that down to a specific TCP port – for example, port 22 – we could use a command such as `lsof -nP -iTCP:22 -sTCP:LISTEN`:

```
root@client1:~# lsof -nP -iTCP:22 -sTCP:LISTEN
COMMAND   PID USER    FD    TYPE DEVICE SIZE/OFF NODE NAME
sshd     1116 root    3u    IPv4  39729       0t0  TCP *:22 (LISTEN)
sshd     1116 root    4u    IPv6  39731       0t0  TCP *:22 (LISTEN)
```

Figure 4.22 – Narrowing the lsof output to TCP port 22 only

If we need to check opened ports specified by port range, `lsof` allows that, by using the `lsof -i` option. For example, let's use that on port range 22 to 1000:

```
root@client1:~# lsof -i :22-1000
COMMAND      PID           USER    FD    TYPE DEVICE SIZE/OFF NODE NAME
sshd        1116           root    3u    IPv4  39729       0t0  TCP *:ssh (LISTEN)
sshd        1116           root    4u    IPv6  39731       0t0  TCP *:ssh (LISTEN)
systemd-r   2838 systemd-resolve  12u    IPv4  46763       0t0  UDP localhost:domain
systemd-r   2838 systemd-resolve  13u    IPv4  46764       0t0  TCP localhost:domain (LISTEN)
```

Figure 4.23 – lsof by port range

Now that we've used some commands on our local machine, let's turn our attention to remote machines and discuss how to find open ports on them, as well as other information that might be necessary. For that, we are going to use the `nmap` command. Let's first use client1 (IP address `192.168.1.1`) to scan `server1` (IP address `192.168.1.254`). `server1` is just a vanilla CentOS 8 installation, as explained in the last recipe of this chapter.

First, let's do a general scan, by using the nmap `192.168.1.254` command:

```
root@client1:~# nmap 192.168.1.254
Starting Nmap 7.91 ( https://nmap.org ) at 2021-08-12 09:46 UTC
Nmap scan report for _gateway (192.168.1.254)
Host is up (0.00060s latency).
Not shown: 998 filtered ports
PORT     STATE  SERVICE
22/tcp   open   ssh
9090/tcp closed zeus-admin
MAC Address: 00:0C:29:2C:B6:5F (VMware)

Nmap done: 1 IP address (1 host up) scanned in 5.20 seconds
```

Figure 4.24 – Using nmap on a single IP address

If we wanted a bit more verbosity, we could've added the -v option before or after the IP address, as well. Still, we can see that the remote IP address has a couple of open TCP ports. Let's try to find some more information, by initiating nmap with the -A option (OS information scan):

```
root@client1:~# nmap -A 192.168.1.254
Starting Nmap 7.91 ( https://nmap.org ) at 2021-08-12 09:49 UTC
Nmap scan report for _gateway (192.168.1.254)
Host is up (0.0011s latency).
Not shown: 998 filtered ports
PORT     STATE  SERVICE  VERSION
22/tcp   open   ssh      OpenSSH 8.0 (protocol 2.0)
| ssh-hostkey:
|   3072 c2:0a:9a:73:6a:fb:65:a7:91:f0:ea:cb:e8:af:d3:e3 (RSA)
|   256 ac:7c:a4:1e:81:1e:ec:4a:80:8d:4b:42:7c:e9:ec:23 (ECDSA)
|_  256 fa:35:e0:aa:37:53:91:e8:36:bd:07:45:53:52:23:25 (ED25519)
9090/tcp closed zeus-admin
MAC Address: 00:0C:29:2C:B6:5F (VMware)
Aggressive OS guesses: Linux 5.1 (98%), Linux 3.10 - 4.11 (97%), Linux 3.2 - 4.9 (96%), Linux 3.16 -
 4.6 (95%), Linux 5.0 - 5.4 (95%), Linux 2.6.32 - 3.13 (95%), Linux 5.4 (94%), Linux 4.10 (93%), Lin
ux 2.6.22 - 2.6.36 (93%), Linux 2.6.39 (93%)
No exact OS matches for host (test conditions non-ideal).
Network Distance: 1 hop

TRACEROUTE
HOP RTT      ADDRESS
1   1.08 ms  _gateway (192.168.1.254)

OS and Service detection performed. Please report any incorrect results at https://nmap.org/submit/
.
Nmap done: 1 IP address (1 host up) scanned in 9.52 seconds
```

Figure 4.25 – More detailed version of the previous nmap session

We can see even more details on this output. If we just wanted to do OS fingerprinting, we could've used the -O option:

```
root@client1:~# nmap -O 192.168.1.254
Starting Nmap 7.91 ( https://nmap.org ) at 2021-08-12 09:52 UTC
Nmap scan report for _gateway (192.168.1.254)
Host is up (0.0013s latency).
Not shown: 998 filtered ports
PORT     STATE  SERVICE
22/tcp   open   ssh
9090/tcp closed zeus-admin
MAC Address: 00:0C:29:2C:B6:5F (VMware)
Device type: general purpose
Running: Linux 5.X
OS CPE: cpe:/o:linux:linux_kernel:5.1
OS details: Linux 5.1
Network Distance: 1 hop

OS detection performed. Please report any incorrect results at https://nmap.org/submit/ .
Nmap done: 1 IP address (1 host up) scanned in 6.99 seconds
```

Figure 4.26 – nmap OS fingerprinting

We could also scan for various other things, such as the following:

- Specific TCP ports – nmap -p T:9090,22 192.168.1.254
- Specific UDP ports – nmap -sU 53 192.168.1.254
- Scan port range – nmap -p 22-2000 192.168.1.254
- Find remote host service versions – nmap -sV 192.168.1.254
- Scan a subnet – nmap 192.168.1.*
- Scan multiple hosts – nmap 192.168.1.252,253,254
- Scan a complete IP range – nmap 192.168.1.1-254

Let's now discuss how these four utilities work and wrap this recipe up.

How it works

netstat, ss, and lsof are kind of similar, yet have their differences. The usual way in which people use netstat is just to check their routing table. But, having said that, by default, netstat is a tool that gives us a list of opened TCP sockets/UDP connections on the network layer. lsof, on the other hand, lists open files (kernel-level functionality), but it's also capable of determining which processes are using those opened files. Keep in mind that, in Unix operating systems, almost everything is a file, which also includes objects such as network sockets. As such, lsof is often used when dealing with security aspects of our Linux systems, as it obviously gives many more technical details when compared to netstat.

ss, as an alternative to netstat, can be used to work with network information and statistics, which makes it kind of similar to netstat. It can be used to get details about network connections, sockets, statistical data, TCP state filtering, connections to and from specific IP addresses, and so on. And, not to be forgotten, ss is quite a bit simpler to use, and when you compare man page sizes of netstat and ss, you'll see the difference there as well.

nmap, on the other hand, is completely different from all of these commands. It's a tool that's aimed much more broadly in terms of functionality – it can scan both local and remote hosts, domains, IP ranges, and ports. It's a regular network scanner, with all the good and the bad that comes with it, as people both love and dislike it – love using it, dislike being the target of it. It works by establishing connections to remote IP addresses and ports, sending them information and gathering output from them to get information. Therefore, it's a perfect tool to do security scanning and auditing as it's able to find open ports and report the fact that there are open ports. It's also heavily used to search for certain security issues.

There's more

If you need to learn more about netstat, lsof, ss, and nmap, make sure that you check the following links:

- nmap documentation: https://nmap.org/docs.html
- netstat man page: https://man7.org/linux/man-pages/man8/ netstat.8.html
- lsof man page: https://man7.org/linux/man-pages/man8/ lsof.8.html
- ss man page: https://man7.org/linux/man-pages/man8/ss.8.html

Configuring /etc/hosts and DNS resolving

Name resolution is an essential part of any operating system, specifically its networking stack. Generally speaking, operating systems have multiple different ways of making DNS queries – usually, it involves some kind of hosts file, caches, and – of course – network interface configuration. Let's go through the configuration capabilities of /etc/hosts and see how that fits in the grand scheme of name resolution.

Getting ready

Keep the CLI1 virtual machine powered on and let's discuss how to work with name resolution in general, using /etc/hosts (a file that we can fill with hostnames and IP addresses for local resolving) and /etc/resolv.conf (a file that determines which DNS servers are being used for network resolving, and which domain is the Linux server a part of) as integral parts of that process. When editing /etc/hosts or /etc/resolv.conf, we have to be logged in as root or use sudo, as this is a system-wide operation that's only allowed to administrative users. The way in which the name resolution process works changed years ago as systemd took over from init and upstart, and introduced a service called systemd-resolved. On top of this, the configuration is different on Ubuntu when compared to CentOS. So, let's dig into all of that and explain what's going on.

How to do it

Let's first take care of Ubuntu, then we'll switch to CentOS. This is the default /etc/resolv.conf file from our Ubuntu CLI1 machine:

```
# This file is managed by man:systemd-resolved(8). Do not edit.
#
# This is a dynamic resolv.conf file for connecting local clients to the
# internal DNS stub resolver of systemd-resolved. This file lists all
# configured search domains.
#
# Run "resolvectl status" to see details about the uplink DNS servers
# currently in use.
#
# Third party programs should typically not access this file directly, but only
# through the symlink at /etc/resolv.conf. To manage man:resolv.conf(5) in a
# different way, replace this symlink by a static file or a different symlink.
#
# See man:systemd-resolved.service(8) for details about the supported modes of
# operation for /etc/resolv.conf.

nameserver 127.0.0.53
options edns0 trust-ad
```

Figure 4.27 – Default /etc/resolv.conf file

As we can see, this file is mostly commented out (the # sign in config files equals a Unix shell-style comment, so these lines are omitted in terms of configuration). We only have two configuration lines, which are a by-product of running the systemd-resolved service (a local service that provides resolving capabilities for DNS, DNS over TLS, DNSSEC, mDNS, and so on), as well as using the netplan service by default:

```
nameserver 127.0.0.53
options edns0 trust-ad
```

There are two approaches to `resolv.conf` configuration:

- We say that we want to stick with systemd-resolved and configure our system that way (and `127.0.0.53` is actually the loopback IP address that systemd-resolved binds to).

- We say that we don't want systemd-resolved and we want to go back to the *old way* of configuring our system, which means installing a package called `resolvconf`. That will give us the capability to configure `/etc/resolv.conf` and `/etc/hosts` as they were always configured and not rely on systemd-resolved making changes to `/etc/resolv.conf` on the fly (most of us usually don't want this).

Let's start with the first approach and then move to the second approach as a lot of us Linux administrators are more prone to using our old-school ways and we find it easier for things to be configured the way they have always been configured since the dawn of Unix time.

If we are using systemd-resolved, we need to mention a couple of files. The first file that we need to mention is `/run/systemd/resolve/stub-resolv.conf` – this is a file that's actually linked to `/etc/resolv.conf` when systemd-resolved is being used. This file is used for maintaining compatibility with old Linux programs that were exclusively using the old way (`/etc/resolv.conf`, `/etc/hosts`) to get access to name resolution information. If we want to permanently set DNS servers to be used, we need to do it via `systemd`. So, let's go to the second file that we need to discuss. It's located in the `/etc/systemd` directory, and it's called `resolved.conf`. At the beginning of this file, there's a `[Resolve]` section that's completely commented out. Let's change it to this:

```
[Resolve]
DNS = 8.8.8.8 8.8.4.4
FallBackDNS = 1.1.1.1
Domains =domain.local
```

The first and second lines set the main and fallback DNS addresses, while the third line sets the default domain that we're querying.

After we do this change, we need to restart the `systemd-resolved` service, which we can do with the following command:

```
systemctl restart systemd-resolved
```

We can check if our changes have been applied by using `systemd-resolve --status`, which should, in accordance with our changes, give us output similar to this one:

```
root@cli1:~# systemd-resolve --status
Global
          LLMNR setting: no
   MulticastDNS setting: no
     DNSOverTLS setting: no
         DNSSEC setting: no
       DNSSEC supported: no
            DNS Servers: 8.8.8.8
                         8.8.4.4
   Fallback DNS Servers: 1.1.1.1
             DNS Domain: domain.local

Link 2 (ens33)
         Current Scopes: DNS
   DefaultRoute setting: yes
          LLMNR setting: yes
   MulticastDNS setting: no
     DNSOverTLS setting: no
         DNSSEC setting: no
       DNSSEC supported: no
            DNS Servers: 8.8.8.8
                         10.0.0.254
                         194.146.109.224
             DNS Domain: domain.local
```

Figure 4.28 – Checking the systemd-resolved status

Let's now check how the DNS cache works – for example, we type in the following commands:

```
nslookup index.hr
nslookup planetf1.com
```

We did this so that we can check the DNS cache, as the DNS cache first needs to be filled with some data at least. If we want to check the state of the `systemd-resolved` cache, we can do it with two commands:

```
killall -USR1 systemd-resolved
journalctl -u systemd-resolved > cache.txt
```

The first command doesn't kill `systemd-resolved` but tells it to write available entries in the DNS cache. The second command exports entries to a file called `cache.txt` (it can be called whatever we want). When we check the content of that file for the string CACHE, we're going to see entries similar to this:

```
Aug 08 12:47:34 cli1 systemd-resolved[4411]: CACHE:
Aug 08 12:47:34 cli1 systemd-resolved[4411]:            index.hr IN A 104.19.141.57
Aug 08 12:47:34 cli1 systemd-resolved[4411]:            index.hr IN A 104.19.140.57
Aug 08 12:47:34 cli1 systemd-resolved[4411]:            index.hr IN A 104.19.139.57
Aug 08 12:47:34 cli1 systemd-resolved[4411]:            index.hr IN A 104.19.138.57
Aug 08 12:47:34 cli1 systemd-resolved[4411]:            index.hr IN A 104.19.142.57
Aug 08 12:47:34 cli1 systemd-resolved[4411]: [Scope protocol=dns]
Aug 08 12:47:34 cli1 systemd-resolved[4411]: CACHE:
Aug 08 12:47:34 cli1 systemd-resolved[4411]:            planetf1.com IN A 65.9.71.3
Aug 08 12:47:34 cli1 systemd-resolved[4411]:            planetf1.com IN A 65.9.71.1
Aug 08 12:47:34 cli1 systemd-resolved[4411]:            planetf1.com IN A 65.9.71.4
Aug 08 12:47:34 cli1 systemd-resolved[4411]:            planetf1.com IN A 65.9.71.3
```

Figure 4.29 – Checking the DNS cache

This is correct – on our testing system, those are two entries that we searched for by using nslookup. If we want to flush the DNS cache, we can use the following command:

```
resolvectl flush-caches
```

If you notice errors with DNS violations in the file, there was a problem during the system installation or upgrade – one that didn't set a symbolic link to resolv.conf properly. As a result of that problem, the symbolic link was created to the wrong file (stub-resolv.conf) instead of the actual file (/run/systemd/resolve/resolv.conf). We can mitigate that issue by using the following commands:

```
mv /etc/resolv.conf /etc/resolv.conf.old
ln -s /run/systemd/resolve/resolv.conf /etc/resolv.conf
systemctl restart systemd-resolved
```

Now, let's try the second approach, which is quite a bit more simple. So, if we wanted to get rid of all of this systemd-resolved configuration and just use a good old administration process via resolv.conf without all of this additional hassle, we could do that easily. So, let's first install the necessary package:

```
apt-get -y install resolvconf
systemctl stop systemd-resolved
systemctl disable systemd-resolved
systemctl mask systemd-resolved
```

Next, let's do a bit of configuration. Let's open the /etc/resolv.conf file and make it look like this (the commented part is not important, start with the nameserver part):

```
# Dynamic resolv.conf(5) file for glibc resolver(3) generated by resolvconf(8)
#     DO NOT EDIT THIS FILE BY HAND -- YOUR CHANGES WILL BE OVERWRITTEN
# 127.0.0.53 is the systemd-resolved stub resolver.
# run "systemd-resolve --status" to see details about the actual nameservers.

nameserver 8.8.8.8
nameserver 8.8.4.4
nameserver 1.1.1.1
search domain.local
```

Figure 4.30 – The resolv.conf configuration

Let's check if this configuration works:

```
root@cli1:~# nslookup packtpub.com
Server:         8.8.8.8
Address:        8.8.8.8#53

Non-authoritative answer:
Name:    packtpub.com
Address: 104.22.0.175
Name:    packtpub.com
Address: 104.22.1.175
Name:    packtpub.com
Address: 172.67.31.83
Name:    packtpub.com
Address: 2606:4700:10::6816:1af
Name:    packtpub.com
Address: 2606:4700:10::6816:af
Name:    packtpub.com
Address: 2606:4700:10::ac43:1f53
```

Figure 4.31 – Checking if DNS resolution works

No problem whatsoever, right? Of course, we used 8.8.8.8, 8.8.4.4, and 1.1.1.1 as examples for DNS servers here – this needs to be configured so that it's valid for the environment where our Linux server is actually running.

Working with the DNS cache requires a bit of extra effort. First, we need to deploy two additional packages – nscd (the service that does the caching), and binutils (this package contains a command called strings, which we'll use to check string content in a binary file):

```
apt-get -y install nscd binutils
strings /var/cache/nscd/hosts
```

The output of the second command should look similar to this:

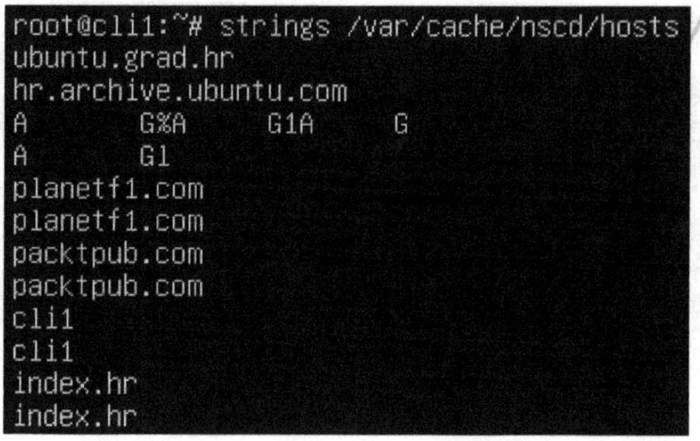

```
root@cli1:~# strings /var/cache/nscd/hosts
ubuntu.grad.hr
hr.archive.ubuntu.com
A        G%A       G1A       G
A        G1
planetf1.com
planetf1.com
packtpub.com
packtpub.com
cli1
cli1
index.hr
index.hr
```

Figure 4.32 – Checking the nscd cache

If we need to clear the nscd hosts cache, we can use the following commands:

```
nscd -i hosts
```

or

```
systemctl restart nscd
```

The first one just clears the hosts table, while the second one restarts the nscd service and, as a part of the process, clears the hosts table.

And that brings us to the hosts table, and – luckily – it works the same on all Linux distributions. If we're in a situation where we just need to add some resolving capabilities without actually building up a DNS server via BIND, dnsmasq, or anything similar to that, using the hosts table seems like a reasonably simple thing to do. Let's say that we need to use temporary resolution for the following two hosts:

server1.domain.local

server2.domain.local

Let's assume that these two servers' IP addresses are 192.168.0.101 and 192.168.0.102. We'd add these entries to the /etc/hosts file by editing it and adding these entries to the bottom of the file:

192.168.0.101 server1.domain.local

192.168.0.102 server2.domain.local

So, our `/etc/hosts` file should look like this:

```
127.0.0.1 localhost
127.0.1.1 cli1

# The following lines are desirable for IPv6 capable hosts
::1      ip6-localhost ip6-loopback
fe00::0 ip6-localnet
ff00::0 ip6-mcastprefix
ff02::1 ip6-allnodes
ff02::2 ip6-allrouters

192.168.0.101    server1.domain.local
192.168.0.102    server2.domain.local
```

Figure 4.33 – /etc/hosts file with additions

If we now use a command such as `ping` to check if these hosts are alive, we will get the following result:

```
root@cli1:~# ping server1.domain.local
PING server1.domain.local (192.168.0.101) 56(84) bytes of data.
^C
--- server1.domain.local ping statistics ---
2 packets transmitted, 0 received, 100% packet loss, time 1026ms

root@cli1:~# ping server2.domain.local
PING server2.domain.local (192.168.0.102) 56(84) bytes of data.
^C
--- server2.domain.local ping statistics ---
1 packets transmitted, 0 received, 100% packet loss, time 0ms
```

Figure 4.34 – ping not working

The `^C` character visible in this output is due to the fact that we used *Ctrl + C* to stop the ping process as these hosts don't actually exist on our network. But that's beside the point – the point of this was to test whether the name resolution works. In other words, does `server1` and `server2.domain.local` resolution work? And it does – we can clearly see that the `ping` command is trying to ping their IP addresses.

We need to briefly discuss the way in which CentOS does these things, as it's a bit different from what Ubuntu does. By default, the latest couple of generations of CentOS use NetworkManager as the default service to configure a network. As a result, `/etc/resolv.conf` gets configured by NetworkManager by default, which is very important to note, especially in the most common use case – when our CentOS machine gets its IP address from the DHCP server. What happens if we need to configure custom DNS servers and we don't want to use the DNS servers that we got from our DHCP server?

There are two basic ways to make sure that everything's configured correctly in CentOS:

- To configure everything via interface files
- To configure everything after the fact, by using the nmcli command

Using configuration files is a hassle here, so let's just do the second thing – configure our DNS entries by using the nmcli command. Let's say that we want to assign 8.8.8.8, 8.8.4.4, and 1.1.1.1 as DNS servers for our CentOS server. Let's check our network interface name first:

```
nmcli con show
```

Our system tells us that it's using the ens33 network interface. Let's modify its settings by typing in the following commands:

```
nmcli con mod ens33 ipv4.ignore-auto-dns yes
nmcli con mod ens33 ipv4.dns "8.8.8.8 8.8.4.4 1.1.1.1"
systemctl restart NetworkManager
```

The key aspect of this configuration is the first line – we're basically telling NetworkManager to quit automatically using the DNS server that it gets from the DHCP server. If we didn't want that, we could've just omitted that specific line.

If we check the contents of our /etc/resolv.conf file, it should now look like this:

Figure 4.35 – /etc/resolv.conf configured correctly

And that's a wrap in terms of configuration – using both Ubuntu and CentOS. Let's now focus on how all of this works *under the hood*.

How it works

There are two concepts that we need to dig into and explain. We need to understand how systemd-resolved works, and, of course, the opposite – how everything works when we remove systemd-resolved from the administrative equation, if you will. Having in mind that there was Linux before systemd and name resolution before systemd-resolved, let's start by explaining how the *old method* (pre-systemd-resolved) worked.

The core concept was called the **Name Service Switch** (**NSS**). The basic idea behind NSS was to connect to various mechanisms – databases, files, services – to provide various services. Services such as authentication (`/etc/passwd`, `/etc/shadow`, and `/etc/group`), network configuration, and, of course, services such as name resolution (`/etc/hosts` and so on). Our focus will be solely on name resolution, which is why we need to discuss a configuration file, `/etc/nsswitch.conf`. Specifically, we will ignore all of the configuration options in that file and focus on one configuration line, which is usually similar to this:

```
hosts:        files dns
```

This configuration line tells our name resolution system *how* to do its job. The `files` option means *check file /etc/hosts*, while the `dns` option means just that – use other network name resolution methods. But the important thing about this line is the *order*, which clearly states *files first, dns second*. This is the reason why – by default – Linux first checks the contents of the `/etc/hosts` file, and then starts issuing network name resolution calls (for example, `nslookup`) to get to the IP address of some server that we're trying to communicate with. We also have capabilities to store these entries in a database, and we can force NSS to access it to read the necessary data. For example, 20 years ago, when Active Directory and other LDAP-based directories weren't used so often, we used to use NIS/NIS+ a lot – to store user and similar data. We were also able to store host data in NIS/NIS+ databases (`hosts.byname` and `hosts.byaddr`). These maps were basically forward and reverse DNS tables, stored within an external service. That's why we can use the configuration option `db` in `nsswitch.conf`, although pretty much nobody uses that nowadays.

When systemd took over name resolution (`systemd-resolved`), things changed, as we described in our last recipe. The whole point of `systemd-resolved` is to be able to better integrate with systemd and to offer support for some use cases that were – realistically – complicated without it. Stuff such as VPN connections, especially corporate ones, were a constant source of problems when using the old-style configuration. `systemd-resolved` tries to get around that stack of problems (and others) by introducing the capability to do split DNS, which is implemented by using DNS routing domains as a way of determining which DNS requests we're actually making. Please don't mistake this for IP-based ideas of subnet routing, VLAN routing, or anything of the sort – those are completely different concepts, based on completely different ideas. We're specifically talking about the DNS routing domain, which is nothing more than a term saying *let's determine which DNS server should be contacted for correct information about your DNS query*. This has nothing to do with the IP aspect of it, which is handled by using standard routing methods.

Having split DNS is nothing new – it's something that a lot of us have been using for a decade or two. In short, split DNS means having some DNS servers assigned to internal connections and other DNS servers assigned to external connections. From an enterprise standpoint, if we connect via VPN connection to our workplace, a part of our DNS queries is aimed at internal infrastructure, while the other part should be headed to the external DNS servers hosted on the internet. Being able to implement this scenario in Linux also isn't something new – we could've easily done this with `BIND` more than a decade ago. But a way to do this as tightly integrated and as automatically as possible, especially on the client side – which is what `systemd-resolved` does – is actually something new.

Let's imagine for a second that we have a Linux VPN server that we're connecting to by using a Linux machine as a VPN client. Let's say that these two systems both have multiple network interfaces in different subnets (a couple of physical network cards and a wireless network adapter for the VPN client). When we connect from our VPN client to the VPN server, how is the VPN client going to determine where to send DNS queries? Yes, it's going to use `resolv.conf`, but still, `resolv.conf` and `systemd-resolved` need to be configured correctly so that a name resolution request gets sent to the correct DNS server. If we have multiple subnets, and multiple domains (a larger enterprise, for example), things can get messy very quickly. This situation gets taken care of via NetworkManager/netplan's interaction with `systemd-resolved`. By using this interaction, we can have different DNS servers assigned to different network interfaces, that are assigned to multiple different domains. And that's a pretty smart way of dealing with potential VPN client problems.

There's more

If we need to learn more about network name resolution, we can check the following links:

- What is DNS?: `https://www.cloudflare.com/learning/dns/what-is-dns/`

- What is DNS?: `https://aws.amazon.com/route53/what-is-dns/`

- NSCD man page, chapter 8: `https://linux.die.net/man/8/nscd`

- `systemd-resolved` man page: `http://manpages.ubuntu.com/manpages/bionic/man8/systemd-resolved.service.8.html`

- `resolved.conf` man page: `https://www.freedesktop.org/software/systemd/man/resolved.conf.html`

Using network diagnostic tools

Diagnosing problems with network connections is an everyday job for a seasoned system engineer. It doesn't necessarily happen because we have problems in our own network, it can be other factors. For example, sometimes our local network works, while the internet connection doesn't. Or, even worse, customers report that some of them are able to access the internet, while some others can't. How do we approach these situations and which tools should we use? That's what we will talk about in this recipe. So, get ready to talk about `ping`, `route`, `netstat`, `tracepath`, and similar commands – that's what they're there for!

Getting ready

Let's install a CentOS virtual machine called `server1` and use our existing clients (an Ubuntu virtual machine called client1 and a CentOS virtual machine called client2) to work on this recipe. We are going to use client1 to simulate a situation where the server on our local network wants to access internal resources and/or the internet by using `server1` as a default gateway. We are going to use client2 to simulate a situation where our local client or wireless client wants to access internal resources and/or the internet by using `server1` as a default gateway. In order for us to be able to do that, we'll temporarily add another network interface to client2, so that we can have two network interfaces in two different subnets to simulate problems in our scenario. The `server1` virtual machine is just going to be a standard CentOS installation, but with four network interfaces. In our scenario, `server1`'s `ens33` network interface is going to be an external network interface, while network interfaces `ens37`, `ens38`, and `ens39` are going to be internal network interfaces.

How to do it

Let's create a scenario here so that we can go through the whole process. For example, our colleagues from the company that we work for are reporting that they have problems accessing both internal resources (the company network) and external resources (the internet). The company that we're discussing has multiple network subnets:

- `192.168.1.0/24` – This one is used for all of the server machines; we'll call this connection profile `network1` when we configure it via `nmcli`.

- `192.168.2.0/24` – This one is used for all of the client machines; we'll call this connection profile `network2` when we configure it via `nmcli`.

- `192.168.3.0/24` – This one is used for company wireless; we'll call this connection profile `network3` when we configure it via `nmcli`.

The fourth network interface of our machine is going to act as our internet connection. As we mentioned in our second recipe in this chapter (*Using firewalld and ufw*), let's configure that virtual machine so that it allows connectivity for all three of these subnets to the internet and work from there.

The first step will obviously be to allow internet access for these three subnets. Let's do that in the simplest fashion, by using `firewalld`. Specifically, we'll do that by adding these interfaces to the public zone. So, we need a couple of standard commands and configuration steps on `server1`:

```
echo "1" > /proc/sys/net/ipv4/ip_forward
nmcli connection add con-name network1 ifname ens37 type
ethernet ip4 192.168.1.254/24
nmcli connection add con-name network2 ifname ens38 type
ethernet ip4 192.168.2.254/24
nmcli connection add con-name network3 ifname ens39 type
ethernet ip4 192.168.3.254/24
```

If our configuration is correct, when we type in the `nmcli con show` command, we should have something like this (depending on how we configured our external network on `ens33` – and in our virtual machine, it's using the `192.168.159.0/24` network):

```
[root@server1 ~]# nmcli con show
NAME        UUID                                    TYPE      DEVICE
ens33       2f2eeb51-a6bc-4237-959d-2dd1bf5dc9f2    ethernet  ens33
network1    5ec4e8f8-8ef5-4739-85c7-8e167a57c084    ethernet  ens37
network2    ada56b63-9f2f-46a0-b62b-14ccfa951ff3    ethernet  ens38
network3    1b41e39a-d8ed-46c0-8d65-b0774d677ac5    ethernet  ens39
```

Figure 4.36 – Checking our NM connection setup

Also, if we check routing information by using the `ip route` command, we should get something similar to this:

```
[root@server1 ~]# ip route show
default via 192.168.159.2 dev ens33 proto dhcp metric 100
192.168.1.0/24 dev ens37 proto kernel scope link src 192.168.1.254 metric 101
192.168.2.0/24 dev ens38 proto kernel scope link src 192.168.2.254 metric 102
192.168.3.0/24 dev ens39 proto kernel scope link src 192.168.3.254 metric 103
192.168.159.0/24 dev ens33 proto kernel scope link src 192.168.159.137 metric 100
```

Figure 4.37 – Checking our routes

So, we have our three subnets, the routes are configured accordingly, now we need to configure `server1` to act as a router. Let's type in the following commands to set our interfaces to specific zones:

```
nmcli connection modify ens33 connection.zone public
nmcli connection modify network1 connection.zone public
nmcli connection modify network2 connection.zone public
nmcli connection modify network3 connection.zone public
firewall-cmd --zone=public --add-masquerade --permanent
firewall-cmd --reload
firewall-cmd --list-all
```

When we type in the last command, we should get output similar to this:

```
[root@server1 ~]# firewall-cmd --list-all
public (active)
  target: default
  icmp-block-inversion: no
  interfaces: ens33 ens37 ens38 ens39
  sources:
  services: cockpit dhcpv6-client ssh
  ports:
  protocols:
  masquerade: yes
  forward-ports:
  source-ports:
  icmp-blocks:
  rich rules:
```

Figure 4.38 – The firewall-cmd --list-all output

On client1, we need to do a bit of reconfiguration as well, as it was initially set up to use DHCP to get the IP address. First, let's install the `traceroute` package by typing in the following command:

```
apt-get -y install traceroute
```

After that, let's configure this Linux virtual machine so that its IP address is `192.168.1.1/24` and apply that configuration. First, we need to edit netplan's configuration file. For simplicity reasons, let's just use the default configuration file, `/etc/netplan/00-installer-config.yaml`. It needs to have the following content applied via the `netplan apply` command:

```
root@client1:/etc/netplan# cat 00-installer-config.yaml
# This is the network config written by 'subiquity'
network:
  ethernets:
    ens33:
      dhcp4: no
      dhcp6: no
      addresses: [192.168.1.1/24]
      gateway4: 192.168.1.254
      nameservers:
        addresses: [8.8.8.8, 8.8.4.4]
  version: 2

root@client1:/etc/netplan# netplan apply
root@client1:/etc/netplan# _
```

Figure 4.39 – The netplan configuration file

Let's now test if internet access from this machine works. As noted on the screenshot shown previously, we're using server1 as the default gateway (192.168.1.254):

```
root@client1:~# ping packtpub.com
PING packtpub.com (104.22.0.175) 56(84) bytes of data.
64 bytes from 104.22.0.175 (104.22.0.175): icmp_seq=1 ttl=127 time=39.2 ms
64 bytes from 104.22.0.175 (104.22.0.175): icmp_seq=2 ttl=127 time=47.2 ms
64 bytes from 104.22.0.175 (104.22.0.175): icmp_seq=3 ttl=127 time=42.3 ms
64 bytes from 104.22.0.175 (104.22.0.175): icmp_seq=4 ttl=127 time=48.4 ms
64 bytes from 104.22.0.175 (104.22.0.175): icmp_seq=5 ttl=127 time=47.5 ms
^C
--- packtpub.com ping statistics ---
5 packets transmitted, 5 received, 0% packet loss, time 4014ms
rtt min/avg/max/mdev = 39.212/44.930/48.367/3.558 ms
root@client1:~# _
```

Figure 4.40 – Checking the configuration works

So, connectivity works. Let's now configure client2. Our CentOS virtual machine called client2 has a network interface called ens39. Let's set it up so that it's a part of the network2 subnet (we defined that subnet on server1). Let's say client2 is going to temporarily use 192.168.2.2/24 as its IP address:

```
nmcli connection add con-name network2 ifname ens39 type
ethernet ipv4.address 192.168.2.2/24 gateway 192.168.2.254
ipv4.dns 8.8.8.8,8.8.4.4
nmcli con reload network2
```

We previously configured `server1` to act as a default gateway, and, as a result, `client1` and `client2` can happily use it as a default gateway and access the external network. We can easily test that by using `ping`. Let's use `client2` as an example:

```
[root@client2 ~]# ping packtpub.com
PING packtpub.com (104.22.1.175) 56(84) bytes of data.
64 bytes from 104.22.1.175 (104.22.1.175): icmp_seq=1 ttl=127 time=7.59 ms
64 bytes from 104.22.1.175 (104.22.1.175): icmp_seq=2 ttl=127 time=7.48 ms
^C
--- packtpub.com ping statistics ---
2 packets transmitted, 2 received, 0% packet loss, time 1002ms
rtt min/avg/max/mdev = 7.475/7.532/7.590/0.104 ms
```

Figure 4.41 – Checking the configuration works after configuration changes

Now that we have verified that everything is configured correctly, let's now check a few different scenarios that might require additional network troubleshooting:

- `ping` to an external host is not working, but external network access works.

- External network access is not working.

- Can't route between two subnets.

- Name resolution not working properly.

Let's start with the first scenario. Usually, this is a firewall configuration setting (we're not calling it a problem on purpose). Let's first ping a site that we want to access:

```
[root@client2 ~]# ping packtpub.com
PING packtpub.com (104.22.0.175) 56(84) bytes of data.
From _gateway (192.168.2.254) icmp_seq=1 Packet filtered
From _gateway (192.168.2.254) icmp_seq=2 Packet filtered
From _gateway (192.168.2.254) icmp_seq=3 Packet filtered
From _gateway (192.168.2.254) icmp_seq=4 Packet filtered
^C
--- packtpub.com ping statistics ---
4 packets transmitted, 0 received, +4 errors, 100% packet loss, time 3055ms
```

Figure 4.42 – Scenario start – ping doesn't work

At the same time, if we try to browse `packtpub.com` from our web browser, that works without any problems:

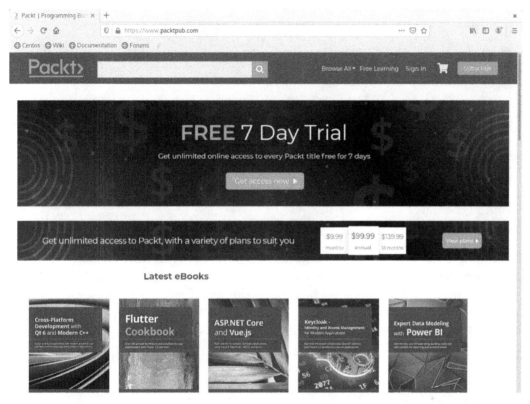

Figure 4.43 – It works in the browser

This type of problem is common – and from the output of `ping`, we can see that the firewall that we're passing on the way from our client2 to `packtpub.com` is filtering `ping` (**ICMP**, or **Internet Control Message Protocol**) traffic. This is nothing to be worried about, although it might be confusing. We need to keep in mind that ping/ICMP traffic has nothing to do with HTTP(S)/TCP traffic and that these protocols can be filtered separately. This is exactly what was done here – ping/ICMP traffic was filtered, while HTTP(S)/TCP traffic wasn't.

Let's pile up additional complexity now and go through a scenario where external network access isn't available. Let's try pinging one of Google's DNSes from client2 and `server1`, just to see the symptoms:

```
[root@client2 ~]# ping 8.8.8.8 -c 5
PING 8.8.8.8 (8.8.8.8) 56(84) bytes of data.
From 172.20.0.63 icmp_seq=3 Destination Host Unreachable
From 172.20.0.63 icmp_seq=5 Destination Host Unreachable

--- 8.8.8.8 ping statistics ---
5 packets transmitted, 0 received, +2 errors, 100% packet loss, time 4119ms

[root@server1 ~]# ping 8.8.8.8 -c 5
PING 8.8.8.8 (8.8.8.8) 56(84) bytes of data.
From 172.20.0.63 icmp_seq=3 Destination Host Unreachable
From 172.20.0.63 icmp_seq=5 Destination Host Unreachable

--- 8.8.8.8 ping statistics ---
5 packets transmitted, 0 received, +2 errors, 100% packet loss, time 4122ms
```

Figure 4.44 – External network access doesn't work

It's one thing if the network client (client2) can't get to the external network. It's a completely different thing if the default gateway (in our case, `server1`) can't get to the external network. That points to a bigger problem, and if we didn't touch the firewall configuration and other network devices, it's probably some kind of a problem either with connectivity to the **Internet Service Provider** (**ISP**), or something on the ISP's end.

We could do a bit more detective work by using additional tools, such as **traceroute** or **tracepath**:

```
[root@client2 ~]# traceroute 8.8.8.8
traceroute to 8.8.8.8 (8.8.8.8), 30 hops max, 60 byte packets
 1  _gateway (192.168.2.254)  1.193 ms  1.090 ms  1.011 ms
 2  192.168.159.2 (192.168.159.2)  0.954 ms  0.886 ms  0.829 ms
 3  * * *
 4  * * *
 5  * * *
 6  *^C
[root@client2 ~]# tracepath 8.8.8.8
 1?: [LOCALHOST]                        pmtu 1500
 1:  _gateway                                         1.255ms
 1:  _gateway                                         0.690ms
 2:  192.168.159.2                                    0.813ms
 3:  no reply
 4:  no reply
 5:  no reply
 6:  no reply
```

Figure 4.45 – Further verification that external network access doesn't work

If we are using external DNS servers, we could even use the `nslookup`, `host`, or `dig` commands to almost conclusively determine that the problem lies with internet access, not our client or server:

```
[root@server1 ~]# cat /etc/resolv.conf
# Generated by NetworkManager
nameserver 8.8.8.8
nameserver 8.8.4.4
nameserver 1.1.1.1
[root@server1 ~]# nslookup packtpub.com
```

Figure 4.46 – resolv.conf is configured correctly; DNS name resolution doesn't work

Let's say that the problem was that a cable connecting from our `server1` to the ISP router broke down. When we change that cable, `ping` should work perfectly, as shown here:

```
[root@server1 ~]# ping 8.8.8.8 -c 5
PING 8.8.8.8 (8.8.8.8) 56(84) bytes of data.
64 bytes from 8.8.8.8: icmp_seq=1 ttl=128 time=70.6 ms
64 bytes from 8.8.8.8: icmp_seq=2 ttl=128 time=74.6 ms
64 bytes from 8.8.8.8: icmp_seq=3 ttl=128 time=72.3 ms
64 bytes from 8.8.8.8: icmp_seq=4 ttl=128 time=68.2 ms
64 bytes from 8.8.8.8: icmp_seq=5 ttl=128 time=63.2 ms

--- 8.8.8.8 ping statistics ---
5 packets transmitted, 5 received, 0% packet loss, time 4014ms
```

Figure 4.47 – The connection works again

Let's now check a scenario where we can't go from one subnet to another subnet. We're going to use network1 and network2 as an example – so, we're going to use client2 (`192.168.2.2/24`) to try to access client1 (`192.168.1.1/24`). As these two hosts are *not* a part of the same Layer 2 network, we have to have some kind of mechanism to *route* traffic between them. Let's check if that routing configuration works properly:

```
[root@client2 ~]# traceroute 192.168.1.1
traceroute to 192.168.1.1 (192.168.1.1), 30 hops max, 60 byte packets
 1  _gateway (192.168.2.254)  0.627 ms  0.556 ms  0.522 ms
```

Figure 4.48 – Routing across subnets working

As previously configured, we allowed the forwarding of traffic between all of the networks on `server1`. We achieved that by allowing masquerading and putting all interfaces in the public `firewalld` zone. Sometimes when we configure our routing devices, we make mistakes. The results of our mistakes might be that two networks can't communicate with each other anymore (usually two VLANs, as we are discussing an internal networking scenario here). Let's see the symptoms:

```
[root@client2 ~]# nmcli connection show network2 | grep -i ipv4.address
ipv4.addresses:                          192.168.2.2/24
[root@client2 ~]# ping 192.168.1.1 -c 5
PING 192.168.1.1 (192.168.1.1) 56(84) bytes of data.

--- 192.168.1.1 ping statistics ---
5 packets transmitted, 0 received, 100% packet loss, time 4109ms

[root@client2 ~]# traceroute 192.168.1.1
traceroute to 192.168.1.1 (192.168.1.1), 30 hops max, 60 byte packets
 1  * * *
 2  * * *
 3  * * *
 4  * * *
 5  * * *
 6  * * *
 7  * * *
 8  * * *
 9  * * *
10  * * *
11  * *^C
```

Figure 4.49 – Routing doesn't work anymore

If we check our routing table by using the `netstat -rn` command, we can see the following information:

```
[root@client2 ~]# netstat -rn
Kernel IP routing table
Destination     Gateway         Genmask         Flags   MSS Window  irtt Iface
0.0.0.0         192.168.2.254   0.0.0.0         UG        0 0          0 ens39
192.168.2.0     0.0.0.0         255.255.255.0   U         0 0          0 ens39
[root@client2 ~]# ping 192.168.2.254
PING 192.168.2.254 (192.168.2.254) 56(84) bytes of data.
64 bytes from 192.168.2.254: icmp_seq=1 ttl=64 time=0.468 ms
^C
--- 192.168.2.254 ping statistics ---
1 packets transmitted, 1 received, 0% packet loss, time 0ms
rtt min/avg/max/mdev = 0.468/0.468/0.468/0.000 ms
```

Figure 4.50 – Checking that routing on our Linux machine is set up properly, which it is

So, our gateway works (we can ping it), but it doesn't forward us correctly to the 192.168.1.0/24 network. Seeing that we configured the 192.168.1.0/24 network to be a local network for server1, it's clear that we have some kind of routing problem here. It could be a firewalld misconfiguration, stopping firewalld as a service, a routing table misconfiguration, or maybe someone played with the /proc filesystem and set the ip_forward flag back to 0. Whatever the case may be, the source of our problem is our default gateway. In larger enterprises, we usually have a networking team taking care of these things, so showing them output from ping, traceroute, and netstat should tell them where the problem is (in their own backyard). We'd usually tell them that they have a VLAN routing problem between VLAN X (subnet 1) and VLAN Y (subnet 2), send them outputs of these previously mentioned commands, and let them work from there.

Let's finish this recipe by talking about a few name resolution issues. These issues can happen because of a service misconfiguration (systemd-resolved, for example), a wrong /etc/resolv.conf configuration, and even an /etc/hosts configuration that we did ourselves. Let's go through a couple of common problems.

First, we are going to edit /etc/resolv.conf on client1 and put some custom DNS servers there. Then, we are going to reboot our client1 Linux virtual machine and see what happens when we check the content of /etc/resolv.conf. This screenshot is pre-reboot (we added two name servers, 8.8.8.8 and 8.8.4.4):

```
nameserver 127.0.0.53
nameserver 8.8.8.8
nameserver 8.8.4.4
options edns0 trust-ad
```

Figure 4.51 – Editing /etc/resolv.conf manually

This next screenshot was taken post-reboot. We can clearly see that the content of this file has been changed:

```
nameserver 127.0.0.53
options edns0 trust-ad
```

Figure 4.52 – /etc/resolv.conf after reboot

This was to be expected – as we described in our previous recipe, changing /etc/resolv.conf on a Linux machine that's running systemd-resolved is always going to end like this. If we want to change DNS settings, we need to do it properly. That means using nmcli in CentOS and, in this case, using netplan configuration on Ubuntu. This might only be a *local* issue, but it can still have a big impact in various scenarios where split-dns is involved.

The next problem is going to be about the opposite – let's say that we installed the `resolvconf` package on our Ubuntu machine, disabled `systemd-resolved`, and configured `/etc/resolv.conf` like this:

```
namserver 8.8.8.8
```

Figure 4.53 – Putting a wrong config option in /etc/resolv.conf

And when we try to resolve something by using the `nslookup`, `host`, or `dig` commands, it ends up nowhere, although our internet connection works, as shown with our manual DNS server configuration in `nslookup`:

```
root@client1:~# nslookup packtpub.com
;; connection timed out; no servers could be reached

root@client1:~# nslookup
> server 8.8.8.8
Default server: 8.8.8.8
Address: 8.8.8.8#53
> packtpub.com
Server:         8.8.8.8
Address:        8.8.8.8#53

Non-authoritative answer:
Name:    packtpub.com
Address: 172.67.31.83
Name:    packtpub.com
Address: 104.22.1.175
Name:    packtpub.com
Address: 104.22.0.175
Name:    packtpub.com
Address: 2606:4700:10::6816:af
Name:    packtpub.com
Address: 2606:4700:10::6816:1af
Name:    packtpub.com
Address: 2606:4700:10::ac43:1f53
```

Figure 4.54 – The network obviously works, but DNS name resolution doesn't,
which points us in the right direction

This clearly points to a wrong DNS server configuration as the internet access works, but we can't resolve a host. It's obvious that the option that we used (`namserver`) is wrong – it has to be `nameserver`. This brings us to the point: we always have to make sure that the syntax of our configuration files – in this case, `resolv.conf` – is correct. Mistakes are easily made if we are making changes by using a text editor, especially when we, for example, ignore red, highlighted fields in vi. If we were using commands to configure this and made an error in the syntax (`nmcli` or `netplan`), we would have an error somewhere, which would be easy to debug.

The last scenario that we are going to work on is a common one for those of us dealing with a public website migration from one provider to another, thus changing the public IP address. When we are configuring these scenarios, oftentimes we need to have the old website running while we test the new website. We could have two IP address entries in our public DNS servers pointing to two different web servers, but that's not what we're after, ever. It would confuse our website visitors and us as well. So, we want to have a quick way of testing the new website until it's fully debugged while offering the general public access to the old one.

Obviously, the simplest thing to do would be to add an entry to /etc/hosts, so that it points to the new website. Then, on the same machine where we made that change, we can debug our new website as much as we need – the public DNS entry still points to the old website, while our local machine goes to the new one.

After the debugging process is done, we need to do a switchover – we need to change the public DNS entries and remove the /etc/hosts entry on our debugging machine. That's an ideal scenario where we can make some mistakes. So, we go to our public DNS provider, change the IP address of our website so that it points to the new IP, and save the configuration. Then, we go to our local debugging machine and remove the /etc/hosts entry pointing to the new website, start a web browser, point it to our website URL, and – lo and behold – we are still being presented with the old website. What is going on here?

The simple fact is public DNS records need a bit of time to become active. It could be a minute, 15 minutes, an hour, a day – depending on how it's configured, but still, it needs time. Also, from various parts of the world – if our website is for an international audience – it could take different amounts of time to synchronize, which is why we have to be armed with patience when dealing with scenarios like these as we are probably going to get some emails about this scenario. We just need to do these types of configuration changes over the weekend when the amount of website visitors is at its lowest, and then sit and wait it out for all of the DNS entries to sync. From the time when we changed the DNS entry until everything is working, it's out of our hands. It's just the way it works.

As you can clearly see from these examples, there are quite a few different scenarios that might come into play as you're administering your Linux servers, clients, and networks.

How it works

All of the commands that we covered in this recipe work on the same idea – we have a networking stack that is either configured correctly or not. If it is, we mostly don't need them, but if something is misconfigured and/or not working properly, which can happen for a variety of external reasons as well, then we need to know how these commands work.

If we are discussing networking generally, there are a few well-known concepts: the configured IP address, netmask, gateway, DNS server(s), and a fully qualified domain name of any given Linux server. Keeping in mind that networks are isolated into multiple subnets and that DNS is a hierarchical structure, if any of these concepts aren't configured correctly, we will have issues with network communication. That's why most of us system engineers take extra care to configure all of these settings correctly, as it's a basis for us to not get permanent headaches, if you will.

When we do `ping`, `traceroute`, and `tracepath`, all of the traffic that we generate by using those utilities either goes to our local network or to non-local networks, which requires routing. On top of routing, firewalls might get in the way – sometimes people configure firewalls with ICMP traffic denied.

Then, even if all of that works as it should, there's the DNS, sitting on top of it like a Jedi master trying to balance the Force. And sometimes, it just seems a bit evil and as if it's bugging us for no reason whatsoever. That's where utilities such as `nslookup`, `host`, and `dig` come in handy – so that we can find out if it's something *lower* in the networking stack, or if it's the DNS. As we discussed in our previous recipe, using `systemd-resolved` changed quite a few things in terms of DNS configuration. We must be extra careful to configure things properly when we're using it – so, using `nmcli` and netplan's config files. We shouldn't just go and start editing files more often than not. That's just going to make more problems.

That being said, when we configure everything correctly and some other device on the network (our network or an external network) is at fault, things can get complicated very quickly. If a device doing network routing (a switch, router, Linux server, firewall, or whatever it might be) isn't configured correctly, we won't be able to communicate between multiple subnets. Imagine trying to go from Paris to Barcelona without knowing the way. There are so many possible ways of going from Paris to Barcelona (which we can equate to routing) that we wouldn't know which way to go. Usually, we start our debugging process by pinging some addresses on our networks (to check that the local network is working properly), then the default gateway, and checking if DNS is available as a service. On a more personal note, over the years, I have seen students and course attendees becoming painfully aware of just how complex DNS is as a system, especially at scale. We have a saying here at our college that students repeat over and over again – *It's always DNS*. So, we need to make sure that we have strong foundations in terms of DNS knowledge and understanding of how routing works. Then everything becomes much, much easier, as combining these two concepts can get insanely complicated. Especially when there are dynamic routing protocols such as BGP, EIGRP, and OSPF involved with split DNS and multiple locations.

And that's a wrap for this chapter. The next chapter is going to be all about using the shell to manage software packages on our Linux systems. We are going to discuss how to use `apt` and `apt-get`, `yum`, and `dnf`, software repositories, and other subjects related to software management. Until then, we bid you adieu!

There's more

If you need to learn more about network debugging, you can check the following links:

- A beginner's guide to network troubleshooting in Linux: `https://www.redhat.com/sysadmin/beginners-guide-network-troubleshooting-linux`

- Five Linux network troubleshooting commands: `https://www.redhat.com/sysadmin/five-network-commands`

5
Using Commands for File, Directory, and Service Management

When working with files and folders, **CLI** is the most common way. It doesn't really matter if we're copying or moving content around, finding content, configuring file and folder security, or doing some basic text-based work – it's all about the command line. When we expand on that model, we usually start using archiving and compression to be more space-efficient. In terms of services, we need to learn some basics of how to manage them, which is commonly done by using the systemctl command. This is exactly what we're going to cover in this chapter, by covering the following recipes:

- Basic file and directory-based commands
- Additional commands for manipulating file/directory security aspects
- Finding files and folders
- Manipulating text files by using commands
- Archiving and compressing files and folders
- Managing services and targets

Technical requirements

For these recipes, we're going to use one Linux machine – in our case, let's use `cli1`. We just need to make sure that it's powered on.

Basic file and directory-based commands

Let's discuss various shell commands that can be used to work with files and directories. In a nutshell, what we're interested in are these commands:

- `ls` – for listing folder contents
- `touch` – for creating an empty text file
- `cd` – for changing directories, both in absolute and relative terms
- `pwd` – for showing the current directory
- `mkdir` and `rm` – for creating and deleting a file or directory
- `cp` and `mv` – for copying or moving a file or a directory
- `ln` – for working with soft and hard links

These commands are some of the most frequently used commands in the everyday life of a system administrator/engineer. Let's see what they are all about.

Getting ready

We need to go through these commands to really understand what happens on the filesystem as we execute them. So, let's make sure that our `cli1` machine is running and let's do it!

How to do it...

Starting with the simplest command of them all, `ls`, it's all about checking the content of a folder. So, if we want to check the content of a directory in a nice, readable format, we can do it like this:

```
root@cli1:~# ls -al /root
total 164
drwx------   7 root root  4096 Jan  9 15:07 .
drwxr-xr-x  20 root root  4096 Mar 14  2021 ..
-rw-------   1 root root 17467 Nov 28 23:37 .bash_history
-rw-r--r--   1 root root  3771 Jul 27 09:39 .bashrc
drwx------   4 root root  4096 Jul 28 11:39 .config
-rw-------   1 root root    47 Sep 12 17:58 .lesshst
drwxr-xr-x   3 root root  4096 Apr 11  2021 .local
-rw-r--r--   1 root root   807 Jul 27 09:39 .profile
-rw-r--r--   1 root root    75 Jul 28 19:11 .selected_editor
drwx------   2 root root  4096 Mar 14  2021 .ssh
-rw-------   1 root root 13268 Jan  9 15:07 .viminfo
-rwxr-xr-x   1 root root   991 Nov 28 18:56 backup1.sh
-rw-r--r--   1 root root   523 Nov 28 19:04 backup2.sh
-rw-r--r--   1 root root 58463 Aug  8 13:01 dns-cache.txt
drwxr-xr-x   2 root root  4096 Mar 16  2021 links
-rw-r--r--   1 root root   122 Nov 28 19:12 logfile.Sun.tar.gz
-rw-r--r--   1 root root    36 Nov 28 19:12 logfile.snar
drwxr-xr-x   3 root root  4096 Nov 28 19:45 snap
-rw-r--r--   1 root root    98 Sep 28 06:19 txtsample.txt
root@cli1:~#
```

Figure 5.1 – Using ls with the most common options

The *nice, readable* part is achieved by using -la options, where the l option stands for *long listing*, and the a option stands for *all files (including the ones starting with a dot)*. We can also see that, by default, the ls command colors its output. For example, folders are colored blue, while files marked in red are archive files (in this case, the tar.gz file). We will go into more detail regarding archive files a bit later, when we start dealing with archiving and compressing files and folders later in this chapter. There are other colors that ls uses in its default output. Here are a couple of examples:

- Green – executable file
- Cyan – symbolic link
- Red with black background – broken link

The ls command can be used in a **recursive** mode, which means starting to list the content of a folder from that folder onward and going into all its subfolders. Here's an example of that, by using a common system directory in Ubuntu called /etc/network, and its recursive option (capital letter R):

```
root@cli1:~# ls -alR /etc/network
/etc/network:
total 20
drwxr-xr-x   5 root root 4096 Aug  8 13:04 .
drwxr-xr-x 107 root root 4096 Jan  9 15:07 ..
drwxr-xr-x   2 root root 4096 Aug  8 13:04 if-down.d
drwxr-xr-x   2 root root 4096 Oct 22  2020 if-pre-up.d
drwxr-xr-x   2 root root 4096 Aug  8 13:04 if-up.d
-rw-r--r--   1 root root    0 Mar 14  2021 interfaces

/etc/network/if-down.d:
total 12
drwxr-xr-x 2 root root 4096 Aug  8 13:04 .
drwxr-xr-x 5 root root 4096 Aug  8 13:04 ..
-rwxr-xr-x 1 root root  256 Aug 19  2020 resolvconf

/etc/network/if-pre-up.d:
total 12
drwxr-xr-x 2 root root 4096 Oct 22  2020 .
drwxr-xr-x 5 root root 4096 Aug  8 13:04 ..
-rwxr-xr-x 1 root root  344 Jun 30  2016 ethtool

/etc/network/if-up.d:
total 16
drwxr-xr-x 2 root root 4096 Aug  8 13:04 .
drwxr-xr-x 5 root root 4096 Aug  8 13:04 ..
-rwxr-xr-x 1 root root 1053 Aug 19  2020 000resolvconf
-rwxr-xr-x 1 root root 1685 Jun 30  2016 ethtool
root@cli1:~#
```

Figure 5.2 – Using ls in recursive mode

By using the R option, we instructed the ls command to do its job in recursive mode.

We can also use ls to display the content of the folder and sort the output by using the last modified time:

```
root@cli1:~# ls -alt
total 164
drwx------   7 root root  4096 Jan  9 15:07 .
-rw-------   1 root root 13268 Jan  9 15:07 .viminfo
-rw-------   1 root root 17467 Nov 28 23:37 .bash_history
drwxr-xr-x   3 root root  4096 Nov 28 19:45 snap
-rw-r--r--   1 root root    36 Nov 28 19:12 logfile.snar
-rw-r--r--   1 root root   122 Nov 28 19:12 logfile.Sun.tar.gz
-rw-r--r--   1 root root   523 Nov 28 19:04 backup2.sh
-rwxr-xr-x   1 root root   991 Nov 28 18:56 backup1.sh
-rw-r--r--   1 root root    98 Sep 28 06:19 txtsample.txt
-rw-------   1 root root    47 Sep 12 17:58 .lesshst
-rw-r--r--   1 root root 58463 Aug  8 13:01 dns-cache.txt
-rw-r--r--   1 root root    75 Jul 28 19:11 .selected_editor
drwx------   4 root root  4096 Jul 28 11:39 .config
-rw-r--r--   1 root root   807 Jul 27 09:39 .profile
-rw-r--r--   1 root root  3771 Jul 27 09:39 .bashrc
drwxr-xr-x   3 root root  4096 Apr 11  2021 .local
drwxr-xr-x   2 root root  4096 Mar 16  2021 links
drwx------   2 root root  4096 Mar 14  2021 .ssh
drwxr-xr-x  20 root root  4096 Mar 14  2021 ..
root@cli1:~#
```

Figure 5.3 – Sorting the ls output according to the last modification time

The next command on our list is touch, and it's a simple one. We use the touch command to create an empty file, like this:

```
root@cli1:~# touch file
root@cli1:~# ls -al file
-rw-r--r-- 1 root root 0 Jan  9 15:33 file
root@cli1:~#
```

Figure 5.4 – Touching a file in Linux means creating an empty file

Following that, it's time to explain two commands that are closely related – cd and pwd. cd, or the *change directory* command, is there so that we can leave one directory in the shell and go to another. In contrast, pwd is a command that tells us what our current directory is. Let's try that out by again using /etc/network as an example:

```
root@cli1:~# cd /etc/network
root@cli1:/etc/network# ls
if-down.d  if-pre-up.d  if-up.d  interfaces
root@cli1:/etc/network# pwd
/etc/network
root@cli1:/etc/network#
```

Figure 5.5 – Using cd and pwd to get our bearings in terms of directories

The next two commands that we're going to deal with are mkdir and rm. We use mkdir to create a directory, while we use the rm command to remove a file or a folder. So, let's show how these commands are used by means of an example. First, we are going to create a directory called temporary. Then, we're going to create two files in that directory called tempfile and tempfile2. After that, we're going to remove tempfile2, and then remove the entire temporary directory with all its contents, recursively. Let's do that now:

```
root@cli1:~# mkdir temporary
root@cli1:~# touch temporary/tempfile
root@cli1:~# ls -al temporary/tempfile
-rw-r--r-- 1 root root 0 Jan  9 15:41 temporary/tempfile
root@cli1:~# touch temporary/tempfile2
root@cli1:~# rm temporary/tempfile2
root@cli1:~# rm -rf temporary
root@cli1:~# ls -al
total 164
drwx------  7 root root  4096 Jan  9 15:42 .
drwxr-xr-x 20 root root  4096 Mar 14  2021 ..
-rw-------  1 root root 17467 Nov 28 23:37 .bash_history
-rw-r--r--  1 root root  3771 Jul 27 09:39 .bashrc
drwx------  4 root root  4096 Jul 28 11:39 .config
-rw-------  1 root root    47 Sep 12 17:58 .lesshst
drwxr-xr-x  3 root root  4096 Apr 11  2021 .local
-rw-r--r--  1 root root   807 Jul 27 09:39 .profile
-rw-r--r--  1 root root    75 Jul 28 19:11 .selected_editor
drwx------  2 root root  4096 Mar 14  2021 .ssh
-rw-------  1 root root 13268 Jan  9 15:07 .viminfo
-rwxr-xr-x  1 root root   991 Nov 28 18:56 backup1.sh
-rw-r--r--  1 root root   523 Nov 28 19:04 backup2.sh
-rw-r--r--  1 root root 58463 Aug  8 13:01 dns-cache.txt
-rw-r--r--  1 root root     0 Jan  9 15:33 file
drwxr-xr-x  2 root root  4096 Mar 16  2021 links
-rw-r--r--  1 root root   122 Nov 28 19:12 logfile.Sun.tar.gz
-rw-r--r--  1 root root    36 Nov 28 19:12 logfile.snar
drwxr-xr-x  3 root root  4096 Nov 28 19:45 snap
-rw-r--r--  1 root root    98 Sep 28 06:19 txtsample.txt
root@cli1:~#
```

Figure 5.6 – Working with mkdir and rm

The next topic of our discussion is the cp and mv commands – they enable us to copy or move files and/or folders where we want to move them. So, let's copy a file and folder (recursively), and then let's move them someplace else. We're going to use the same example as with mkdir and rm, but we're going to adjust the example slightly to fit the purpose. Specifically, we're going to create a directory with two files, but this time these files are going to be in a subdirectory. Then, we're going to add additional files to the first folder, after which we're going to copy and move a single file to a new location, and then a folder to another location. Let's see how that's going to work:

```
root@clil:~# mkdir temporary
root@clil:~# mkdir temporary/tempdir2
root@clil:~# touch temporary/tempdir2/file1
root@clil:~# touch temporary/tempdir2/file2
root@clil:~# touch temporary/file3
root@clil:~# touch temporary/file4
root@clil:~# mkdir newlocation
root@clil:~# cp temporary/tempdir2/file1 newlocation/
root@clil:~# mv temporary/tempdir2/file2 newlocation/
root@clil:~# mkdir newlocation2
root@clil:~# cp -R temporary/tempdir2 newlocation2
root@clil:~# mv temporary/tempdir2 newlocation
root@clil:~#
```

Figure 5.7 – Copying and moving files and folders

The first two commands can be aggregated into one by executing mkdir -p temporary/tempdir2.

This will create both of these directories in one command.

And finally, let's discuss the ln command, which can be used to create hard links (pointers to the file content) and soft links (pointers to the file/directory name, usually referred to as shortcuts). For hard links, we just use the ln command without any additional options, while soft links require us to use the ln command with the -s option. Let's create an example to drive the point of this home, and we'll explain how it works as soon as we're done with this example. The scenario is going to include the following:

- Copying a file with a bit of content to a new location so that we have a source file that's going to be used for hard and soft links

- Creating a hard link

- Creating a soft link

- Deleting the original file, and then checking what happens with the soft link and the hard link

- Copying the hard link to the original file, then checking what happens with soft link and hard link

This is how it's done:

```
root@cli1:~# cp /usr/share/dict/words .
root@cli1:~# ln words hardlink
root@cli1:~# ln -s words softlink
root@cli1:~# ls -al words hardlink softlink
-rw-r--r-- 2 root root 976256 Jan  9 18:45 hardlink
lrwxrwxrwx 1 root root      5 Jan  9 18:45 softlink -> words
-rw-r--r-- 2 root root 976256 Jan  9 18:45 words
root@cli1:~# rm words
root@cli1:~# ls -al words hardlink softlink
ls: cannot access 'words': No such file or directory
-rw-r--r-- 1 root root 976256 Jan  9 18:45 hardlink
lrwxrwxrwx 1 root root      5 Jan  9 18:45 softlink -> words
root@cli1:~# cp hardlink words
root@cli1:~# ls -al words hardlink softlink
-rw-r--r-- 1 root root 976256 Jan  9 18:45 hardlink
lrwxrwxrwx 1 root root      5 Jan  9 18:45 softlink -> words
-rw-r--r-- 1 root root 976256 Jan  9 18:45 words
root@cli1:~#
```

Figure 5.8 – Hard link and soft link operations

Now that we have gone through all the predetermined scenarios, let's explain some of the concepts behind these commands so that we can understand what happens on the filesystem as we execute them.

How it works...

Some of these commands are very straightforward and don't necessarily require further explanation, such as ls, touch, cd, mkdir, and pwd. But others do require a bit of background and technical explanation, especially ln. So, let's build on that premise and go through how rm, cp, mv, and ln work.

First, to cover basic file-related commands that we use the most often – rm, cp, and mv. These commands are straightforward, as we use them to remove files or folders (rm), copy files or folders (cp), or move files or folders (mv). Please note *remove* and *move* in our description of rm and mv, as those two things are different – removing is about deleting, while moving is about placing something someplace else. This sometimes confuses novice users, although it shouldn't.

However, the most technically demanding command from the bunch is ln, which requires us to explain what soft links are, and what hard links are. So, let's do that first.

Soft links are what we usually refer to as *shortcuts*, similarly to what we can do in Windows – create a shortcut to a file or a folder. As it's obvious from the picture of our scenario, when we removed the original file, the soft link stopped working. The reason for this is simple – soft links point to a *file or folder name*. If we delete a file or folder to which a soft link is pointing, that effectively means that the soft link points to nothing. And that's the reason why, in our scenario, we had a soft link turning red in color.

Hard links are a completely different concept. They don't point to a filename – they point to *file content*. When using hard links, try to think of them as two files pointing to the same content. If we delete the original file, the file content is still there, as that's the way modern filesystems work – they don't waste time deleting content, especially if another file is pointing to the same content. That would introduce a lot of latency into the process of deleting files if the files are big. A file delete operation therefore just deletes a pointer (filename) to the file content from a filesystem table. The filesystem takes care of the rest – when the time comes, if that file content is no longer used, the filesystem is going to use it to write new content over it.

We could've deduced this much from checking the original scenario, where we can clearly see the *difference in size* between soft and hard links – the soft link points to a filename and is therefore related to the size of a filename (the number of characters in the filename). As we explained, the hard link points to the file content, which is why the hard link has the same size as the file that it's pointing to.

There are two fundamental differences between these two concepts. Having in mind that the hard links point to *file content*, it's logical that they have two limitations – they can't point to a directory (just a file), and they can't go across partitions. So, if we have a disk partition mounted to `/directory` and another disk partition mounted to `/home directory`. We can't go to `/home directory` and create a hard link that points to a file that's located in `/partition`. One partition can't see the content of another partition, which is an important security concept. It also precludes any chance that hard links across partitions are going to work.

The next recipe is going to go much deeper into file-directory security concepts, as we're going to discuss permissions, special permissions, and **Access Control Lists (ACLs)**. These concepts are core concepts of IT security and something that we deal with daily. So, let's go through a scenario related to that next.

See also

If you need more information about these commands, we suggest that you visit these links:

- `ls` man page: `https://man7.org/linux/man-pages/man1/ls.1.html`
- `touch` man page: `https://man7.org/linux/man-pages/man1/touch.1.html`
- `cd` man page: `https://linuxcommand.org/lc3_man_pages/cdh.html`
- `pwd` man page: `https://man7.org/linux/man-pages/man1/pwd.1.html`
- `mkdir` man page: `https://man7.org/linux/man-pages/man1/mkdir.1.html`
- `rm` man page: `https://man7.org/linux/man-pages/man1/rm.1.html`
- `cp` man page: `https://man7.org/linux/man-pages/man1/cp.1.html`
- `mv` man page: `https://linux.die.net/man/1/mv`
- `ln` man page: `https://man7.org/linux/man-pages/man1/ln.1.html`

Additional commands for manipulating file/directory security aspects

In this recipe, we're going to use our users – Jack, Joe, Jill, and Sarah – to create a specific scenario to explain permissions, ACLs, and umask usage. A short explanation of these concepts is as follows: permissions are used to control access to files and folders in read, write, and execute mode. As they're limited in granularity, a concept of ACL was developed, to be able to manage permissions on a more finely grained level. Umask is a variable that pre-determines which permissions are going to be assigned to a newly created file or directory.

The recipe will go like this:

- We need to create a collaborative directory for our students located at `/share/students`
- We need to create a collaborative directory for our professors located at `/share/professors`
- Members of the student group need to have access to `/share/students` to collaborate on project files

- Members of the student group can create new files, which need to be group-owned by group students, in their /share/students folder

- One member of the student group can't use the rm command to delete other members' files in their /share/students folder

- One member of the student group must have permission to edit other members' files in their /share/students folder

- Professors need to have read-write access to all the student files, and all of the newly created student files

- Only professors have access to their shared folder, /share/professors, where they can delete each other's files, read them, and edit them.

Let's make this recipe happen.

Getting ready

Let's use our cli2 machine (CentOS) for this recipe, so make sure that it's powered on.

How to do it...

We're going to first create our users and groups by using the useradd and groupadd commands by using a scenario. Let's say that our task is as follows:

- Create four users called jack, joe, jill, and sarah

- Create two user groups called profs and pupils

- Reconfigure the jack and jill user accounts to be members of the profs group

- Reconfigure the joe and sarah user accounts to be members of the students group

- Assign a standard password to all the accounts (we're going to use P@ckT2021 for this purpose)

- Configure user accounts so that they must change their password when next logging in

- Set specific expiry data for the professors' user group – the minimum days before the password change should be set to 15, the maximum days before a forced password change should be set to 30, the warning regarding a password change needs to start a week before it expires, and the expiry date for accounts should be set to 2023/01/01 (January 1, 2023)

- Set specific expiry data for the students' user group – the minimum days before the password change should be set to 7, the maximum days before a forced password change should be set to 30, the warning regarding a password change needs to start 10 days before it expires, and the expiry date for accounts should be set to 2022/09/01 (September 1, 2022)

- Modify the profs group to be called professors, and the pupils group to be called students

> **Note**
> There are a lot of commands in this recipe, so make sure that you refer to the *How it works…* section of the recipe to understand everything about the new commands that we haven't used before.

The first task is to create user accounts, with their unique home directories and the Bash shell as the default shell:

```
useradd -m -s /bin/bash jack
useradd -m -s /bin/bash joe
useradd -m -s /bin/bash jill
useradd -m -s /bin/bash sarah
```

This will create entries for these four users in the /etc/passwd file (where most of the users' information is stored – username, user ID, group ID, default home directory, and default shell), and the /etc/shadow file (where users' passwords and aging information are stored).

Then, we need to create groups:

```
groupadd profs
groupadd pupils
```

This will create entries for these groups in the /etc/group file, where the system keeps all of the system groups.

The next step is to manage the membership of the user groups, for both the professors and student user groups. Before we do that, we need to be aware of one fact. There are two distinctive local group types – *primary group* and *supplementary group*. The *primary group* is important in terms of being the key parameter used when creating new files and directories, as the users' primary group will be used by default for that (there are exceptions, as we'll mention in recipe #4 in this chapter, about umask, permissions, and ACLs).

The *supplementary group* is important when dealing with sharing files and folders and related scenarios and exceptions. This is what's usually used for some additional settings for more advanced scenarios. These scenarios are going to be explained partially in the aforementioned recipe #4 in this chapter, as well as in recipes regarding NFS and Samba in *Chapter 9, An Introduction to Shell Scripting*.

Primary and supplementary groups are stored in /etc/group file.

Now that we've gotten that out of the way, let's modify our users' settings so that they belong to the *supplementary* groups as assigned by the scenario:

```
usermod -G profs jack
usermod -G profs jill
usermod -G pupils joe
usermod -G pupils sarah
```

Let's now check how that changes the /etc/group file:

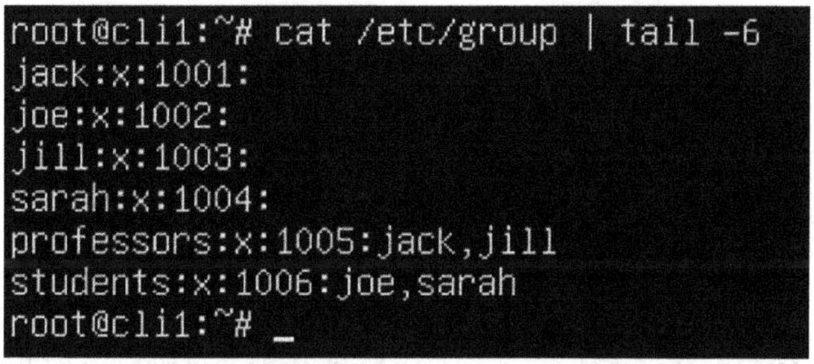

```
root@cli1:~# cat /etc/group | tail -6
jack:x:1001:
joe:x:1002:
jill:x:1003:
sarah:x:1004:
professors:x:1005:jack,jill
students:x:1006:joe,sarah
root@cli1:~# _
```

Figure 5.9 – Entries in the /etc/group file

The first four entries in the /etc/group file were actually created when we used the useradd command to create these user accounts. The next two entries (except for the last part, after the : sign), were created by groupadd commands, while entries after the : sign were created after the usermod commands.

Let's now set their initial password and set a forced password change when next logging in. We can do it in a couple of different ways, but let's learn the more *programmatic* approach to doing this by echoing a string and using it as the plaintext password for a user account:

```
echo "jack:P@ckT2021" | chpasswd
echo "joe:P@ckT2021" | chpasswd
echo "jill:P@ckT2021" | chpasswd
echo "sarah:P@ckT2021" | chpasswd
```

This is not necessarily something that we should recommend doing as it would leave these commands in the command history. We're just using this as an example.

The echo part – without the rest of the command – would just type P@ckT2021 to a terminal, like this:

```
echo "P@ckT2021"
P@ckT2021
```

In CentOS and similar distributions, we could use the passwd command with the --stdin parameter, which would mean that we want to add a password for the user account via standard input (keyboard, variable, ...). In Ubuntu, this is not available. So, we can echo the username:P@ckT2021 string to shell and pipe that to the chpasswd command, which achieves just that purpose – instead of outputting the string to our terminal, the chpasswd command uses it as standard input into itself.

Let's set the expiry data for professors and students. For this purpose, we need to learn how to use the chage command and some of its parameters (-m, -M, -W, -E):

- If we use the -m parameter, this means that we want to assign the minimum number of days before a password change is allowed
- If we use the -M parameter, this means that we want to assign the maximum number of days before a password change is forced
- If we use the -W parameter, this means that we want to set the number of warning days prior to password expiration, which, in turn, means that the shell is going to start throwing us messages about needing to change our password before it expires
- If we use the -E parameter, this means that we want to set account expiration to a certain date (in YYYY-MM-DD format)

Let's now translate that into commands:

```
chage -m 15 -M 30 -W 7 -E 2023-01-01 jack
chage -m 15 -M 30 -W 7 -E 2023-01-01 jill
chage -m 7 -M 30 -W 10 -E 2022-09-01 joe
chage -m 7 -M 30 -W 10 -E 2022-09-01 sarah
```

And finally, let's modify the groups to their final settings by modifying the group name from professors to profs and from students to pupils:

```
groupmod -n professors profs
groupmod -n students pupils
```

These commands will only change group names, not their other data (such as group ID), which is going to be reflected in our users' information as well:

```
root@cli1:~/links# id jack
uid=1001(jack) gid=1001(jack) groups=1001(jack),1005(professors)
root@cli1:~/links# id jill
uid=1003(jill) gid=1003(jill) groups=1003(jill),1005(professors)
root@cli1:~/links# id sarah
uid=1004(sarah) gid=1004(sarah) groups=1004(sarah),1006(students)
root@cli1:~/links# id joe
uid=1002(joe) gid=1002(joe) groups=1002(joe),1006(students)
root@cli1:~/links#
```

Figure 5.10 – Checking created users' settings

As we can see, `jack` and `jill` are members of a group that's now called `professors`, while `joe` and `sarah` are now members of a group called `students`.

We deliberately left the `userdel` and `groupdel` commands for last, as they come with some caveats and shouldn't be used lightly. Let's create a user called `temp` and a group called `temporary`, and then let's delete them:

```
useradd temp
groupadd temporary
userdel temp
groupdel temporary
```

This will work just fine. The thing is, because we used the `userdel` command without any parameters, it will leave the user's home directory intact. Since users' home directories are usually stored in the `/home` directory, by default, this means that the `/home/temp` directory is still going to be there. When deleting users, this is sometimes something we want – to delete a user, but not to delete their files. If you specifically want to delete a user account and all the data from that user account, use the `userdel -r username` command. But think twice before doing it!

Furthermore, we obviously need to create a bunch of directories and files and change a whole stack of permissions and ACLs. As a general note, the `chmod` command changes permissions, while the `setfacl` command modifies ACLs. This is the correct way to do it:

```
mkdir -p /share/students
mkdir /share/professors
chgrp students /share/students
chmod 3775 /share/students
```

```
setfacl -m g:professors:rwx /share/students
chgrp professors /share/professors
chmod 2770 /share/professors/
setfacl -m d:g:professors:rwx /share/students/
```

Let's now test this to check whether it works. First, we're going to log in as our two students from the first recipe (joe and sarah) and create a couple of files. Then we're going to use joe's account to try to delete Sarah's files, and vice versa, so we should first use su to log in as joe, su - joe, and type in the root password.

Let's see how that works out:

```
[joe@localhost ~]$ cd /share/students/
[joe@localhost students]$ touch myfile1 myfile2
[joe@localhost students]$ logout
[root@localhost share]# su - sarah
[sarah@localhost ~]$ cd /share/students/
[sarah@localhost students]$ touch myfile3 myfile4
[sarah@localhost students]$ ls -al
total 0
drwxrwsr-t+ 2 root    students 66 Dec  6 20:58 .
drwxr-xr-x. 4 root    root     40 Dec  6 20:51 ..
-rw-rw-r--+ 1 joe     students  0 Dec  6 20:58 myfile1
-rw-rw-r--+ 1 joe     students  0 Dec  6 20:58 myfile2
-rw-rw-r--+ 1 sarah   students  0 Dec  6 20:58 myfile3
-rw-rw-r--+ 1 sarah   students  0 Dec  6 20:58 myfile4
[sarah@localhost students]$ rm myfile1
rm: cannot remove 'myfile1': Operation not permitted
[sarah@localhost students]$ logout
[root@localhost share]# su - joe
Last login: Sun Dec  6 20:58:16 CET 2020 on tty1
[joe@localhost ~]$ cd /share/students
[joe@localhost students]$ rm myfile3
rm: cannot remove 'myfile3': Operation not permitted
```

Figure 5.11 – The scenario works flawlessly from the students' perspective

Part of our scenario required us to be able to edit each other's files, while not being able to delete them outright. Let's test that now:

```
[joe@localhost students]$ echo "Joe adds a bit of content to Jill's file" >> myfile3
[joe@localhost students]$ ls -al myfile3
-rw-rw-r--+ 1 sarah students 41 Dec  6 21:09 myfile3
[joe@localhost students]$ cat myfile3
Joe adds a bit of content to Jill's file
[joe@localhost students]$ rm myfile3
rm: cannot remove 'myfile3': Operation not permitted
[joe@localhost students]$
```

Figure 5.12 – Changing the content of the file works, while removing the file doesn't

There's a reason why we picked this type of scenario – this is a real-life scenario that file server administrators often encounter. It's basically the best of both worlds – collaboration works, but users can't delete each other's files by accident. Therefore, this recipe covers some of the most common things that happen on a file server, such as one user deleting another user's file by accident (the key point here being the *lack of intention* to delete a file). It has happened to all of us. On the other hand, changing a file's content is something that we can only do *intentionally, consciously*. This is also something that we can easily track by using filesystem auditing and file attributes, if we set up our system that way.

Let's now review things from the professor's perspective. We'll use the `jill` account for this purpose:

```
[jill@localhost ~]$ cd /share/students/
[jill@localhost students]$ ls -al
total 4
drwxrwsr-t+ 2 root   students 66 Dec  6 21:12 .
drwxr-xr-x. 4 root   root     40 Dec  6 20:51 ..
-rw-rw-r--+ 1 joe    students  0 Dec  6 20:58 myfile1
-rw-rw-r--+ 1 joe    students  0 Dec  6 20:58 myfile2
-rw-rw-r--+ 1 sarah students 41 Dec  6 21:12 myfile3
-rw-rw-r--+ 1 sarah students  0 Dec  6 20:58 myfile4
[jill@localhost students]$ echo "Prof.Jill looked through this file" >> myfile3
[jill@localhost students]$ ls -al myfile3
-rw-rw-r--+ 1 sarah students 76 Dec  6 21:16 myfile3
[jill@localhost students]$ cat myfile3
Joe adds a bit of content to Jill's file
Prof.Jill looked through this file
[jill@localhost students]$
```

Figure 5.13 – Checking whether our configuration works for Jill

We also need to check whether the professors' share works. Let's test it:

```
[jill@localhost ~]$ cd /share/professors/
[jill@localhost professors]$ touch prof1 prof2
[jill@localhost professors]$ logout
[root@localhost students]# su - jack
Last login: Sun Dec  6 21:21:06 CET 2020 on tty1
[jack@localhost ~]$ cd /share/professors/
[jack@localhost professors]$ touch prof3 prof4
[jack@localhost professors]$ echo "jack tests appending to jill's file" >> prof1
[jack@localhost professors]$ logout
[root@localhost students]# su - jill
Last login: Sun Dec  6 21:22:14 CET 2020 on tty1
[jill@localhost ~]$ cd /share/professors/
[jill@localhost professors]$ echo "jill tests appending to jack's file" >> prof3
[jill@localhost professors]$ ls -al
total 8
drwxrws---. 2 root professors 58 Dec  6 21:22 .
drwxr-xr-x. 4 root root       40 Dec  6 20:51 ..
-rw-rw-r--. 1 jill professors 36 Dec  6 21:23 prof1
-rw-rw-r--. 1 jill professors  0 Dec  6 21:22 prof2
-rw-rw-r--. 1 jack professors 36 Dec  6 21:23 prof3
-rw-rw-r--. 1 jack professors  0 Dec  6 21:22 prof4
[jill@localhost professors]$ rm prof4
[jill@localhost professors]$ _
```

Figure 5.14 – From the professors' standpoint, their share works as requested

Let's try to use a student's account to get into the professors' shared folder:

```
[root@localhost students]# su - joe
Last login: Sun Dec  6 21:12:27 CET 2020 on tty1
[joe@localhost ~]$ cd /share/professors
-bash: cd: /share/professors: Permission denied
[joe@localhost ~]$
```

Figure 5.15 – A student tries to get to the professors' share and is denied access

We can also see that these files created by users get the 664 default permission. That's what umask is all about. Check how umask works in the *How it works…* section of this recipe.

So, the whole scenario works, but how exactly does it work? Let's check that out now.

How it works...

Before we get to the detailed explanations of these commands, let's just cover the basics and describe the commands that we have used:

- useradd – the command that's used to create a local user account

- usermod – the command that's used to modify a local user account

- `userdel` – the command that's used to delete a local user account
- `groupadd` – the command that's used to create a local group
- `groupmod` – the command that's used to modify a local group
- `groupdel` – the command that's used to delete a local group
- `passwd` – the command that's most often used to assign passwords to user accounts, but it can be used for some other scenarios (for example, locking user accounts)
- `chage` – the command that's used to manage user password expiry
- `chgrp` – the command that changes the group ownership of a file or folder
- `chmod` – the command that changes the permission of a file or folder
- `setfacl` – the command that changes the ACL of a file or folder

Now that we have discussed these commands, let's explain the details.

Every file or directory on a Linux filesystem has a number of attributes:

- Permissions
- Ownership
- File size
- Date of creation
- File/directory name

We will focus on permissions and ownership in this recipe as that's the core of this specific recipe. When we issue a command such as `ls -al` in the `/share/students` directory, this is what we get:

```
[root@localhost students]# ls -al
total 4
drwxrwsr-t+ 2 root    students 66 Dec   6 21:12 .
drwxr-xr-x. 4 root    root     40 Dec   6 20:51 ..
-rw-rw-r--+ 1 joe     students  0 Dec   6 20:58 myfile1
-rw-rw-r--+ 1 joe     students  0 Dec   6 20:58 myfile2
-rw-rw-r--+ 1 sarah   students 76 Dec   6 21:16 myfile3
-rw-rw-r--+ 1 sarah   students  0 Dec   6 20:58 myfile4
```

Let's now use the `myfile1` output as an example. Reading from left to right, the `-rw-rw-r--+` part is related to permissions on that specific file. The second part (`joe`, followed by `students`) is related to ownership of that specific file.

Let's parse through this a bit, going with permissions first:

- The first – means that this is a file – this field is used for the type of content.

- The first `rw-` means that we have read and write permission for the file owner (`joe`) – which we refer to as the user class (u).

- The second `rw-` means that we have read and write permission for the group owner (`students`) – which we refer to as the group class (g).

- R—means that all the other users have just read permissions – we refer to this class as `others` (o).

- + at the end means that we have an active ACL on this specific file (to be discussed |a bit later in this explanation).

We can assign numerical values (weights) to these permissions. The read permission has a weight of 4 (2^2), the write permission has a weight of 2 (2^1), and the execute permission has a weight of 1 (2^0).

So, if we wanted to assign all permissions to all classes of users (user owner, group owner, others), we'd use the following:

```
chmod 777 file_name
```

Why? If we add 4+2+1, that equals 7. That means read + write + execute. And we can use that on all three classes – u, g, and o – so that gives us 777. The first 7 refers to u (user owner), the second one to g (group owner), and the third one to everyone else (others). That simplifies the management of permissions significantly.

If we're talking about files, the meaning of these permissions is straightforward – read means read, write means write, delete, and modify, and execute means the ability to start the file.

With directories, it gets a bit trickier. Default permissions that we need in order to be able to read directory content and pass through it (folder traverse) are read and execute permissions. Write permission on a directory and read permission is needed to list the contents of the directory, while x permission is needed to traverse the directory (being able to go into subfolders of that directory). Write permission means write, delete, and modify on the folder level for the files in that folder (unless there are explicit denies, for example, set by ACL).

As you can clearly see in the command output, files have two types of ownership:

- **user owner** – in this specific example, `joe`
- **group owner** – in this specific example, `students`

What does that mean?

It means that a *user called Joe* owns that file. At the same time, it means that a *group called students* own that file from the group perspective.

Let's now add to that discussion by talking about the second line of this output, specifically:

```
drwxrwsr-t+ 2 root   students 66 Dec  6 21:12 .
```

The same principles apply, it's just that we need to discuss a couple of new settings. We can clearly see a couple of letters that we didn't mention previously – s in the group ownership class, and t in the others class. What is that all about?

The thing is, there are additional, special permissions on top of r(ead), w(rite), and (e) x(ecute). These are used for special use cases:

- sticky bit – we set this special permission on a folder level. When enabled on a folder level, it's there to protect from accidental file deletion from our scenario. For example, myfile1 is owned by the user joe. Although sarah is the member of the same group (students), which group-owns the file, she still can't delete that file. That's what the sticky bit is all about.

- setgid – we set this special permission on the folder level as well. When set on the folder level, this special permission means that all of the newly created files (after setgid was set) are going to take their group ownership from the parent folder. In our scenario, that means that all of the newly created files are group-owned by the student group, as requested by the scenario. This is why we used the chgrp command on the folder level to set folder ownership to students.

- setuid – almost never used now, as it's a security risk. It used to be used on files a bit, specifically, so that when files are started by a non-owner user, it seems that the user owning the file started it (similar to Run As IDEA in Windows).

These permissions are also set by the chmod command, like the first number. That's why our chmod command had four numbers instead of three – the first number is all about special permissions. In general, when we use a three-digit number with chmod, it expands that to include a zero from the left side.

Going back to our recipe, we issued the following command:

```
chmod 3775 /share/students
```

That means that we used chmod to set the sticky bit and setgid (1+2 equals 3 on the first digit) on the folder, as well as rwx for user and group owners (77) and rx for others (5).

The next, more complicated part relates to ACLs. ACLs are most commonly used to take care of *exceptions* (regular ACLs) or *permission inheritance* (default ACLs). Let's describe them in a bit more detail. We used the following command:

```
setfacl -m g:professors:rwx /share/students
```

This means that we want to modify (the -m parameter) the ACL on the directory called /share/students. And we want to modify it so that the group called professors gets rwx (read-write-execute) permissions on that directory. You can clearly see why we said that ACLs are most commonly used to treat exceptions. Our scenario required that the /share/students folder has group ownership of students. We can't assign two users to be owners of a directory (there can be only one, (c) according to the Highlander movie). So, there's no direct way for us to do that, which means we have to use something else to create an exception. That's where ACL comes in.

We could've done this differently (not that we should have). We could have issued two user-based ACLs for both members of our professors' group. Those two commands would be as follows:

```
setfacl -m u:jack:rwx /share/students
setfacl -m u:jill:rwx /share/students
```

The trouble with that approach is really simple to understand. Let's say that we add five more professors to our system. We then need to issue five more setfacl commands to set the same ACLs to them. It's just easier to use a group and add users to a group. It's a well-known concept that everyone uses on all of the operating systems used today.

If we wanted to set explicit deny ACL for others, we could have used this command:

```
setfacl -m o:--- /share/students
```

This way, we make sure that all of the members of that others class don't get access to the folder.

The second setfacl command that we used was as follows:

```
setfacl -m d:g:professors:rwx /share/students/
```

This command sets a *default ACL*, which is a completely different concept to regular ACL we described just before this. Default ACLs are used so that every newly created file or folder under a directory (in this case, /share/students) automatically inherits permissions from the parent folder as set by the default ACL. In our scenario, this command means that every single file or folder that gets created after we set this default ACL is going to have an ACL set to g:professors:rwx.

Clearly, you can see how ACLs and default ACLs are useful, as without them, we'd have way less scope to configure more advanced, finely-grained scenarios for data access.

Let's now discuss the last important aspect of this scenario – default file permissions. We mentioned in the recipe that we need to cover the subject of umask. Let's do that now.

If we check the output from one of the previous screenshots, we can see this:

```
-rw-rw-r--. 1 jill professors  0 Dec  6 21:22 prof2
-rw-rw-r--. 1 jack professors 36 Dec  6 21:23 prof3
```

The question is, why are default permissions rw-rw-r--?

The answer to this question is called umask (user mask).

As a concept, umask is used specifically for that – to set default permissions for newly created files and directories. It can be set by shell configuration files, by a user profile, or by a command. Let's use the umask command to explain how it does what it does:

```
[jill@localhost professors]$ umask
0002
[jill@localhost professors]$ touch prof4
[jill@localhost professors]$ umask 0022
[jill@localhost professors]$ touch prof5
[jill@localhost professors]$ umask 0222
[jill@localhost professors]$ touch prof6
[jill@localhost professors]$ ls -al
total 8
drwxrws---. 2 root professors 84 Dec  6 22:32 .
drwxr-xr-x. 4 root root       40 Dec  6 20:51 ..
-rw-rw-r--. 1 jill professors 36 Dec  6 21:23 prof1
-rw-rw-r--. 1 jill professors  0 Dec  6 21:22 prof2
-rw-rw-r--. 1 jack professors 36 Dec  6 21:23 prof3
-rw-rw-r--. 1 jill professors  0 Dec  6 22:32 prof4
-rw-r--r--. 1 jill professors  0 Dec  6 22:32 prof5
-r--r--r--. 1 jill professors  0 Dec  6 22:32 prof6
```

You can clearly see that as we change the umask variable for a user, the default permissions for newly created files change. When we used an umask value of 0002, the prof4 file was created with permission 664. When we used an umask value of 0022, the prof5 file was created with permission 644. Lastly, when we used an umask value of 0222, the prof6 file was created with permission 444. We could also ignore the leading zero when assigning umask in accordance with the same principle we used for the chmod command.

Mask for files is set to 666, and for directories, it's set to 777. So, if we want to calculate default permissions for newly created files or folders, we just need to subtract the umask value from these values (666 or 777) and get default permissions for files (or folders).

If we don't resort to manual configuration, all users' umask values are set by the /etc/profile file, which is loaded by default when a user logs in. In that file, there's an if statement that looks like this:

```
if [ $UID -gt 199 ] && [ "`/usr/bin/id -gn`" = "`/usr/bin/id
-un`" ]; then
    umask 002
else
    umask 022
fi
```

Basically, what this if-then-else structure does is, for all the UIDs that are greater than 199, umask is set to 002, otherwise, it's set to 022. That's why regular users have umask 002, while the root user has umask 022 (the root's UID is 0).

Let's now move to our next recipe, which is all about using commands to manipulate text files – commands including cat, cut, more, less, head, and tail.

See also

If you need more information about file permissions, special permissions, or ACLs, we suggest that you visit these links:

- Managing file permissions: https://access.redhat.com/documentation/en-us/red_hat_enterprise_linux/8/html/configuring_basic_system_settings/assembly_managing-file-permissions_configuring-basic-system-settings

- Linux permissions – SUID, SGID, and sticky bit: `https://www.redhat.com/sysadmin/suid-sgid-sticky-bit`

- An introduction to Linux ACLs: `https://www.redhat.com/sysadmin/linux-access-control-lists`

Manipulating text files by using commands

Let's now switch our attention to learning about commands that enable us to manipulate text files – just for output reasons `head`, `tail`, `more`, `less`, `cat`. Some other commands related to the same concepts are going to be covered in later chapters, such as *Chapter 8, Using the Command Line to Find, Extract, and Manipulate Text Content*, where we discuss more advanced scenarios with text files, such as merging, cutting, and using regular expressions with `grep` and `sed` to manipulate text content.

Getting ready

We still need the same virtual machines as with our previous recipes.

How to do it...

Let's start by using `head` and `tail` command, commands that can be used to show the beginning and end of a text file. For example, let's use the `/root/.bashrc` file:

```
[student@cli1 22:28] head /root/.bashrc
# ~/.bashrc: executed by bash(1) for non-login shells.
# see /usr/share/doc/bash/examples/startup-files (in the
package bash-doc)
# for examples


# If not running interactively, don't do anything
case $- in
    *i*) ;;
      *) return;;
esac
```

Let's now check the tail end of the same file:

```
[student@cli1 22:29] tail /root/.bashrc
   if [ -f /usr/share/bash-completion/bash_completion ]; then
      . /usr/share/bash-completion/bash_completion
   elif [ -f /etc/bash_completion ]; then
      . /etc/bash_completion
   fi
fi
PS1="\e[0;31m[\u@\H \A] \e[0m"
export VISUAL=nano
export EDITOR=nano
```

Unlike that, more and less are used to just display output, but in a page-by-page formatted fashion, making a long output much more humanly readable. So, when we execute the following command:

```
less /root/.bashrc
```

Or we can execute this command:

```
more /root/.bashrc
```

The expected output from these commands is something like this:

```
# ~/.bashrc: executed by bash(1) for non-login shells.
# see /usr/share/doc/bash/examples/startup-files (in the package bash-doc)
# for examples

# If not running interactively, don't do anything
case $- in
    *i*) ;;
      *) return;;
esac

# don't put duplicate lines or lines starting with space in the history.
# See bash(1) for more options
HISTCONTROL=ignoreboth

# append to the history file, don't overwrite it
shopt -s histappend

# for setting history length see HISTSIZE and HISTFILESIZE in bash(1)
HISTSIZE=1000
HISTFILESIZE=2000

# check the window size after each command and, if necessary,
# update the values of LINES and COLUMNS.
shopt -s checkwinsize

# If set, the pattern "**" used in a pathname expansion context will
# match all files and zero or more directories and subdirectories.
#shopt -s globstar

# make less more friendly for non-text input files, see lesspipe(1)
[ -x /usr/bin/lesspipe ] && eval "$(SHELL=/bin/sh lesspipe)"

# set variable identifying the chroot you work in (used in the prompt below)
if [ -z "${debian_chroot:-}" ] && [ -r /etc/debian_chroot ]; then
    debian_chroot=$(cat /etc/debian_chroot)
fi

# set a fancy prompt (non-color, unless we know we "want" color)
case "$TERM" in
    xterm-color|*-256color) color_prompt=yes;;
esac

# uncomment for a colored prompt, if the terminal has the capability; turned
# off by default to not distract the user: the focus in a terminal window
/root/.bashrc
```

Figure 5.16 – Using the more and less commands looks very similar

to this – page-by-page viewing of text content

Cat (the command, not a feline) is completely opposite to the *discipline* of more or less – it just displays the whole file content, without any stoppages. This is cool when a text file is short, but mostly useless if that file is long, and is one of the most common reasons why we use more or less commands. So, let's pick a short file and cat it, for example, the /root/.profile file:

```
root@cli1:~# cat .profile
# ~/.profile: executed by the command interpreter for login shells.
# This file is not read by bash(1), if ~/.bash_profile or ~/.bash_login
# exists.
# see /usr/share/doc/bash/examples/startup-files for examples.
# the files are located in the bash-doc package.

# the default umask is set in /etc/profile; for setting the umask
# for ssh logins, install and configure the libpam-umask package.
#umask 022

# if running bash
if [ -n "$BASH_VERSION" ]; then
    # include .bashrc if it exists
    if [ -f "$HOME/.bashrc" ]; then
        . "$HOME/.bashrc"
    fi
fi

# set PATH so it includes user's private bin if it exists
if [ -d "$HOME/bin" ] ; then
    PATH="$HOME/bin:$PATH"
fi

# set PATH so it includes user's private bin if it exists
if [ -d "$HOME/.local/bin" ] ; then
    PATH="$HOME/.local/bin:$PATH"
fi
root@cli1:~#
```

Figure 5.17 – Using the cat command on an appropriate file – a text file
that's not too big to fit on one terminal page

Cat can be used for one more thing, which is to combine multiple text files into one. This operation is often used when combining multiple log files into one to concatenate them. We're going to discuss this scenario later in this book, in *Chapter 8, Using the Command Line to Find, Extract, and Manipulate Text Content.*

How it works...

`more` and `less` are page viewers – they enable us to display content page by page. As we can see in the last line of our example, using these commands didn't *finish* – the command stopped showing the file content after displaying one page of that file. Now it's interactively waiting for us to either continue listing file content page by page, do something else (for example, searching by using the / sign), or quit by using the q key.

The `head` and `tail` commands are named appropriately – they show the head (beginning) and the tail (end) of a text file. They can also be used with a variety of options to further parametrize the output that we want. For example, if we execute the following command:

```
tail -n 15 /root/.bashrc
```

We're going to get the last 15 lines of that file. The same can be done with the `head` command.

Our next topic of discussion is using the `find` command to find files and folders. Let's deal with that first and then move on to the next recipes, which are going to involve archiving, compression, and dealing with services via `systemctl`.

There's more...

If we need to learn more about these commands, we can check the following links:

- `head` man page: `https://man7.org/linux/man-pages/man1/head.1.html`
- `tail` man page: `https://man7.org/linux/man-pages/man1/tail.1.html`
- `more` man page: `https://man7.org/linux/man-pages/man1/more.1.html`
- `less` man page: `https://man7.org/linux/man-pages/man1/less.1.html`
- `cat` man page: `https://man7.org/linux/man-pages/man1/cat.1.html`

Finding files and folders

Our next topic of the day is to learn to use the `find` command, an incredibly useful command. It can be used in a variety of different ways – to find files and folders according to specific criteria (permissions, ownership, modified date, and others), but also to *prepare data* to be further manipulated *after* the `find` command. We'll go through some examples of both principles in this recipe.

Getting ready

We need to leave our `cli1` virtual machine running. If it's not powered on, we need to power it back on.

How to do it...

Let's use a couple of examples to explain how the `find` command works. Here are some examples that we're going to use:

- Finding files in the / directory that have permission `2755`
- Finding files in the / directory owned by the user `jill`
- Finding files in the / directory owned by a group student
- Finding files in the / directory with a specific name (for example, `network`)
- Finding files of a specific type (for example, all files with the `php` extension)
- Finding all empty directories
- Finding files that have been modified in the past two hours (120 minutes)
- Finding files that are 100-200 MB in size

Let's make these scenarios happen:

```
find / -type f -perm 2755
find / -type f -user jill
find / -type f -group student
find / -type f -name network
find / -type f -name "*.php"
find / -type d -empty
find / -mmin -120
find / -size +100M -size -200M
```

To further drive home the importance of this command, let's use it, as we mentioned previously, to *prepare* data to do something *after* we find the necessary content. For example, let's find all files that have the `.avi` extension on the whole filesystem and remove them:

```
find / -type f -name "*.avi" -exec rm -f {} \;
```

This command finds all files with the `.avi` extension, puts them in an array, and removes them, one by one, by using the `rm -f` command. This is very useful if you have users who are abusing corporate resources for unnecessary content.

How it works...

Finding files and folders by name is something that we often do. For example, let's say that we have a rudimentary shell script that performs a backup and that it uses specific criteria to create a list of files that it's going to copy to a pre-determined backup folder. If we're doing this kind of operation on a large production server with hundreds of users, chances are that there will be a lot of new files daily. Using the `find` command makes a lot of sense in these sorts of scenarios.

Most commonly, we use the `find` command to locate files (`-type f` option) or folders (`-type d` option), and then we narrow our search by using more criteria. Criteria such as modification date, user or group ownership, permissions – there are a lot of options available. If we look at the `find` command man page, we'll quickly become painfully aware of how many options and advanced scenarios can be covered by using the `find` command. This is why there's a common way of using `find`, which is to start with something such as a file type or extension, and then narrow it down further by using other options that we mentioned. If we start with that in mind, we'll get to our results quickly.

The next recipe on our plate is related to archiving and compressing files and folders. So, let's learn how to use `tar` and its sidekicks, `gzip`, `bzip`, `xzip`, and commands alike.

There's more...

If you need to learn more about Bash reserved variables and PS variables, refer to the Find command man page: `https://man7.org/linux/man-pages/man1/find.1.html`.

Archiving and compressing files and folders

Being efficient in terms of how we use disk space is nothing new – it's always been around. Yes, we're at a point in history where large capacity hard drives and other media is available, but that doesn't mean we can be reckless about it. This is the reason why we've been using archiving and compressing for decades, and it's a topic that we're going to cover now as well.

Getting ready

We need to make sure that our `cli1` machine is ready to be used, which will make our work on this recipe easy.

How to do it...

Let's go through another scenario-based example to cover all the necessary topics. So, this is what we're going to do in the first part of our recipe:

- Create a `tar` archive with the current folder content
- Create a `tar.gz` compressed archive with the current folder content
- Create a `tar.bz2` compressed archive with the current folder content
- Create a `tar.xz` compressed archive with the current folder content

In the second part of our recipe, we're going to extract these archives:

- Extract a `tar`, `tar.gz`, `tar.bz2`, or `tar.xz` archive
- Extract a `tar`, `tar.gz`, `tar.bz2`, or `tar.xz` archive to a specific folder (let's say `/tmp/extract`)

Let's say that we're located in the `/root` directory, and that we want to save all of our archives to the `/tmp` directory. This is how we'd do the first part of our scenario:

```
cd /root
tar cfp /tmp/root.tar .
```

If we were creating the `tar.gz` archive, we'd do this:

```
tar cfpz /tmp/root.tar.gz .
```

If we were creating the `tar.bz2` archive, we'd do this:

```
tar cfpj /tmp/root.tar.bz2 .
```

If we were creating the `tar.xz` archive, we'd do this:

```
tar cfpJ /tmp/root.tar.xz .
```

The second part of our scenario begins by opening an archive. We just need to change one `tar` parameter when compared to the first set of examples in our scenario and ditch the last part of our command (`.` for the current directory). So, we'd need to do this (don't do this in the real world; this is merely for illustration purposes):

```
tar fpx /tmp/root.tar
```

Alternatively, we could need to do this:

```
tar zfpx /tmp/root.tar.gz
```

Or we could do this:

```
tar jfpx /tmp/root.tar.bz2
```

Or we could do this:

```
tar Jfpx /tmp/root.tar.xz
```

The problem with this is the location of the output – where's the output of this extraction process going to go? So, the correct way to do this would be as follows:

```
cd /tmp
mkdir /tmp/extract
tar fpx /tmp/root.tar -C /tmp/extract
```

Alternatively, we could do this:

```
tar zfpx /tmp/root.tar.gz -C /tmp/extract
```

Or we could do this:

```
tar jfpx /tmp/root.tar.bz2 -C /tmp/extract
```

Or we could do this:

```
tar Jfpx /tmp/root.tar.xz -C /tmp/extract
```

Again, this depends on the archive type.

Tar has a myriad of other available options, for example, for manipulating ACLs and SELinux contexts such as the following:

- --acls – Use ACLs when creating an archive
- --no-acls – Ignore ACLs when creating an archive
- --selinux – Use SELinux contexts when creating an archive
- --noselinux – Ignore SELinux contexts when creating an archive

We cannot stress how important it is to check the corresponding man page if we're looking for something specific. So, we need to make sure that we do.

How it works...

tar, or Tape ARchiver, has been around for decades now. Its original use case included archiving content on tape, which is how it got its name. Archiving, as the manual states, means storing multiple files in a single file. All the other options that we use are additional options that have been added over the past 40+ years, given that it was introduced way back in 1979.

In terms of the parameters used in our examples, we have the following:

- c – Create an archive
- x – Extract an archive
- f – Or --file, to select an output archive filename
- p – Option to preserve permissions
- C – Select an output folder
- z – Use gzip to compress the tar archive
- j – Use bzip2 to compress the tar archive
- J – Use xzip to compress the tar archive

These are the most frequently used tar parameters, which is why we specifically selected those for our recipe.

This concludes our tar recipe, and we're ready to move on to the final recipe in this chapter, which is all about managing services by using the systemctl command. Let's work on that for a bit.

There's more...

If you want to learn more about the `tar` command, make sure that you refer to the `tar` command man page: `https://man7.org/linux/man-pages/man1/tar.1.html`.

Managing services and targets

Managing services tends to be something that we need to do at times. For example, when we install a new piece of software that comes bundled as a service, we need to be able to manage it so that it can work properly. This is what we're going to work on as we go through this recipe. We're also going to give a short description of how `systemctl` configuration files work, but without going on a 100 mile-long journey, as it's the recipe that's the focus. However, we will make sure that we provide you with additional links where you can learn a whole lot more about `systemd` as it's a big topic and an important one.

Getting ready

We're going to use our `cli2` CentOS for this recipe, just so that it doesn't feel left out.

How to do it...

The basic idea behind managing services in practical terms is to have services start either at the point in time when we want them to be started, or to have them available after we boot our Linux server.

The management of services and targets became quite a bit easier over the past couple of releases of any Linux distribution. If you've been using CentOS for an extended period, you will probably remember upstart, init, and all those beautiful things that will remain buried deep in our not-so-fond memories. In terms of the management of services, both from the administrative and *development* perspective (we'll get to that in a sec), things became much easier with CentOS 7. CentOS 8 follows the same path. There are, and always will be, differing opinions regarding the whole idea of `systemd`, but that's not the subject at hand here. So, let's focus on services and targets. First, we'll log in as root and type some commands, starting with the following:

```
systemctl set-default multi-user.target
```

This is going to switch our default boot target to multi-user, which means that our CentOS machine is going to be booted to text mode by default. So, after we reboot the machine, it's going to be started in text mode. Then, we're going to switch to using text mode instantly:

```
systemctl isolate multi-user.target
```

This is going to kill all the GUI processes, check the service delta between `graphical.target` and `multi-user.target`, and do its magic.

The next thing that we're going to do is, we're going to pick a service (for example, `sshd`), and use the `systemctl` command to manage it – both momentarily (manage its state at the time of command execution) and permanently (manage what happens with the `sshd` service during system boot). Let's type in these commands:

```
systemctl stop sshd.service
systemctl status sshd.service
```

The result of these two commands is going to be as follows:

```
[root@localhost ~]# systemctl stop sshd.service
[root@localhost ~]# systemctl status sshd.service
● sshd.service - OpenSSH server daemon
   Loaded: loaded (/usr/lib/systemd/system/sshd.service; enabled; vendor preset>
   Active: inactive (dead) since Tue 2020-12-08 00:05:47 CET; 4s ago
     Docs: man:sshd(8)
           man:sshd_config(5)
  Process: 1104 ExecStart=/usr/sbin/sshd -D $OPTIONS $CRYPTO_POLICY (code=exite>
 Main PID: 1104 (code=exited, status=0/SUCCESS)

Dec 08 00:00:26 localhost.localdomain systemd[1]: Starting OpenSSH server daemo>
Dec 08 00:00:26 localhost.localdomain sshd[1104]: Server listening on 0.0.0.0 p>
Dec 08 00:00:26 localhost.localdomain sshd[1104]: Server listening on :: port 2>
Dec 08 00:00:26 localhost.localdomain systemd[1]: Started OpenSSH server daemon.
Dec 08 00:05:47 localhost.localdomain systemd[1]: Stopping OpenSSH server daemo>
Dec 08 00:05:47 localhost.localdomain sshd[1104]: Received signal 15; terminati>
Dec 08 00:05:47 localhost.localdomain systemd[1]: sshd.service: Succeeded.
Dec 08 00:05:47 localhost.localdomain systemd[1]: Stopped OpenSSH server daemon.
lines 1-16/16 (END)
```

Figure 5.18 – Using systemctl to manage a service – in this case, the SSH service

It's telling us that `sshd.service` is disabled – Active: inactive (dead). We will enable it and check its state by typing the following command:

```
systemctl start sshd.service
systemctl status sshd.service
```

Let's check the result of these last two commands:

```
[root@localhost ~]# systemctl start sshd.service
[root@localhost ~]# systemctl status sshd.service
● sshd.service - OpenSSH server daemon
   Loaded: loaded (/usr/lib/systemd/system/sshd.service; enabled; vendor preset>
   Active: active (running) since Tue 2020-12-08 00:07:36 CET; 4s ago
     Docs: man:sshd(8)
           man:sshd_config(5)
 Main PID: 3590 (sshd)
    Tasks: 1 (limit: 12116)
   Memory: 1.7M
   CGroup: /system.slice/sshd.service
           └─3590 /usr/sbin/sshd -D -oCiphers=aes256-gcm@openssh.com,chacha20-p>

Dec 08 00:07:36 localhost.localdomain systemd[1]: Starting OpenSSH server daemo>
Dec 08 00:07:36 localhost.localdomain sshd[3590]: Server listening on 0.0.0.0 p>
Dec 08 00:07:36 localhost.localdomain sshd[3590]: Server listening on :: port 2>
Dec 08 00:07:36 localhost.localdomain systemd[1]: Started OpenSSH server daemon.
lines 1-15/15 (END)
```

Figure 5.19 – Checking the state of the SSH service following a configuration change

We can see that `sshd.service` is now active and ready to accept network connections.

There's another aspect of this, which is to configure a service so that it's enabled on system boot. If we use `sshd.service` as an example:

```
systemctl enable sshd.service
```

Also, if we don't want `sshd.service` to be enabled on system boot, we can do the opposite:

```
systemctl disable sshd.service
```

When we deploy a new service, we can start and enable it at the same time. For example, let's say that we just installed the `sshd` service from a package. Let's enable and start it by using one command:

```
systemctl enable --now sshd.service
```

Of course, that pre-supposes that we know the name of any given service, which isn't always the case. Let's learn how to overcome this problem in a text mode-driven way before we wrap this scenario up with a summary table to make things easier.

If we want to list all of the available services, we can use the following command, as `systemd` and `systemctl` *know* more objects than just services (not the topic of this scenario):

```
systemctl list-units --type=service
-----
.......
```

```
.... part of the output ommited ....
systemd-journal-flush.service loaded active exited  Flush
Journal to Persistent Storage
systemd-journald.service      loaded active running Journal
Service
systemd-logind.service        loaded active running Login
Service
systemd-machined.service      loaded active running Virtual
Machine and Container Registration Service
systemd-modules-load.service  loaded active exited  Load Kernel
Modules
systemd-random-seed.service   loaded active exited  Load/Save
Random Seed
systemd-remount-fs.service    loaded active exited  Remount
Root and Kernel File Systems
systemd-sysctl.service        loaded active exited  Apply
Kernel Variables
systemd-sysusers.service      loaded active exited  Create
System Users
systemd-tmpfiles-setup-dev.service    loaded active exited
Create Static Device Nodes in /dev
systemd-tmpfiles-setup.service        loaded active exited
Create Volatile Files and Directories
systemd-udev-settle.service   loaded active exited  udev Wait
for Complete Device Initialization
systemd-udev-trigger.service  loaded active exited  udev
Coldplug all Devices
systemd-udevd.service         loaded active running udev Kernel
Device Manager
systemd-update-done.service   loaded active exited  Update is
Completed
systemd-update-utmp.service   loaded active exited  Update UTMP
about System Boot/Shutdown
systemd-user-sessions.service loaded active exited  Permit User
Sessions
tuned.service                 loaded active running Dynamic
System Tuning Daemon
udisks2.service               loaded active running Disk
Manager
```

upower.service power management	loaded active	running	Daemon for
user-runtime-dir@0.service mount wrapper	loaded active	exited	/run/user/0
user-runtime-dir@42.service user/42 mount wrapper	loaded active	exited	/run/
user@0.service Manager for UID 0	loaded active	running	User
user@42.service Manager for UID 42	loaded active	running	User
vdo.service services	loaded active	exited	VDO volume
vgauthd.service Service for open-vm-tools	loaded active	running	VGAuth
vmtoolsd.service virtual machines hosted on VMware	loaded active	running	Service for
wpa_supplicant.service supplicant	loaded active	running	WPA

When we were going through one of our previous recipes, we discussed vdo. We can clearly see the vdo service listed here. Remember that we started it by using the systemctl command?

If we want to check the list of all enabled services, we can use the following command:

```
systemctl list-units --type=service --state=enabled
```

If we need a list of services running currently, execute the following command:

```
systemctl list-units --type=service --state=running
```

The systemctl command – because of how it works and the related config files (covered in the *How it works…* section a bit later) – can also list service dependencies. For example, the sshd service needs some other services to be started so that it can work. Let's list the sshd dependencies:

```
systemctl list-dependencies sshd
```

Therefore, let's just create a table with some of the most common service names so that we can manage this problem more efficiently:

Service	systemd service name	systemd short service name
cups	cups.service	cups
firewalld	firewalld.service	firewalld
NetworkManager	NetworkManager.service	NetworkManager
rsyslog	rsyslog.service	rsyslog
sshd	sshd.service	sshd
vdo	vdo.service	vdo
postfix	postfix.service	postfix
cron	cron.service	cron
nginx	nginx.service	nginx
Apache	httpd.service (some distributions use apache2.service)	httpd
PosgreSQL	postgresql.service	postgresql
Redis	redis.service	redis
MySQL	mysql.service (or mariadb.service, depending on version)	mysql or mariadb

Table 5.1 – A table with details about services and systemd service names

We can use the short names of these services (without .service) as well as the *Tab* key to use Bash shell completion to scroll through options and service names in systemctl. We can also mask systemd services, and therefore make them invisible to the system, by linking them from the service start up perspective to /dev/null:

```
systemctl mask cups.service
Created symlink /etc/systemd/system/cups.service → /dev/null.
```

Here, we can get an idea of how all of this works. The systemctl command obviously uses some configuration files to do its job. Let's now discuss how and what's done.

We could write books about systemd, but having this specific scenario in mind, we need to stick to the job at hand. We used the systemctl command to manage a service – right now (start/stop/restart), and permanently (enable/disable).

How it works...

From the perspective of purely managing services, the systemctl command looks by checking into its configuration files. So, let's check the anatomy of a service file for systemd, again, using sshd as an example:

```
[Unit]
Description=OpenSSH server daemon
Documentation=man:sshd(8) man:sshd_config(5)
After=network.target sshd-keygen.target
Wants=sshd-keygen.target

[Service]
Type=notify
EnvironmentFile=-/etc/crypto-policies/back-ends/opensshserver.config
EnvironmentFile=-/etc/sysconfig/sshd
ExecStart=/usr/sbin/sshd -D $OPTIONS $CRYPTO_POLICY
ExecReload=/bin/kill -HUP $MAINPID
KillMode=process
Restart=on-failure
RestartSec=42s

[Install]
WantedBy=multi-user.target
```

Figure 5.20 – Systemd configuration file – in this example, SSHD

It's almost readable without a lot of explanation, which is one of the big differences between these service files and what was used in the past.

The first part, which starts with the [Unit] section, is related to the general settings of the service – with a description and man pages for documentation being the first part of that. Then a statement tells us the order; that is, after which services should this specific service be started. Wants is related to dependencies – in this case, which targets need to be enabled for this service to be successfully started.

The [Service] section is a bit trickier, as it tells us basic configuration details and start up options (the EnvironmentFile option), which commands should be used for starting and reloading the service, how to kill the service, and details related to restarting. Restarting is for selecting if the service will be restarted when there's a timeout, kill, or exit from the process service. RestartSec is about sleep time before the service restart.

The [Install] section is a bit more global and related to how systemd works with this unit. WantedBy is used in the sense of creating additional dependencies between this specific service and other services, completely opposite to what [Unit] statements do.

This is why, when we change or create new systemd unit/service/whatever files, we have to use the `systemctl daemon-reload` command. That command specifies that `systemctl` goes through all the config files and treats them as *yes, the administrator might have changed something in any of these files, but this is on purpose and OK.*

There's more...

Bearing in mind the importance of systemd and its internals, as promised, let's provide our readers with some additional content related to systemd:

- Introduction to systemd: `https://access.redhat.com/documentation/en-us/red_hat_enterprise_linux/8/html/configuring_basic_system_settings/introduction-to-systemd_configuring-basic-system-settings`

- Managing system services with systemctl: `https://access.redhat.com/documentation/en-us/red_hat_enterprise_linux/8/html/configuring_basic_system_settings/managing-system-services-with-systemctl_configuring-basic-system-settings`

- Working with systemd targets: `https://access.redhat.com/documentation/en-us/red_hat_enterprise_linux/8/html/configuring_basic_system_settings/working-with-systemd-targets_configuring-basic-system-settings`

- Working with systemd unit files: `https://access.redhat.com/documentation/en-us/red_hat_enterprise_linux/8/html/configuring_basic_system_settings/assembly_working-with-systemd-unit-files_configuring-basic-system-settings`

- Optimizing systemd to shorten the boot time: `https://access.redhat.com/documentation/en-us/red_hat_enterprise_linux/8/html/configuring_basic_system_settings/optimizing-systemd-to-shorten-the-boot-time_configuring-basic-system-settings`

- Demystifying systemd: `https://www.youtube.com/watch?v=tY9GYsoxeLg`

6
Shell-Based Software Management

Copying content over a network is usually done as a manual process – for example, we just use `scp` or `FTP` to transfer a file and that's that. But what happens if we need to make this process a permanent one? We then need to figure out a way to perform file/directory synchronization, which is what `rsync` is all about. That being said, with all of the security-related incidents in the past 5+ years, it's always a good idea to implement some kind of encryption, so using `ssh` and `scp` seems like a reasonable approach. And that's exactly what we are going to do.

In this chapter, we are going to learn about the following topics:

- Using `dnf` and `apt` for package management
- Using additional repositories, streams, and profiles
- Creating custom repositories
- Compiling third-party software

Technical requirements

For these recipes, we're going to use two Linux machines – we can use the `cli1` and `cli2` virtual machines from our previous recipes. These recipes are doable both on CentOS and/or Ubuntu, so there is no reason to use separate virtual machines for these scenarios.

So, let's start our virtual machines, and let's get cracking!

Using dnf and apt for package management

Packages and package groups are different ways of deploying software to our CentOS and Ubuntu virtual machines. A package is nothing more than a stack of files that can be installed on our machine in an automated fashion, without our manual input. Package groups are more of a RedHat/CentOS concept. Just like the term suggests, they are a way of grouping packages into larger groups so that we can use these groups to install multiple packages without manually specifying every single package from the group. Let's learn how to use them to our benefit, specifically, for deployment purposes.

Getting ready

Let's continue using our `cli1` and `cli2` machines for this one, so make sure that they're powered on and ready to go. We are going to use `cli1` for the `apt` part of this recipe, and `cli2` for the `yum`/`dnf` part, as `cli1` is Ubuntu-based and `cli2` is CentOS-based.

How to do it...

Let's start with the basics of `yum` and `dnf` for CentOS on cli2. Let's list all the available packages on the system:

```
yum list
```

The output should look like this (abbreviated):

```
xorg-x11-xinit-session.x86_64       1.3.4-18.el8                    AppStream
xorg-x11-xkb-utils.x86_64           7.7-27.el8                      AppStream
xorriso.x86_64                      1.4.8-4.el8                     AppStream
xrestop.x86_64                      0.4-21.el8                      AppStream
xsane.x86_64                        0.999-30.el8                    AppStream
xsane-common.x86_64                 0.999-30.el8                    AppStream
xsane-gimp.x86_64                   0.999-30.el8                    AppStream
xterm.x86_64                        331-1.el8                       AppStream
xterm-resize.x86_64                 331-1.el8                       AppStream
xz-devel.i686                       5.2.4-3.el8                     BaseOS
xz-devel.x86_64                     5.2.4-3.el8                     BaseOS
xz-libs.i686                        5.2.4-3.el8                     BaseOS
yajl.i686                           2.1.0-10.el8                    AppStream
yajl.x86_64                         2.1.0-10.el8                    AppStream
yelp.x86_64                         2:3.28.1-3.el8                  AppStream
yelp-libs.i686                      2:3.28.1-3.el8                  AppStream
yelp-libs.x86_64                    2:3.28.1-3.el8                  AppStream
yelp-tools.noarch                   3.28.0-3.el8                    AppStream
yelp-xsl.noarch                     3.28.0-2.el8                    AppStream
yp-tools.x86_64                     4.2.3-1.el8                     AppStream
ypbind.x86_64                       3:2.5-2.el8                     AppStream
ypserv.x86_64                       4.0-6.20170331git5bfba76.el8    AppStream
yum.noarch                          4.2.17-7.el8_2                  BaseOS
yum-utils.noarch                    4.0.12-4.el8_2                  BaseOS
zenity.x86_64                       3.28.1-1.el8                    AppStream
zip.x86_64                          3.0-23.el8                      BaseOS
zlib.i686                           1.2.11-16.el8_2                 BaseOS
zlib.x86_64                         1.2.11-16.el8_2                 BaseOS
zlib-devel.i686                     1.2.11-16.el8_2                 BaseOS
zlib-devel.x86_64                   1.2.11-16.el8_2                 BaseOS
zsh.x86_64                          5.5.1-6.el8_1.2                 BaseOS
zsh-html.noarch                     5.5.1-6.el8_1.2                 AppStream
zstd.x86_64                         1.4.2-2.el8                     AppStream
zziplib.i686                        0.13.68-8.el8                   AppStream
zziplib.x86_64                      0.13.68-8.el8                   AppStream
zziplib-utils.x86_64                0.13.68-8.el8                   AppStream
```

Figure 6.1 – Shortened yum list output

We've shortened this screenshot in *Figure 6.1* as it contains thousands of packages. There are three columns in this output. Going from left to right, the first column is the package name, the second column is the package version, and the third one is the package repository where that specific package is located.

If we want to find out more details about a package, we can use yum info (or dnf info), for example:

```
[root@localhost ~]# dnf info NetworkManager
Last metadata expiration check: 1:16:19 ago on Wed 02 Dec 2020 02:29:33 PM CET.
Installed Packages
Name         : NetworkManager
Epoch        : 1
Version      : 1.22.8
Release      : 4.el8
Architecture : x86_64
Size         : 8.6 M
Source       : NetworkManager-1.22.8-4.el8.src.rpm
Repository   : @System
From repo    : anaconda
Summary      : Network connection manager and user applications
URL          : http://www.gnome.org/projects/NetworkManager/
License      : GPLv2+ and LGPLv2+
Description  : NetworkManager is a system service that manages network interfaces and
             : connections based on user or automatic configuration. It supports
             : Ethernet, Bridge, Bond, VLAN, Team, InfiniBand, Wi-Fi, mobile broadband
             : (WWAN), PPPoE and other devices, and supports a variety of different VPN
             : services.
```

Figure 6.2 – Getting information regarding a package

By using this command, we get much more information about the package. Also, please note that we didn't use x86_64 in the package name, as it's not necessary. Bearing in mind the fact that we're using a 64-bit distribution, it becomes understandable that using *architecture* in the package name is almost always unnecessary.

Let's now install a package, for example, mc (Midnight Commander):

```
[root@localhost ~]# dnf -y install mc
Last metadata expiration check: 1:28:45 ago on Wed 02 Dec 2020 08:59:37 AM CET.
Dependencies resolved.
================================================================================
 Package              Architecture     Version            Repository      Size
================================================================================
Installing:
 mc                   x86_64           1:4.8.19-9.el8      AppStream       1.9 M
Installing dependencies:
 perl-File-Temp       noarch           0.230.600-1.el8     BaseOS          63 k

Transaction Summary
================================================================================
Install  2 Packages

Total download size: 2.0 M
Installed size: 6.9 M
Downloading Packages:
(1/2): perl-File-Temp-0.230.600-1.el8.noarch.rpm        2.2 MB/s |  63 kB   00:00
(2/2): mc-4.8.19-9.el8.x86_64.rpm                       8.2 MB/s | 1.9 MB   00:00
--------------------------------------------------------------------------------
Total                                                   5.8 MB/s | 2.0 MB   00:00
Running transaction check
Transaction check succeeded.
Running transaction test
Transaction test succeeded.
Running transaction
  Preparing        :                                                         1/1
  Installing       : perl-File-Temp-0.230.600-1.el8.noarch                   1/2
  Installing       : mc-1:4.8.19-9.el8.x86_64                                2/2
  Running scriptlet: mc-1:4.8.19-9.el8.x86_64                                2/2
  Verifying        : mc-1:4.8.19-9.el8.x86_64                                1/2
  Verifying        : perl-File-Temp-0.230.600-1.el8.noarch                   2/2
Installed products updated.

Installed:
  mc-1:4.8.19-9.el8.x86_64                perl-File-Temp-0.230.600-1.el8.noarch

Complete!
[root@localhost ~]#
```

Figure 6.3 – Installing a package

The beauty of Linux package systems is evident here. It's not only about the fact that a package gets installed without hassle – dependencies get installed by default, as well, and that's really useful. Back in the days when we only had the rpm command to install packages in CentOS, it was much more difficult to resolve dependencies. We had to deploy them before deploying the package that we wanted to deploy, and in a specific order, which complicated the deployment process.

We can remove that package by using the following command:

```
dnf -y remove mc
```

If we want to find which package installed a specific file, we can use the `yum provides` or `dnf provides` command:

```
[root@localhost ~]# yum provides /usr/bin/ls
Last metadata expiration check: 1:01:04 ago on Wed 02 Dec 2020 02:29:33 PM CET.
coreutils-8.30-6.el8_1.1.x86_64 : A set of basic GNU tools commonly used in shell scripts
Repo        : @System
Matched from:
Filename    : /usr/bin/ls

coreutils-8.30-7.el8_2.1.x86_64 : A set of basic GNU tools commonly used in shell scripts
Repo        : BaseOS
Matched from:
Filename    : /usr/bin/ls

coreutils-single-8.30-7.el8_2.1.x86_64 : coreutils multicall binary
Repo        : BaseOS
Matched from:
Filename    : /usr/bin/ls
```

Figure 6.4 – Checking which package installed a specific file

If we need to find package dependencies (which package depends on which package), we can use the following command:

```
[root@localhost ~]# yum deplist bash
Last metadata expiration check: 1:03:10 ago on Wed 02 Dec 2020 02:29:33 PM CET.
package: bash-4.4.19-10.el8.x86_64
  dependency: filesystem >= 3
   provider: filesystem-3.8-2.el8.x86_64
  dependency: libc.so.6(GLIBC_2.15)(64bit)
   provider: glibc-2.28-101.el8.x86_64
  dependency: libdl.so.2()(64bit)
   provider: glibc-2.28-101.el8.x86_64
  dependency: libdl.so.2(GLIBC_2.2.5)(64bit)
   provider: glibc-2.28-101.el8.x86_64
  dependency: libtinfo.so.6()(64bit)
   provider: ncurses-libs-6.1-7.20180224.el8.x86_64
  dependency: rtld(GNU_HASH)
   provider: glibc-2.28-101.el8.i686
   provider: glibc-2.28-101.el8.x86_64
[root@localhost ~]#
```

Figure 6.5 – Checking package dependencies

We used bash in this example, but we could have used any package name for this query.

We can also use dnf and yum to download and install packages locally. Let's say that we want to download and install the joe editor locally. This is how we'd do it:

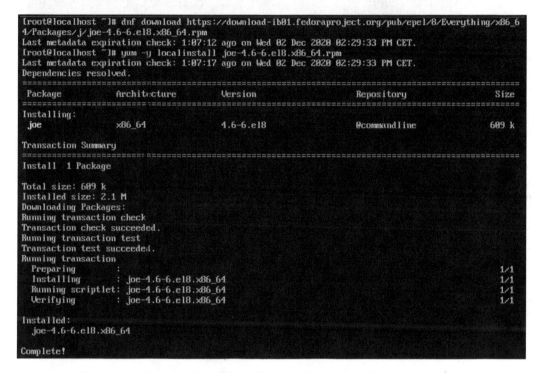

```
[root@localhost ~]# dnf download https://download-ib01.fedoraproject.org/pub/epel/8/Everything/x86_6
4/Packages/j/joe-4.6-6.el8.x86_64.rpm
Last metadata expiration check: 1:07:12 ago on Wed 02 Dec 2020 02:29:33 PM CET.
[root@localhost ~]# yum -y localinstall joe-4.6-6.el8.x86_64.rpm
Last metadata expiration check: 1:07:17 ago on Wed 02 Dec 2020 02:29:33 PM CET.
Dependencies resolved.
================================================================================
 Package          Architecture        Version            Repository        Size
================================================================================
Installing:
 joe              x86_64              4.6-6.el8           @commandline      609 k

Transaction Summary
================================================================================
Install  1 Package

Total size: 609 k
Installed size: 2.1 M
Downloading Packages:
Running transaction check
Transaction check succeeded.
Running transaction test
Transaction test succeeded.
Running transaction
  Preparing        :                                                        1/1
  Installing       : joe-4.6-6.el8.x86_64                                   1/1
  Running scriptlet: joe-4.6-6.el8.x86_64                                   1/1
  Verifying        : joe-4.6-6.el8.x86_64                                   1/1

Installed:
  joe-4.6-6.el8.x86_64

Complete!
```

Figure 6.6 – Downloading and installing a package manually from a local disk

We can, of course, search for packages by using the `yum search` or `dnf search` command:

```
[root@localhost ~]# yum search php
Last metadata expiration check: 1:09:24 ago on Wed 02 Dec 2020 02:29:33 PM CET.
============================== Name & Summary Matched: php ==============================
php.x86_64 : PHP scripting language for creating dynamic web sites
php-common.x86_64 : Common files for PHP
php-fpm.x86_64 : PHP FastCGI Process Manager
php-dbg.x86_64 : The interactive PHP debugger
php-cli.x86_64 : Command-line interface for PHP
php-pgsql.x86_64 : A PostgreSQL database module for PHP
php-devel.x86_64 : Files needed for building PHP extensions
php-xml.x86_64 : A module for PHP applications which use XML
php-ldap.x86_64 : A module for PHP applications that use LDAP
php-json.x86_64 : JavaScript Object Notation extension for PHP
php-embedded.x86_64 : PHP library for embedding in applications
php-enchant.x86_64 : Enchant spelling extension for PHP applications
php-pear.noarch : PHP Extension and Application Repository framework
php-intl.x86_64 : Internationalization extension for PHP applications
php-odbc.x86_64 : A module for PHP applications that use ODBC databases
php-dba.x86_64 : A database abstraction layer module for PHP applications
php-pdo.x86_64 : A database access abstraction module for PHP applications
php-soap.x86_64 : A module for PHP applications that use the SOAP protocol
php-gmp.x86_64 : A module for PHP applications for using the GNU MP library
php-mysqlnd.x86_64 : A module for PHP applications that use MySQL databases
php-process.x86_64 : Modules for PHP script using system process interfaces
php-bcmath.x86_64 : A module for PHP applications for using the bcmath library
php-recode.x86_64 : A module for PHP applications for using the recode library
php-gd.x86_64 : A module for PHP applications for using the gd graphics library
php-snmp.x86_64 : A module for PHP applications that query SNMP-managed devices
php-xmlrpc.x86_64 : A module for PHP applications which use the XML-RPC protocol
php-mbstring.x86_64 : A module for PHP applications which need multi-byte string handling
================================== Name Matched: php ==================================
php-opcache.x86_64 : The Zend OPcache
php-pecl-zip.x86_64 : A ZIP archive management extension
php-pecl-apcu.x86_64 : APC User Cache
php-pecl-apcu-devel.x86_64 : APCu developer files (header)
```

Figure 6.7 – Using the yum/dnf search command

Sometimes, the list of these packages is going to be quite long, so additional filtering might be required.

Let's now talk a bit about package groups, starting with the `dnf grouplist` command:

```
[root@localhost ~]# dnf grouplist
Last metadata expiration check: 1:35:37 ago on Wed 02 Dec 2020 08:59:37 AM CET.
Available Environment Groups:
   Server with GUI
   Minimal Install
   Workstation
   Virtualization Host
   Custom Operating System
Installed Environment Groups:
   Server
Installed Groups:
   Container Management
   Headless Management
Available Groups:
   .NET Core Development
   RPM Development Tools
   Development Tools
   Graphical Administration Tools
   Legacy UNIX Compatibility
   Network Servers
   Scientific Support
   Security Tools
   Smart Card Support
   System Tools
[root@localhost ~]# 
```

Figure 6.8 – Using dnf group list commands gives us a list of package groups

The output of that command is going to give us the names of **package groups** that we can use for much *larger* package deployments. For example, let's check what's going to happen if we install the **Development Tools** package group by issuing the following command:

```
dnf groupinstall "Development Tools"
```

This command will ask us whether we want to download and deploy more than 100 packages. If we answer *yes*, that's exactly what's going to happen (the screenshot is rendered smaller on purpose, just to show the end of the command output):

```
xorg-x11-font-utils                 x86_64 1:7.5-40.el8              AppStream 103 k
xorg-x11-fonts-ISO8859-1-100dpi     noarch 7.5-19.el8               AppStream 1.1 M
xorg-x11-server-utils               x86_64 7.7-27.el8               AppStream 198 k
zstd                                x86_64 1.4.2-2.el8              AppStream 385 k
Installing weak dependencies:
gcc-gdb-plugin                      x86_64 8.3.1-5.el8.0.2          AppStream 117 k
kernel-devel                        x86_64 4.18.0-193.28.1.el8_2    BaseOS     15 M
perl-IO-Socket-IP                   noarch 0.39-5.el8              AppStream  47 k
perl-IO-Socket-SSL                  noarch 2.066-4.el8             AppStream 297 k
perl-Mozilla-CA                     noarch 20160104-7.el8          AppStream  15 k
Enabling module streams:
javapackages-runtime                       201801
Installing Groups:
Development Tools

Transaction Summary
================================================================================
Install   161 Packages

Total download size: 185 M
Installed size: 602 M
Is this ok [y/N]:
```

Figure 6.9 – Installing a package group

As we can see, having the ability to deploy package groups greatly increases the speed of package deployment.

The next step in our process is to cover everything that we've covered in this recipe, but to deliver it on Ubuntu. So, let's switch to our `cli1` machine and start from scratch. First, let's describe a couple of commands that we're interested in:

- `apt-get` or `apt`: Commands used to install, remove, upgrade, and update packages

- `apt-cache`: Mostly used to search and find information about packages

Let's now learn to use them. First, we are going to discuss regular operations – installing, removing, purging, updating, and upgrading. Let's install a package, for example, `mc`:

```
root@cli1:~# apt-get install mc
Reading package lists... Done
Building dependency tree
Reading state information... Done
Suggested packages:
  arj catdvi | texlive-binaries dbview djvulibre-bin epub-utils genisoimage gv imagemagick
  libaspell-dev links | w3m | lynx odt2txt poppler-utils python python-boto python-tz xpdf
  | pdf-viewer zip
The following NEW packages will be installed:
  mc
0 upgraded, 1 newly installed, 0 to remove and 1 not upgraded.
Need to get 0 B/486 kB of archives.
After this operation, 1530 kB of additional disk space will be used.
Selecting previously unselected package mc.
(Reading database ... 129957 files and directories currently installed.)
Preparing to unpack .../mc_3%3a4.8.25-1_amd64.deb ...
Unpacking mc (3:4.8.25-1) ...
Setting up mc (3:4.8.25-1) ...
Processing triggers for mime-support (3.64ubuntu1) ...
root@cli1:~# _
```

Figure 6.10 – Using apt-get to install a package

Now, let's remove it:

```
root@cli1:~# apt-get -y remove mc
Reading package lists... Done
Building dependency tree
Reading state information... Done
The following packages were automatically installed and are no longer required:
  libssh2-1 mc-data unzip
Use 'apt autoremove' to remove them.
The following packages will be REMOVED:
  mc
0 upgraded, 0 newly installed, 1 to remove and 1 not upgraded.
After this operation, 1530 kB disk space will be freed.
(Reading database ... 130055 files and directories currently installed.)
Removing mc (3:4.8.25-1) ...
Processing triggers for mime-support (3.64ubuntu1) ...
root@cli1:~# _
```

Figure 6.11 – Using apt-get to remove a package

We can see that we have a standard situation – the package *was* removed, but some of its dependencies weren't. We can do that, as well, by using the `apt-get` `autoremove` command:

```
root@cli1:~# apt-get autoremove
Reading package lists... Done
Building dependency tree
Reading state information... Done
The following packages will be REMOVED:
  libssh2-1 mc-data unzip
0 upgraded, 0 newly installed, 3 to remove and 1 not upgraded.
After this operation, 7137 kB disk space will be freed.
Do you want to continue? [Y/n] y
(Reading database ... 129966 files and directories currently installed.)
Removing libssh2-1:amd64 (1.8.0-2.1build1) ...
Removing mc-data (3:4.8.25-1) ...
Removing unzip (6.0-25ubuntu1) ...
Processing triggers for hicolor-icon-theme (0.17-2) ...
Processing triggers for libc-bin (2.32-0ubuntu3) ...
Processing triggers for man-db (2.9.3-2) ...
Processing triggers for mime-support (3.64ubuntu1) ...
root@cli1:~# _
```

Figure 6.12 – Removing packages that are no longer needed

This is very useful as we're reducing the attack surface of our server (for security breaches) by removing unnecessary software packages.

Now let's check what happens if we use the `update` option:

```
root@cli1:~# apt-get -y update
Hit:1 http://hr.archive.ubuntu.com/ubuntu groovy InRelease
Hit:2 http://hr.archive.ubuntu.com/ubuntu groovy-updates InRelease
Hit:3 http://hr.archive.ubuntu.com/ubuntu groovy-backports InRelease
Hit:4 http://hr.archive.ubuntu.com/ubuntu groovy-security InRelease
Reading package lists... Done
root@cli1:~#
```

Figure 6.13 – Updating repository and package info

As we can see, `apt` refreshed its package list before the upgrade process could happen – these steps are mostly used in sequence – `update` followed by `upgrade`:

```
root@cli1:~# apt-get -y upgrade
Reading package lists... Done
Building dependency tree
Reading state information... Done
Calculating upgrade... Done
The following packages have been kept back:
    ubuntu-advantage-tools
0 upgraded, 0 newly installed, 0 to remove and 1 not upgraded.
root@cli1:~# _
```

Figure 6.14 – Upgrading available packages – this time, no upgrades are necessary

Interestingly enough, no packages were installed, which is a very rare situation, to be honest. Usually, we'd have at least a few packages to be upgraded.

> **Note**
>
> Before we get into the topic of doing dist-upgrade, we are absolutely *NOT* recommending this for a production server. Using dist-upgrade and do-release-upgrade is something that we *can* do, but shouldn't. Migration is always a better idea, however much time it might take.

Let's now push this situation to the extreme by trying to do dist-upgrade, followed by do-release-upgrade. What the dist-upgrade apt option does is simple in theory – it tries to prepare our current distribution so that it's possible to upgrade it to the latest one in the branch. At first, it might just be getting a couple of new packages. Usually, these packages contain new repositories and information about locations, from which apt will upgrade our distribution to the latest one. Here's an example:

```
root@cli1:~# apt-get dist-upgrade
Reading package lists... Done
Building dependency tree
Reading state information... Done
Calculating upgrade... Done
The following NEW packages will be installed:
    distro-info
The following packages will be upgraded:
    ubuntu-advantage-tools
1 upgraded, 1 newly installed, 0 to remove and 0 not upgraded.
Need to get 855 kB of archives.
After this operation, 2979 kB of additional disk space will be used.
Do you want to continue? [Y/n]
```

Figure 6.15 – Using dist-upgrade to get information about new distribution versions

The next step after that one is to use do-release-upgrade, a standalone command that's not an apt subcommand. We need to remember that this is not an apt option (there's no apt do-release-upgrade, it's just do-release-upgrade). After executing it, our system is going to ask us whether we want to continue with the distribution release upgrade:

```
Checking package manager
Reading package lists... Done
Building dependency tree
Reading state information... Done

Calculating the changes

Calculating the changes

Do you want to start the upgrade?


1 package is going to be removed. 56 new packages are going to be
installed. 559 packages are going to be upgraded.

You have to download a total of 516 M. This download will take about
1 minute with your connection.

Installing the upgrade can take several hours. Once the download has
finished, the process cannot be canceled.

Continue [yN]  Details [d]_
```

Figure 6.16 – Using do-release-upgrade, not to be recommended in production environments

If we confirm, the process is going to start, and it's going to take a while. The end result should be an Ubuntu machine that's fully updated to the latest version, with all of the latest package versions. Remember, we specifically mentioned that this shouldn't be done in production – it's just an extreme example of using apt capabilities to do a system-wide package upgrade. Hundreds, perhaps thousands of packages, will get updated if we do something like that, and the process isn't reversible, so it carries a lot of risk. Try it out on some test virtual machine just for practice. If successful, this procedure will upgrade to the latest Ubuntu version. The end result, at the time of writing, looks like this:

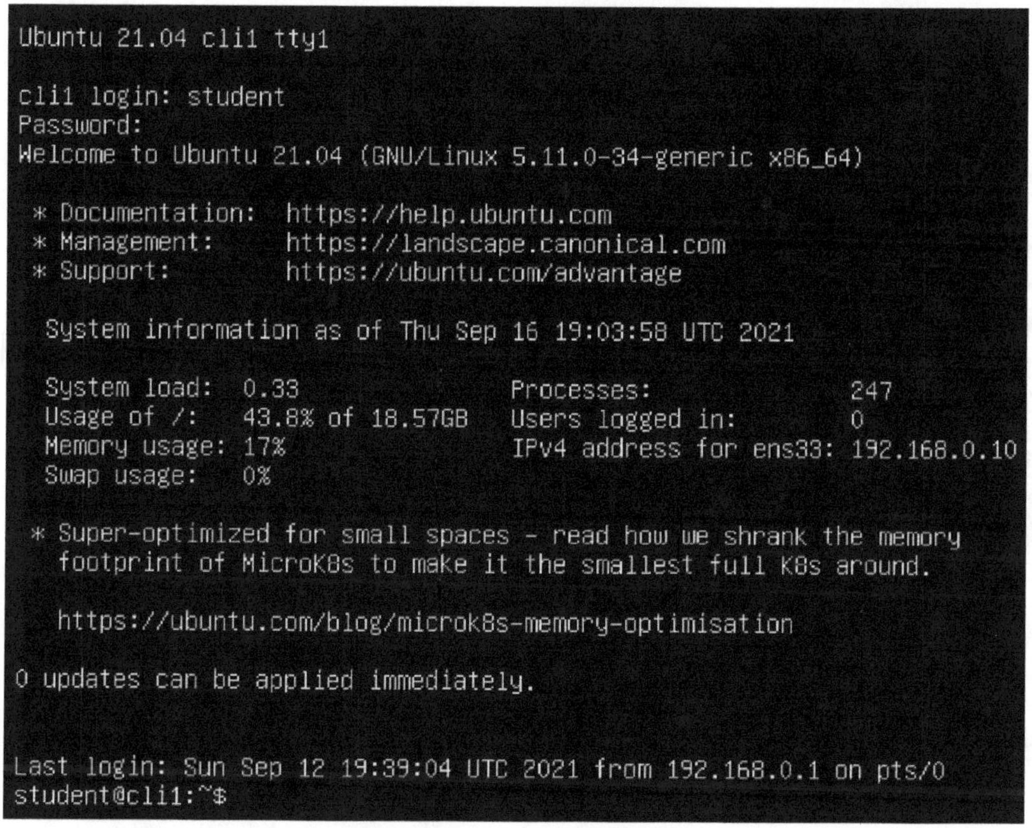

```
Ubuntu 21.04 cli1 tty1

cli1 login: student
Password:
Welcome to Ubuntu 21.04 (GNU/Linux 5.11.0-34-generic x86_64)

 * Documentation:  https://help.ubuntu.com
 * Management:      https://landscape.canonical.com
 * Support:         https://ubuntu.com/advantage

  System information as of Thu Sep 16 19:03:58 UTC 2021

  System load:   0.33              Processes:             247
  Usage of /:    43.8% of 18.57GB  Users logged in:       0
  Memory usage:  17%               IPv4 address for ens33: 192.168.0.10
  Swap usage:    0%

 * Super-optimized for small spaces - read how we shrank the memory
   footprint of MicroK8s to make it the smallest full K8s around.

   https://ubuntu.com/blog/microk8s-memory-optimisation

0 updates can be applied immediately.


Last login: Sun Sep 12 19:39:04 UTC 2021 from 192.168.0.1 on pts/0
student@cli1:~$
```

Figure 6.17 – The end result of do-release-upgrade, and in our experience, we got lucky this time!

Notice that our Ubuntu machine was upgraded to the latest (21.04) version.

There are a few more important `apt` commands – for example, for a package search, we can use the `apt-cache showpkg package_name` command. Let's use it, for example, on `mc`, a package we installed previously:

```
root@cli1:~# apt-cache showpkg mc | more;
Package: mc
Versions:
3:4.8.26-1 (/var/lib/apt/lists/hr.archive.ubuntu.com_ubuntu_dists_hirsute_universe_binary-amd64_Pack
ages)
 Description Language:
             File: /var/lib/apt/lists/hr.archive.ubuntu.com_ubuntu_dists_hirsute_universe_binary
-amd64_Packages
             MD5: 252a5c5aeeb7425db45357d4ab8aa55f
 Description Language: en
             File: /var/lib/apt/lists/hr.archive.ubuntu.com_ubuntu_dists_hirsute_universe_i18n_T
ranslation-en
             MD5: 252a5c5aeeb7425db45357d4ab8aa55f
 Description Language:
             File: /var/lib/dpkg/status
             MD5: 252a5c5aeeb7425db45357d4ab8aa55f

3:4.8.25-1 (/var/lib/dpkg/status)
 Description Language:
             File: /var/lib/apt/lists/hr.archive.ubuntu.com_ubuntu_dists_hirsute_universe_binary
-amd64_Packages
             MD5: 252a5c5aeeb7425db45357d4ab8aa55f
 Description Language: en
             File: /var/lib/apt/lists/hr.archive.ubuntu.com_ubuntu_dists_hirsute_universe_i18n_T
ranslation-en
             MD5: 252a5c5aeeb7425db45357d4ab8aa55f
 Description Language:
             File: /var/lib/dpkg/status
             MD5: 252a5c5aeeb7425db45357d4ab8aa55f
```

Figure 6.18 – Using apt-get to get package info

There is a somewhat shorter version of the same thing if we use the `apt` command:

```
root@cli1:~# apt show mc
Package: mc
Version: 3:4.8.26-1
Priority: optional
Section: universe/utils
Origin: Ubuntu
Maintainer: Ubuntu Developers <ubuntu-devel-discuss@lists.ubuntu.com>
Original-Maintainer: Dmitry Smirnov <onlyjob@debian.org>
Bugs: https://bugs.launchpad.net/ubuntu/+filebug
Installed-Size: 1542 kB
Provides: mcedit
Depends: libc6 (>= 2.15), libext2fs2 (>= 1.37), libglib2.0-0 (>= 2.59.2), libgpm2 (>= 1.20.7), libsl
ang2 (>= 2.2.4), libssh2-1 (>= 1.2.8), mc-data (= 3:4.8.26-1)
Recommends: mime-support, perl, unzip, sensible-utils
Suggests: arj, bzip2, catdvi | texlive-binaries, dbview, djvulibre-bin, epub-utils, file, genisoimag
e, gv, imagemagick, libaspell-dev, links | w3m | lynx, odt2txt, poppler-utils, python, python-boto,
python-tz, unar, wimtools, xpdf | pdf-viewer, zip
Homepage: https://www.midnight-commander.org
Download-Size: 489 kB
APT-Sources: http://hr.archive.ubuntu.com/ubuntu hirsute/universe amd64 Packages
Description: Midnight Commander - a powerful file manager
 GNU Midnight Commander is a text-mode full-screen file manager. It
 uses a two panel interface and a subshell for command execution. It
 includes an internal editor with syntax highlighting and an internal
 viewer with support for binary files. Also included is Virtual
 Filesystem (VFS), that allows files on remote systems (e.g. FTP, SSH
 servers) and files inside archives to be manipulated like real files.

N: There is 1 additional record. Please use the '-a' switch to see it
```

Figure 6.19 – Using apt to get package info – somewhat shorter and more concise

If we need to add repositories, we can use the `add-apt-repository` command. Let's say that we want to add an unofficial repository, such as **Personal Package Archives (PPA)**, that is hosted on Launchpad. Generally speaking, we should only add *reputable* repositories, and not just *any* repository just because it has a certain package that we might need. We are going to use an example here – let's say that we need to install the latest PHP 7.4 version on our Ubuntu machine. We can do it like this:

```
apt-get install software-properties-common
add-apt-repository ppa:ondrej/php
apt-get update
apt-get install -y php7.4
```

This should be the result if we started `php` from the shell:

```
root@cli1:~# php -v
PHP 7.4.9 (cli) (built: Jul  5 2021 13:33:00) ( NTS )
Copyright (c) The PHP Group
Zend Engine v3.4.0, Copyright (c) Zend Technologies
    with Zend OPcache v7.4.9, Copyright (c), by Zend Technologies
root@cli1:~#
```

Figure 6.20 – The end result of us using the ppa repository to deploy the latest release of PHP 7.4

This covers all the necessary commands we need for both Ubuntu and CentOS. Let's now explain some background information about where some of the more important information is stored – for both CentOS (dnf/yum) and Ubuntu (apt/apt-get).

How it works...

yum and dnf work in tandem with files located in /etc/yum.repos.d repository files, as well as the /etc/yum.conf configuration file. We covered repository files, so let's now discuss /etc/yum.conf and a couple of important configuration options that we can use from it. This is a global configuration file for dnf and yum commands.

There are a couple of really useful configuration items that we can manage in it. Let's just illustrate that point by using two commonly used examples. Let's add these two options to it:

```
exclude: kernel* open-vm*
gpgcheck=0
```

By using these two commands, we instructed yum/dnf to exclude all kernel and open-vm packages (by name) in any kind of operation, such as a yum update (which updates all packages on the machine). gpgcheck=0 sets a global policy that tells yum and dnf *not* to use GPG key checking when working with packages. This can also be managed in /etc/yum.repos.d, as discussed in our recipe.

Ubuntu has a very similar principle; it's just that directories are different as well as the file structure, somewhat. The most important information regarding the software repository location is kept in the /etc/apt directory, specifically, in the /etc/apt/sources.list file. Here's an excerpt:

```
deb http://hr.archive.ubuntu.com/ubuntu groovy universe
# deb-src http://hr.archive.ubuntu.com/ubuntu groovy universe
deb http://hr.archive.ubuntu.com/ubuntu groovy-updates universe
# deb-src http://hr.archive.ubuntu.com/ubuntu groovy-updates universe

## N.B. software from this repository is ENTIRELY UNSUPPORTED by the Ubuntu
## team, and may not be under a free licence. Please satisfy yourself as to
## your rights to use the software. Also, please note that software in
## multiverse WILL NOT receive any review or updates from the Ubuntu
## security team.
deb http://hr.archive.ubuntu.com/ubuntu groovy multiverse
# deb-src http://hr.archive.ubuntu.com/ubuntu groovy multiverse
deb http://hr.archive.ubuntu.com/ubuntu groovy-updates multiverse
# deb-src http://hr.archive.ubuntu.com/ubuntu groovy-updates multiverse

## N.B. software from this repository may not have been tested as
## extensively as that contained in the main release, although it includes
## newer versions of some applications which may provide useful features.
## Also, please note that software in backports WILL NOT receive any review
## or updates from the Ubuntu security team.
deb http://hr.archive.ubuntu.com/ubuntu groovy-backports main restricted universe multiverse
# deb-src http://hr.archive.ubuntu.com/ubuntu groovy-backports main restricted universe multiverse
```

Figure 6.21 – Main apt configuration file called sources.list

The general structure is simple enough. The second part of our `apt` equation is located in the `/etc/apt/sources.list.d` directory. A couple of steps ago, we added the PPA repository, and, sure enough, we have a configuration file for that repository configuration file there, called `ondrej-ubuntu-php-groovy.list`:

```
root@cli1:/etc/apt/sources.list.d# cat ondrej-ubuntu-php-groovy.list
deb http://ppa.launchpad.net/ondrej/php/ubuntu/ groovy main
# deb-src http://ppa.launchpad.net/ondrej/php/ubuntu/ groovy main
root@cli1:/etc/apt/sources.list.d#
```

Figure 6.22 – Additional apt configuration file located at /etc/apt/sources.list.d

That covers our package and package groups recipe. Let's now move on to the next recipe, which is about using modules and module streams.

There's more...

If you need more information about networking in CentOS and Ubuntu, make sure that you check out the following resources:

- **Yum cheat sheet**: `https://access.redhat.com/sites/default/files/attachments/rh_yum_cheatsheet_1214_jcs_print-1.pdf`

- **Yum to DNF cheat sheet**: `https://fedoraproject.org/wiki/Yum_to_DNF_Cheatsheet`

- **Apt cheat sheet**: `https://packagecloud.io/blog/apt-cheat-sheet/`

Using additional repositories, streams, and profiles

Repositories are the most important objects/locations to manage as they provide us with packages and package groups that we can install on our CentOS machine. Let's now learn how to manage repositories by using `yum-config-manager` and `dnf`. Also, let's get to know some configuration files that are key for this process.

Adding to the idea of package groups, which group packages into larger groups, `dnf` introduced the idea of additional modularity. It's all about package organization – we want to have simple ways of deploying software – runtimes, applications, bits and pieces of software. These concepts also enable us to have control over *versions of software* that we want to install, which is really handy. For example, let's say that you need to deploy PHP 7.2 and 7.3 on the machine. Doing that manually isn't going to be much fun. As we're going to demonstrate by using an example, this is much more easily done if we use a module stream.

Profiles act as quasi-repositories, without actually being repositories, within the **AppStream** repository. This concept enables us to additionally filter what we install. Just as an example, the httpd module has a couple of profiles (minimal, devel, common). The minimal profile means just the minimum number of packages that need to be installed for httpd to work. Unlike that, common is a default profile that's ready for production and additionally treated in terms of security (hardened).

Getting ready

Start the cli2 virtual machine created in the previous recipes. We're going to use it to work with streams and profiles on our CentOS machine.

How to do it...

To manage repositories, we have to learn to use two commands – yum-config-manager and dnf. Also, we need to look into the /etc/yum.conf file, as well as the /etc/yum.repos.d directory. Yum.conf gives us global yum command configuration options, and the /etc/yum.repos.d directory contains configuration files with repository locations.

How it works...

Let's look at yum-config-manager first. This command was introduced in Red Hat Enterprise Linux/CentOS 7 to easily add additional repositories to your Red Hat Enterprise Linux or CentOS machine. And it does just that – it lets us skip the whole manual repository configuration and get straight to business. If we didn't have this command, we would need to learn the configuration file options for /etc/yum.repos.d directory files.

If we go to the first virtual machine that we installed (source), and list the content of the /etc/yum.repos.d directory, this is what we'll get:

```
[root@localhost yum.repos.d]# ls -al
total 68
drwxr-xr-x.  2 root root 4096 Dec   1 23:01 .
drwxr-xr-x. 80 root root 8192 Dec   1 23:21 ..
-rw-r--r--.  1 root root  731 Jun   3 03:02 CentOS-AppStream.repo
-rw-r--r--.  1 root root  712 Jun   3 03:02 CentOS-Base.repo
-rw-r--r--.  1 root root  798 Jun   3 03:02 CentOS-centosplus.repo
-rw-r--r--.  1 root root 1043 Jun   3 03:02 CentOS-CR.repo
-rw-r--r--.  1 root root  668 Jun   3 03:02 CentOS-Debuginfo.repo
-rw-r--r--.  1 root root  743 Jun   3 03:02 CentOS-Devel.repo
-rw-r--r--.  1 root root  756 Jun   3 03:02 CentOS-Extras.repo
-rw-r--r--.  1 root root  338 Jun   3 03:02 CentOS-fasttrack.repo
-rw-r--r--.  1 root root  738 Jun   3 03:02 CentOS-HA.repo
-rw-r--r--.  1 root root  928 Jun   3 03:02 CentOS-Media.repo
-rw-r--r--.  1 root root  736 Jun   3 03:02 CentOS-PowerTools.repo
-rw-r--r--.  1 root root 1382 Jun   3 03:02 CentOS-Sources.repo
-rw-r--r--.  1 root root   74 Jun   3 03:02 CentOS-Vault.repo
```

Figure 6.23 – /etc/yum.repos.d directory content

Let's say that we want to add a custom repository, url, to our machine. We can do this in three different ways. The first approach involves using yum-package-manager, and that tool requires the yum-utils package (url is the location of the repository that we want to use):

```
yum -y install yum-utils
yum-config-manager --add-repo url
yum-config-manager --enable repo
```

We can also check the list of currently configured repositories by using the following command:

```
yum repolist all
```

If we need to find the list of currently disabled (unused) repositories, we can use the following command:

```
yum repolist enabled
```

If we need to enable a disabled repository, we can use the following (repository_id is a parameter that you can get from the yum repolist all command):

```
yum-config-manager --enable repository_id
```

The most obvious problem with using `yum-config-manager` is the fact that there are some parameters that we can't assign via that command itself. This is where the manual editing of `/etc/yum.repos.d` configuration files comes in handy.

This command is being phased out little by little and redirected to its new `dnf` counterparts (`dnf config-manager`), just like `yum` is being used in parallel with the `dnf` command. If we want to do the same job by using `dnf` tools, we can do this:

```
dnf config-manager --add-repo url
```

That will create a new configuration file in the `/etc/yum.repos.d` directory with the repository definition and enable it by default.

Our next step is going to be to learn a bit about these configuration files, as they're really not all that difficult to understand. Let's use a repository configuration file to explain their concept, for example, `/etc/yum.repos.d/CentOS-Sources.repo`:

```
[BaseOS-source]
name=CentOS-$releasever - BaseOS Sources
baseurl=http://vault.centos.org/$contentdir/$releasever/BaseOS/Source/
gpgcheck=1
enabled=0
gpgkey=file:///etc/pki/rpm-gpg/RPM-GPG-KEY-centosofficial
```

Figure 6.24 – Part of the /etc/yum.repos.d/CentOS-Sources.repo file

Let's explain these configuration parameters:

- [BaseOS-Source] is the repository ID. This is what we use in `yum-config-manager` or `dnf` to reference repositories.

- The `name` parameter is a description of that repository.

- The `baseurl` parameter describes *where* the location of this repository is, and it can use a variety of different options – `http`, `https`, `ftp`, or `file`. If we create a *local* repository (mount it somewhere on our CentOS machine), then the `file` statement will be used to access it.

- The `gpgcheck` parameter tells `yum`/`dnf` whether or not to check the gpg key against the package signatures. If it's 1, that means that checking is mandatory.

- The `enabled` parameter tells `yum`/`dnf` whether this repository is enabled, which means whether `dnf`/`yum` can use it to get packages. We can also use `yum --enablerepo` to enable a certain defined repository by name, or `yum --disablerepo` to do the opposite.

- The `gpgkey` parameter tells `yum`/`dnf` where the gpg key for `gpgcheck` is located.

Let's now move on to the idea of streams and profiles, the logical next step in our recipe.

After we log in to the source machine, let's use an example to describe what streams and profiles are all about. So, let's use a module stream and profile to remove and re-deploy httpd. In the first step, we're going to use the following command:

```
dnf -y remove @httpd
```

After the process is complete, let's do the opposite:

```
dnf -y install @httpd
```

Let's now check the output of the second command:

```
[root@localhost ~]# dnf -y install @httpd
Last metadata expiration check: 2:40:01 ago on Wed 02 Dec 2020 08:28:43 AM CET.
Dependencies resolved.
================================================================================
 Package      Arch       Version                        Repository    Size
================================================================================
Installing group/module packages:
 mod_ssl      x86_64     1:2.4.37-21.module_el8.2.0+494+1df74eae   AppStream    132 k
Installing dependencies:
 sscg         x86_64     2.3.3-14.el8                   AppStream     49 k
Installing module profiles:
 httpd/common


Transaction Summary
================================================================================
Install  2 Packages

Total download size: 181 k
Installed size: 360 k
Downloading Packages:
(1/2): sscg-2.3.3-14.el8.x86_64.rpm               1.9 MB/s |  49 kB     00:00
(2/2): mod_ssl-2.4.37-21.module_el8.2.0+494+1df    4.2 MB/s | 132 kB     00:00
--------------------------------------------------------------------------------
Total                                             592 kB/s | 181 kB     00:00
Running transaction check
Transaction check succeeded.
Running transaction test
Transaction test succeeded.
Running transaction
  Preparing        :                                                    1/1
  Installing       : sscg-2.3.3-14.el8.x86_64                           1/2
  Installing       : mod_ssl-1:2.4.37-21.module_el8.2.0+494+1df74eae.x86    2/2
  Running scriptlet: mod_ssl-1:2.4.37-21.module_el8.2.0+494+1df74eae.x86    2/2
  Verifying        : mod_ssl-1:2.4.37-21.module_el8.2.0+494+1df74eae.x86    1/2
  Verifying        : sscg-2.3.3-14.el8.x86_64                           2/2

Installed:
  mod_ssl-1:2.4.37-21.module_el8.2.0+494+1df74eae.x86_64
  sscg-2.3.3-14.el8.x86_64
```

Figure 6.25 – Using streams and profiles

We can see that the deployment process automatically defaulted to using the `httpd/common` profile and the default stream (**AppStream**) repository.

Let's do another example. We can check the list of all available modules by using the following command:

```
dnf module list
```

This will give us the following result:

```
[root@localhost ~]# dnf module list
Last metadata expiration check: 0:41:00 ago on Wed 02 Dec 2020 02:29:33 PM CET.
CentOS-8 - AppStream
Name                    Stream        Profiles        Summary
389-ds                  1.4                           389 Directory Server (base)
ant                     1.10 [d]      common [d]      Java build tool
container-tools         rhel8 [d]     common [d]      Common tools and dependencies for container runtimes
container-tools         1.0           common [d]      Common tools and dependencies for container runtimes
container-tools         2.0           common [d]      Common tools and dependencies for container runtimes
freeradius              3.0 [d]       server [d]      High-performance and highly configurable free RADIUS
                                                      server
gimp                    2.8 [d]       common [d], dev gimp module
                                      el
go-toolset              rhel8 [d]     common [d]      Go
httpd                   2.4 [d][e] common [d] [i], Apache HTTP Server
                                      devel, minimal
idm                     DL1           adtrust, client The Red Hat Enterprise Linux Identity Management sys
                                      , common [d], d tem module
                                      ns, server
idm                     client [d] common [d]      RHEL IdM long term support client module
inkscape                0.92.3 [d] common [d]      Vector-based drawing program using SVG
javapackages-runtime 201801 [d] common [d]      Basic runtime utilities to support Java applications
jmc                     rhel8 [d]     common [d], cor Java Mission Control is a profiling and diagnostics
                                      e               tool for the Hotspot JVM
libselinux-python       2.8           common          Python 2 bindings for libselinux
llvm-toolset            rhel8 [d]     common [d]      LLVM
mailman                 2.1 [d]       common [d]      Electronic mail discussion and e-newsletter lists ma
                                                      naging software
mariadb                 10.3 [d]      client, galera, MariaDB Module
                                      server [d]
```

Figure 6.26 – dnf module list, with versions and profiles; abridged output

Let's say that we want to install `container-tools` version 2.0. We can do it this way:

```
[root@localhost ~]# dnf module install container-tools:2.0
Last metadata expiration check: 0:53:58 ago on Wed 02 Dec 2020 02:29:33 PM CET.
Dependencies resolved.
================================================================================
 Package                    Arch    Version                         Repo    Size
================================================================================
Installing group/module packages:
 buildah                    x86_64  1.11.6-7.module_el8.2.0+458+dab581ed  AppStream  8.5 M
 cockpit-podman             noarch  11-1.module_el8.2.0+304+65a3c2ac     AppStream  1.0 M
 conmon                     x86_64  2:2.0.6-1.module_el8.2.0+304+65a3c2ac  AppStream  37 k
 container-selinux          noarch  2:2.124.0-1.module_el8.2.0+304+65a3c2ac  AppStream  47 k
 containernetworking-plugins x86_64 0.8.3-4.module_el8.2.0+304+65a3c2ac  AppStream  20 M
 criu                       x86_64  3.12-9.module_el8.2.0+304+65a3c2ac   AppStream  482 k
 fuse-overlayfs             x86_64  0.7.2-5.module_el8.2.0+304+65a3c2ac  AppStream  60 k
 podman                     x86_64  1.6.4-15.module_el8.2.0+465+f9348e8f  AppStream  12 M
 python-podman-api          noarch  1.2.0-0.2.gitd0a45fe.module_el8.2.0+304+65a3c2ac
                                                                         AppStream  43 k
 runc                       x86_64  1.0.0-64.rc10.module_el8.2.0+304+65a3c2ac  AppStream  2.7 M
 skopeo                     x86_64  1:0.1.40-9.module_el8.2.0+378+19727d8d  AppStream  5.8 M
 slirp4netns                x86_64  0.4.2-3.git21fdece.module_el8.2.0+304+65a3c2ac
                                                                         AppStream  88 k
 toolbox                    noarch  0.0.7-1.module_el8.2.0+304+65a3c2ac  AppStream  15 k
 udica                      noarch  0.2.1-2.module_el8.2.0+304+65a3c2ac  AppStream  48 k
Installing dependencies:
 PackageKit-glib            x86_64  1.1.12-4.el8                         AppStream  141 k
 cairo                      x86_64  1.15.12-3.el8                        AppStream  721 k
 cairo-gobject              x86_64  1.15.12-3.el8                        AppStream  33 k
 checkpolicy                x86_64  2.9-1.el8                            BaseOS     348 k
 cockpit-bridge             x86_64  211.3-1.el8                          BaseOS     618 k
 cockpit-system             noarch  211.3-1.el8                          BaseOS     2.3 M
 containers-common          x86_64  1:0.1.40-9.module_el8.2.0+378+19727d8d  AppStream  50 k
 dnf-plugin-subscription-manager x86_64 1.26.20-1.el8_2                 BaseOS     278 k
 fontconfig                 x86_64  2.13.1-3.el8                         BaseOS     275 k
```

Figure 6.27 – Deploying a specific module version by using the dnf command

As you can see, the result of this action is going to be quite extensive. Sometimes, when we deploy a set of packages from modules and streams, our machine is going to deploy hundreds of packages. So, be prepared to wait for a bit of time to see whether it happens.

In one of our examples, we deployed the httpd package by using the default profile and stream. Every one of these streams can have multiple profiles for our convenience. If the stream has multiple profiles, one of them can be used as the default one (and marked as such). This is not mandatory, but it's a good practice.

In terms of modules, there are 60+ modules available already, with various versions of Python, PHP, PostgreSQL, nginx, and so on, to name a few commonly used services. We can use a module from that list to deploy it. Also, the output of the command gives us details about profiles, which we can then use to deploy a specific module profile.

By using these capabilities, we can modularize our approach to deploy specific packages. The overall idea of modularization via streams and profiles is a good one, although it's a bit clunky and unfinished in terms of upgrades. That being said, it's something that is going to be around in the future, so it's a worthwhile investment of our time to learn about it.

We're done with advanced repository management for the time being. Let's now learn how to create custom repositories.

Creating custom repositories

Sometimes, it's necessary to create your own private repository of packages. Whatever the reason might be – no internet access, low deployment speed – it's a completely normal usage model that's often used all over the world. We are going to show examples for both CentOS and Ubuntu so that we cover everything necessary for most Linux administrators. Let's roll up our sleeves and start!

Getting ready

Keep the `cli1` virtual machine powered on and let's continue using our shell. Let's make sure that the necessary packages are installed by using our standard commands. So, let's use this command:

```
dnf -y install vsftpd createrepo lftp
```

That should be all in terms of preparation, so let's do it.

How to do it...

Setting up a custom CentOS repository is actually quite a simple affair. The first step involves downloading some packages. We are going to download a few of them and place them in the same directory. Then, we are going to make that directory available via the network by using `vsftpd`. A more detailed explanation of `vsftpd` can be found in the next chapter of this book, which is about network-based file synchronization. Here, we are just going to do a *Formula 1* qualifying lap through `vsftpd` to create a repository.

Let's say that we want to create a local repository (hosted on our `cli2` machine) that's going to have two packages in it – the `joe` editor and `desktop-backgrounds-basic`. We need to put them in a directory, `/var/ftp/pub/repository`, so that they're nice and handy inside the `vsftpd` folder structure. We could do it like this:

```
[root@cli2 ~]# mkdir /var/ftp/pub/repository
[root@cli2 ~]# cd /var/ftp/pub/repository/
[root@cli2 repository]# wget https://dl.fedoraproject.org/pub/epel/8/Everything/x86_64/Packages/j/joe-4.6-6.el8.x86_64.rpm https:/
/dl.fedoraproject.org/pub/epel/8/Everything/x86_64/Packages/d/desktop-backgrounds-basic-31.0.0_1.el8.noarch.rpm
--2021-09-17 10:26:44--  https://dl.fedoraproject.org/pub/epel/8/Everything/x86_64/Packages/j/joe-4.6-6.el8.x86_64.rpm
Resolving dl.fedoraproject.org (dl.fedoraproject.org)... 38.145.60.23, 38.145.60.24, 38.145.60.22
Connecting to dl.fedoraproject.org (dl.fedoraproject.org)|38.145.60.23|:443... connected.
HTTP request sent, awaiting response... 200 OK
Length: 623376 (609K) [application/x-rpm]
Saving to: 'joe-4.6-6.el8.x86_64.rpm'

joe-4.6-6.el8.x86_64.rpm        100%[===================================================================>] 608.77K   532KB/s    in 1.1s

2021-09-17 10:26:46 (532 KB/s) - 'joe-4.6-6.el8.x86_64.rpm' saved [623376/623376]

--2021-09-17 10:26:46--  https://dl.fedoraproject.org/pub/epel/8/Everything/x86_64/Packages/d/desktop-backgrounds-basic-31.0.0-1.e
l8.noarch.rpm
Reusing existing connection to dl.fedoraproject.org:443.
HTTP request sent, awaiting response... 200 OK
Length: 2709820 (2.6M) [application/x-rpm]
Saving to: 'desktop-backgrounds-basic-31.0.0-1.el8.noarch.rpm'

desktop-backgrounds-basic-31.0.0 100%[===================================================================>]   2.58M   813KB/s    in 3.3s

2021-09-17 10:26:50 (813 KB/s) - 'desktop-backgrounds-basic-31.0.0-1.el8.noarch.rpm' saved [2709820/2709820]

FINISHED --2021-09-17 10:26:50--
Total wall clock time: 5.4s
Downloaded: 2 files, 3.2M in 4.4s (740 KB/s)
[root@cli2 repository]#
```

Figure 6.28 – Downloading a few packages and getting ready for repository configuration

Since we already installed the `createrepo` package in the introduction to this recipe, we just need to use the `createrepo` command to create the necessary inventory information:

```
[root@cli2 repository]# cd
[root@cli2 ~]# createrepo /var/ftp/pub/repository
Directory walk started
Directory walk done - 2 packages
Temporary output repo path: /var/ftp/pub/repository/.repodata/
Preparing sqlite DBs
Pool started (with 5 workers)
Pool finished
[root@cli2 ~]# ls -al /var/ftp/pub/repository/
total 3264
drwxr-xr-x. 3 root root     111 Sep 17 10:29 .
drwxr-xr-x. 3 root root      24 Sep 17 10:26 ..
-rw-r--r--. 1 root root 2709820 Nov 20  2019 desktop-backgrounds-basic-31.0.0-1.el8.noarch.rpm
-rw-r--r--. 1 root root  623376 Sep 24  2019 joe-4.6-6.el8.x86_64.rpm
drwxr-xr-x. 2 root root    4096 Sep 17 10:29 repodata
[root@cli2 ~]# ls -al /var/ftp/pub/repository/repodata/
total 32
drwxr-xr-x. 2 root root 4096 Sep 17 10:29 .
drwxr-xr-x. 3 root root  111 Sep 17 10:29 ..
-rw-r--r--. 1 root root 2831 Sep 17 10:29 4b407b8e828ab1e5f852bb4d0c17c05bccaff09260ef8aedee983bc16354c35c-filelists.sqlite.bz2
-rw-r--r--. 1 root root 1801 Sep 17 10:29 635f7b4158aa72ec40e6681d3a59c565eab9835ddbbaf64ee01324131343e2d7-filelists.xml.gz
-rw-r--r--. 1 root root 1324 Sep 17 10:29 90a9c3396772c86ba9fdc54b38a012d426c251077a107f16a6c2ddac841ce10f-primary.xml.gz
-rw-r--r--. 1 root root 1688 Sep 17 10:29 9a8a690e05f49997854271f4cc380b8862d271b2097711bd92ecf01db1837979-other.sqlite.bz2
-rw-r--r--. 1 root root 3530 Sep 17 10:29 a2b9c77c25253dc65212282bd73318f0a4a2b8986382a6b2c9bbf7acab65b46d-primary.sqlite.bz2
-rw-r--r--. 1 root root  911 Sep 17 10:29 e4bf5e3f96f65cbb2e352a8bbfb0d15cf99a386f7224e4955d6934cf53afa929-other.xml.gz
-rw-r--r--. 1 root root 3076 Sep 17 10:29 repomd.xml
[root@cli2 ~]#
```

Figure 6.29 – Creating a repository out of a directory with RPM packages

The next step is to allow this directory to be used via `vsftpd`. Again, we have already installed `vsftpd` and, by default, we just need to change one option in its configuration file to allow anonymous FTP. Let's open the configuration file, `/etc/vsftpd/vsftpd.conf`, and locate the offending option:

```
anonymous_enable=NO
```

And change it to the following:

```
anonymous_enable=YES
```

We can now start and enable the service:

```
systemctl restart vsftpd
systemctl enable vsftpd
```

Then, let's try to log in to it to verify whether everything is ready:

```
[root@cli2 ~]# lftp localhost
lftp localhost:~> dir
drwxr-xr-x    3 0        0              24 Nov 16  2020 pub
lftp localhost:/> cd pub
lftp localhost:/pub> cd repository/
lftp localhost:/pub/repository> dir
-rw-r--r--    1 0        0         2709820 Nov 20  2019 desktop-backgrounds-basic-31.0.0-1.el8.noarch.rpm
-rw-r--r--    1 0        0          623376 Sep 24  2019 joe-4.6-6.el8.x86_64.rpm
drwxr-xr-x    2 0        0            4096 Sep 17 08:29 repodata
lftp localhost:/pub/repository>
```

Figure 6.30 – Checking whether the vsftpd configuration works

Everything is now ready from the service perspective. Now we just need to explain to yum/dnf that they need to use this as a repository. So, let's head to the /etc/yum.repos.d directory and create a repository configuration file there. Let's say we'll call it localrepo.repo. The name is irrelevant, it's just that it needs to have the .repo extension. Let's add the following options to it and save it:

```
[MyLocalRepo]
name=My Local Package Repository
baseurl=ftp://localhost/pub/repository
enabled=yes
gpgcheck=no
```

Let's verify whether this repository definition now works. We need to use yum or dnf for that, with the repolist keyword:

```
[root@cli2 yum.repos.d]# dnf repolist
repo id                              repo name
MyLocalRepo                          My Local Package Repository
appstream                            CentOS Linux 8 - AppStream
baseos                               CentOS Linux 8 - BaseOS
extras                               CentOS Linux 8 - Extras
[root@cli2 yum.repos.d]#
```

Figure 6.31 – Checking whether our repository is correctly configured via its repo config file

As we can see, MyLocalRepo is defined and ready to be used. Let's test it by trying to install the desktop-backgrounds-basic package:

```
yum -y install desktop-backgrounds-basic
```

This should be the result:

```
[root@cli2 repository]# yum -y install desktop-backgrounds-basic
Last metadata expiration check: 0:01:56 ago on Fri 17 Sep 2021 11:04:55 AM CEST.
Dependencies resolved.
================================================================================
 Package                    Architecture   Version          Repository      Size
================================================================================
Installing:
 desktop-backgrounds-basic  noarch         31.0.0-1.el8     MyLocalRepo     2.6 M

Transaction Summary
================================================================================
Install  1 Package

Total download size: 2.6 M
Installed size: 2.7 M
Downloading Packages:
desktop-backgrounds-basic-31.0.0-1.el8.noarch.rpm        166 MB/s | 2.6 MB   00:00
--------------------------------------------------------------------------------
Total                                                    156 MB/s | 2.6 MB   00:00
Running transaction check
Transaction check succeeded.
Running transaction test
Transaction test succeeded.
Running transaction
  Preparing        :                                                         1/1
  Installing       : desktop-backgrounds-basic-31.0.0-1.el8.noarch           1/1
  Verifying        : desktop-backgrounds-basic-31.0.0-1.el8.noarch           1/1
Installed products updated.

Installed:
  desktop-backgrounds-basic-31.0.0-1.el8.noarch

Complete!
[root@cli2 repository]#
```

Figure 6.32 – Installing a package from our custom repository

We can clearly see the relevant information here – the repository used was called `MyLocalRepo`, so both `vsftpd` and our repository configuration file work without any problems.

> **Note**
>
> Ubuntu tends to be much, much richer in terms of custom repositories; for example, repositories that are hosted on launchpad.net, and so on. It also tends to be a bit more internet-reliant than CentOS, but, that being said, it's easy enough to create repositories on either one of these distributions.

Let's dive into a short explanation about how this all works in CentOS, and then it's time for another recipe!

How it works...

There are two aspects to this recipe:

- Understanding how creating a repository works
- Understanding how using a *custom-created* repository works from the service and `yum/dnf` perspective

As always, we need to understand both of these concepts so that we can make them work for us. Let's start with the repository creation part.

Obviously, when creating a repository, we have to have some packages for that repository. So, a logical first step would always be to either create some packages or download them. There's just one key point to be made here – there *might* be problems if you download some packages *without* their dependencies. We deliberately chose two packages that don't have any dependencies so that we can have something to start our work with. That problem can be solved in either of the following two ways:

- We download all of the necessary dependencies for the packages that we're creating a repository for.

- We set up our repositories so that some other repository has all of the necessary dependencies for the packages we're creating our repository for.

Generally speaking, we'll go a long way in solving this problem by just enabling the EPEL repository for our CentOS version, so, generally speaking, we should install the EPEL rpm as it's going to help us with dependencies for almost anything:

```
yum -y install epel-release
```

Then, it all just becomes a matter of creating a directory, placing packages there, and using createrepo to create the necessary XML files so that the directory with packages can be used as a repository. Without createrepo, we are going to get an error, so we should always install it and use it prior to using our custom repository.

The second aspect is related to a wider picture – that is, how to make this repository available to other machines on the network and how to configure those machines to use it. That's why we strategically selected vsftpd as a delivery service, as its configuration for this scenario is really easy. We could have used the Apache web server as well, but seeing that our next chapter is related to vsftpd, we thought it would be a fun way to get an introduction to vsftpd out of the way by seeing it in action.

A part of this process is to work on repo files from the repository client perspective – that is, all of the machines that are going to be using our custom repository. It's just a couple of configuration lines that cover the repository's unique name and description, location, and some general settings, such as whether that repository is enabled on a client and if we're using signed packages and verifying their signature. Usually, people tend to skip over this, although it's quite important. If we enable the gpgcheck option, we need to install a gpg key that a repository is using to sign its packages, as well. We can do that with the gpg --import file_name.gpg command, after we download the gpg file.

Let's now get ready for the last part of this chapter, which is all about compiling software from the source code. We're going to use some familiar, usual suspects to do that and learn how to do it along the way.

There's more...

If we need to learn more about `vsftpd`, make sure that you check the following links:

- **yum:** `https://access.redhat.com/documentation/en-us/red_hat_enterprise_linux/6/html/deployment_guide/ch-yum`

- **Install Anonymous FTP server on CentOS 8**: `https://www.centlinux.com/2020/02/install-anonymous-ftp-server-on-centos-8.html`

- **Create Local Repos**: `https://wiki.centos.org/HowTos/CreateLocalRepos`

- **Configuring yum and yum repositories**: `https://access.redhat.com/documentation/en-us/red_hat_enterprise_linux/6/html/deployment_guide/sec-configuring_yum_and_yum_repositories`

Compiling third-party software

Sometimes, a package for a certain application is just not available – either nobody bothered to create it, or that application is so old that it's obsolete and nobody wants to do it. Either way, if an application is useful to us, there's no reason why we shouldn't try to find its source code and compile it.

Compiling software from source code can sometimes be like dark magic, and we have a good example coming up very soon. Sometimes it works without any real effort, and we are going to show you an example of that, too. The main distinction between those two scenarios seems to be the all-important dependencies and their version. Also, there's a lot of software for Linux that needs to be compiled in a specific sequence. A perfect example of that is the LAMP stack. After installing Linux, if you want to compile Apache, MySQL, and PHP, you had better do it in the correct order. Otherwise, your keyboard might find its way to the garbage can sooner than you planned. Let's see what we can do about this so that it doesn't happen.

Getting ready

We can use any machine for this recipe, but the most common scenario is a default installation of some Linux distribution with a lot of packages missing. So, let's install a fresh Ubuntu machine, and let's call it `compile1`, just for fun. So, this one is going to be just a Vanilla Ubuntu installation that will need all of the configurations in order for the compilation process to work.

How to do it...

We'll start with an easy example of a package that's very easy to compile and won't give us a massive headache. Let's compile the `joe` editor and show you what we're talking about. We'll start with the usual procedure:

```
apt-get -y update
apt-get -y upgrade
```

Just to be on the safe side, to get our machine ready for the compilation process, let's use this command to install a large selection of packages:

```
apt-get -y install autoconf g++ subversion linux-source linux-
headers-'uname -r' build-essential tofrodos git-core subversion
dos2unix make gcc automake cmake checkinstall git-core dpkg-
dev fakeroot pbuilder dh-make debhelper devscripts patchutils
quilt git-buildpackage pristine-tar git yasm checkinstall cvs
mercurial
```

As a result, our Ubuntu machine should now be ready for any compilation effort. Let's now download `joe` source:

```
wget https://kumisystems.dl.sourceforge.net/project/joe-editor/
JOE%20sources/joe-4.6/joe-4.6.tar.gz
```

We prefer to keep things tidy in the root's home directory, so let's just create a folder for compilation purposes. Let's call it source and move joe source there, and then open its tar.gz file with the source code:

```
mkdir source
mv joe-4.6.tar.gz source
cd source
tar zfpx joe-4.6.tar.gz
```

The last command (tar) is going to open another subfolder (joe-4.6) with all the necessary files for the compilation process located there. So, let's change the directory to joe-4.6 and start the configuration process:

```
cd joe-4.6
./configure
```

If everything goes well, we should have something like this as a result (shortened for formatting reasons):

```
checking for aspell... ispell
checking that generated files are newer than configure... done
configure: creating ./config.status
config.status: creating Makefile
config.status: creating joe/Makefile
config.status: creating joe/util/Makefile
config.status: creating rc/Makefile
config.status: creating man/Makefile
config.status: creating man/ru/Makefile
config.status: creating syntax/Makefile
config.status: creating po/Makefile
config.status: creating colors/Makefile
config.status: creating charmaps/klingon
config.status: creating desktop/Makefile
config.status: creating joe/autoconf.h
config.status: executing depfiles commands
root@compile1:~/source/joe-4.6# _
```

Figure 6.33 – Configuration step concluded successfully

The configuration process has finished successfully. Let's now continue with the actual compilation process, for which we need the `make` command (hence the reason why we installed all of those packages, `make` being one of them), and we can use some additional options to speed the process up. My Ubuntu machine has four processors, so we can use `make -j4` to speed the process up (so that the compilation process takes all of the available cores, not just one). After a couple of seconds, the compilation process should finish similar to this:

```
gcc -DHAVE_CONFIG_H -I.  -DJOERC="\"/usr/local/etc/joe/\"" -DJOEDATA="\"/usr/local/share/joe/\""  -
g -O2 -MT unicat-9.0.0.o -MD -MP -MF .deps/unicat-9.0.0.Tpo -c -o unicat-9.0.0.o unicat-9.0.0.c
mv -f .deps/options.Tpo .deps/options.Po
mv -f .deps/unicat-9.0.0.Tpo .deps/unicat-9.0.0.Po
mv -f .deps/colors.Tpo .deps/colors.Po
mv -f .deps/cclass.Tpo .deps/cclass.Po
gcc  -g -O2   -o joe b.o blocks.o bw.o cmd.o hash.o help.o kbd.o macro.o main.o menu.o path.o poshis
t.o pw.o queue.o qw.o rc.o regex.o scrn.o tab.o termcap.o tty.o tw.o ublock.o uedit.o uerror.o ufile
.o uformat.o uisrch.o umath.o undo.o usearch.o ushell.o utag.o va.o vfile.o vs.o w.o utils.o syntax.
o utf8.o selinux.o charmap.o mouse.o lattr.o gettext.o builtin.o builtins.o vt.o mmenu.o state.o opt
ions.o unicode.o cclass.o frag.o colors.o unicat-9.0.0.o  -lm -lutil
make[3]: Leaving directory '/root/source/joe-4.6/joe'
make[2]: Leaving directory '/root/source/joe-4.6/joe'
make[1]: Leaving directory '/root/source/joe-4.6/joe'
Making all in colors
make[1]: Entering directory '/root/source/joe-4.6/colors'
make[1]: Nothing to be done for 'all'.
make[1]: Leaving directory '/root/source/joe-4.6/colors'
Making all in desktop
make[1]: Entering directory '/root/source/joe-4.6/desktop'
make[1]: Nothing to be done for 'all'.
make[1]: Leaving directory '/root/source/joe-4.6/desktop'
make[1]: Entering directory '/root/source/joe-4.6'
make[1]: Nothing to be done for 'all-am'.
make[1]: Leaving directory '/root/source/joe-4.6'
root@compile1:~/source/joe-4.6# _
```

Figure 6.34 – Compilation process also concluded successfully

The final step in this process is to install our compiled application. We do that by using the following command:

```
make install
```

After this command finishes its job and installs `joe` system-wide, we should be able to start `joe` from the command line and edit our files. We can also create a `deb` package out of this installation by using the `checkinstall` package. When we run it, it's going to ask us for a package description. We can type in something like `Joe editor v4.6` and be done with it. At the end of this process, we're going to get a `deb` package with installation files that are required to deploy `joe` on other Ubuntu servers.

That wasn't so bad, was it? Yes, we had a couple of steps to do, but overall, it was a very simple process.

Now let's do another example that's the complete opposite of what we'd call a *very simple process*. Let's try to compile the Apache web server. We're going to use the latest version at the time of writing (2.4.49), located at `https://dlcdn.apache.org//httpd/httpd-2.4.49.tar.gz`, by using the same procedure – download the source to our source directory, open the source archive, and start with the configuration process. Let's see what happens:

```
root@compile1:~/source/httpd-2.4.49# ./configure
checking for chosen layout... Apache
checking for working mkdir -p... yes
checking for grep that handles long lines and -e... /usr/bin/grep
checking for egrep... /usr/bin/grep -E
checking build system type... x86_64-pc-linux-gnu
checking host system type... x86_64-pc-linux-gnu
checking target system type... x86_64-pc-linux-gnu
configure:
configure: Configuring Apache Portable Runtime library...
configure:
checking for APR... no
configure: error: APR not found.  Please read the documentation.
root@compile1:~/source/httpd-2.4.49#
```

Figure 6.35 – configure script in action – Missing dependencies – Example 1

Oops! Not going to happen. So then, we go to Dr. Google and check what to do if the message received is *APR is not found*. We'll end up finding some articles that state that we should install some additional packages, so let's do that:

```
apt-get -y install libapr1-dev libaprutil1-dev
```

Try to run the `configure` script again, and check the results:

```
configure:
configure: Configuring Apache Portable Runtime Utility library...
configure:
checking for APR-util... yes
checking for gcc... x86_64-linux-gnu-gcc
checking whether the C compiler works... yes
checking for C compiler default output file name... a.out
checking for suffix of executables...
checking whether we are cross compiling... no
checking for suffix of object files... o
checking whether we are using the GNU C compiler... yes
checking whether x86_64-linux-gnu-gcc accepts -g... yes
checking for x86_64-linux-gnu-gcc option to accept ISO C89... none needed
checking how to run the C preprocessor... x86_64-linux-gnu-gcc -E
checking for x86_64-linux-gnu-gcc option to accept ISO C99... none needed
checking for pcre-config... false
configure: error: pcre-config for libpcre not found. PCRE is required and available from http://pcre
.org/
```

Figure 6.36 – configure script in action – Missing dependencies – Example 2

Another package seems to be missing. And when we – just as an example – try to find a package such as `libpcre` in the `apt` cache, this is what we're going to get:

```
root@compile1:~/source/httpd-2.4.49# apt-cache search libpcre
libpcre16-3 - Old Perl 5 Compatible Regular Expression Library - 16 bit runtime files
libpcre2-16-0 - New Perl Compatible Regular Expression Library - 16 bit runtime files
libpcre2-32-0 - New Perl Compatible Regular Expression Library - 32 bit runtime files
libpcre2-8-0 - New Perl Compatible Regular Expression Library- 8 bit runtime files
libpcre2-dev - New Perl Compatible Regular Expression Library - development files
libpcre2-posix2 - New Perl Compatible Regular Expression Library - posix-compatible runtime files
libpcre3 - Old Perl 5 Compatible Regular Expression Library - runtime files
libpcre3-dbg - Old Perl 5 Compatible Regular Expression Library - debug symbols
libpcre3-dev - Old Perl 5 Compatible Regular Expression Library - development files
libpcre32-3 - Old Perl 5 Compatible Regular Expression Library - 32 bit runtime files
libpcrecpp0v5 - Old Perl 5 Compatible Regular Expression Library - C++ runtime files
clisp-module-pcre - GNU CLISP module that adds libpcre support
libhyperscan5 - High-performance regular expression matching library
libpcre++-dev - C++ wrapper class for pcre (development)
libpcre++0v5 - C++ wrapper class for pcre (runtime)
libpcre-ocaml - OCaml bindings for PCRE (runtime)
libpcre-ocaml-dev - OCaml bindings for PCRE (Perl Compatible Regular Expression)
pcre2-utils - New Perl Compatible Regular Expression Library - utilities
pcregrep - grep utility that uses perl 5 compatible regexes.
root@compile1:~/source/httpd-2.4.49#
```

Figure 6.37 – Trying to figure out which package is missing

Now the question becomes how to know which packages to install from this list? What usually happens is that people lose patience and write a command such as this:

```
apt-get -y install *pcre*
```

And that's going to install more than 200 packages on our machine. If we're security-conscious, that's really not the way to go. It's easy for people like us as we've done this a thousand times, but for normal people, this gets really frustrating really quick. So, let's now install the required package and its dependencies:

```
apt-get -y install libpcre3-dev
```

Before we do the actual configuration/compiling, we do have to mention one thing. Nowadays, a lot of the app code is shared via concepts such as Git. Most of these repositories are hosted by app coders, and usually have additional instructions for dependencies and how to deploy them. However, if we download a source code from a non-Git-like resource, we usually get more information about compiling that source code in files such as INSTALL after we extract the source archive. So, we need to make sure that we check these resources prior to trying to compile an app from some source code.

Run the rest of our procedure in a serial fashion:

```
./configure; make; make install
```

Luckily, there will be no more questions, as we can see in the following screenshot:

```
Installing configuration files
mkdir /usr/local/apache2/conf
mkdir /usr/local/apache2/conf/extra
mkdir /usr/local/apache2/conf/original
mkdir /usr/local/apache2/conf/original/extra
Installing HTML documents
mkdir /usr/local/apache2/htdocs
Installing error documents
mkdir /usr/local/apache2/error
Installing icons
mkdir /usr/local/apache2/icons
mkdir /usr/local/apache2/logs
Installing CGIs
mkdir /usr/local/apache2/cgi-bin
Installing header files
mkdir /usr/local/apache2/include
Installing build system files
mkdir /usr/local/apache2/build
Installing man pages and online manual
mkdir /usr/local/apache2/man
mkdir /usr/local/apache2/man/man1
mkdir /usr/local/apache2/man/man8
mkdir /usr/local/apache2/manual
make[1]: Leaving directory '/root/source/httpd-2.4.49'
root@compile1:~/source/httpd-2.4.49#
```

Figure 6.38 – Compilation and installation completed successfully

We deliberately chose a package that's *a bit* annoying, but not *over-the-top* annoying. There are applications out there that can make us spend hours and hours figuring out all of the dependencies so that we can compile a single package.

How it works...

Now that we have got the step-by-step process out of the way, let's discuss the specifics of how all of this works and fits together. It's pretty obvious that there are multiple steps to the process and that each and every one of them is significant. And it is – one can't be done without the other. So, let's now discuss all of the commands that we used and describe how they work.

The first phase in our compilation process starts with the `./configure` command. It's actually not a command *per se*; it's a `shell` script that almost all source code packages have. This script is there to make sure that the environment is ready for the compilation process – check included files, libraries, dependencies, everything needed for the source code compilation process. It checks for the necessary compiler and its libraries to make sure that the stage is set up for the next part of the process. It also writes down some configuration files that are going to be used by `make` when the build process starts.

The next part of the process involves using the `make` command. By using the configuration files created by the `configure` script and other files, it starts compiling source code. One of these files is called `Makefile`, and it contains a lot of information about what `make` needs to do – which files to compile and how, which compiler flags to use, how to link all of the compiled code into the resulting binaries, and more besides.

The last part of the process is not compiling a source code *per se* – it's about installing the compiled code on our Linux machine. By using relevant information in the configuration files, `make install` installs all of the files necessary for our command to work – libraries, binaries, man pages, documentation, and so on. If the compilation process from the previous part concludes successfully, installation is just about making sure that the compiled application is available to be used.

That was the last recipe in this chapter. The next chapter is about network-based file synchronization, and as part of those recipes, we are going to go much deeper into the inner workings of `vsftpd`, which we just kind of touched on in this chapter without giving it much time or space. Also, we are going to discuss `ssh` and `scp`, two ways of securely connecting to servers and transferring files between servers, and `rsync`, a file synchronization methodology. Stay tuned for the next chapter.

There's more...

If you need to learn more about `vsftpd`, make sure that you check out the following links:

- **How to compile and run C/C++ code in Linux**: `https://www.cyberciti.biz/faq/howto-compile-and-run-c-cplusplus-code-in-linux/`

- **Compiling things on Ubuntu the easy way**: `https://help.ubuntu.com/community/CompilingEasyHowTo`

7
Network-Based File Synchronization

Copying content over a network is usually done manually. For example, we just use SCP or FTP to transfer a file and that's that. But what happens if we need to make this process a permanent one? We then need to figure out a way to do file/directory synchronization, which is what `rsync` is all about. That being said, with all of the security-related incidents in the past few years, it's always a good idea to implement some kind of encryption, so using SSH and SCP seems like a reasonable approach, and that's exactly what we are going to do.

In this chapter, we are going to cover the following topics:

- Learning how to use SSH and SCP
- Learning how to use `rsync`
- Using `vsftpd`

Technical requirements

For these recipes, we're going to use two Linux machines – we can use the `client1` and `gui1` virtual machines from our previous chapters. These recipes will work on both CentOS and Ubuntu, so there is no reason to use separate virtual machines for these scenarios.

So, let's start our virtual machines and let's get cracking!

Learning how to use SSH and SCP

Back in the 1990s, it was a pretty natural thing to use the `Telnet`, `rlogin`, and FTP protocols. Come to think of it, using (anonymous) FTP is still done a lot. Bearing in mind that most local networks in the 1990s were based around network hubs (not switches) and the fact that all of these protocols are plain-text protocols that are easy to eavesdrop on via network sniffers, it really isn't all that strange that we're not using these devices and/or protocols as much anymore. As book authors, we haven't heard of anyone using rlogin since the late 1990s, although Telnet is still widely used to configure network devices (mostly switches and routers). This is the reason why SSH was developed (as a Telnet/rlogin replacement), and, along with SSH, SCP was developed (as a replacement for FTP). To put things into perspective, the first version of SSH was released in the mid-1990s. Let's see how it works.

Getting ready

We just need one Ubuntu and one CentOS machine for this recipe. Let's say we are going to use `cli1` and `cli2` to master these commands.

How to do it...

Our first scenario is going to be connecting from one machine to another by using SSH. We are going to presume that we don't have all of the necessary packages installed – just enough to cover our bases. We know that there are a lot of IT people out there who try to install the smallest number of packages possible on their servers/containers, so these extra steps shouldn't be much of a problem.

On an Ubuntu-based machine, we can do it like this:

```
apt-get -y install libssh-4 openssh-client openssh-server
openssh-sftp-server ssh-import-id
```

On a CentOS machine, we can do it like this:

```
dnf install openssh-server
```

For both of them, we need to start the service and enable it if we want to use it permanently:

```
systemctl start sshd
systemctl enable sshd
```

As a replacement for insecure technologies such as Telnet, rlogin, and FTP, SSH is pretty straightforward to use. We just need to learn the basic syntax. Let's say that we want to log in *from* a user called student on Linux machine cli1 *to* a user called student on Linux machine cli2. As we're logging in from a user called student to a user called student, there are two ways to do that. Here's the first:

```
student@cli1:~$ ssh student@cli2
```

And here's the second:

```
student@cli1:~$ ssh cli2
```

The reason is simple: if we're logging in to the same user that we're using on our source Linux machine, we don't need to explicitly say which account we're logging in to.

If we, however, wanted to go from the student user on cli1 to some other user on cli2, then we have to use the remote username as a parameter. Again, we can do it in two ways. Here's the first:

```
student@cli1:~$ ssh remoteuser@cli2
```

And here's the second:

```
student@cli1:~$ ssh -l remoteuser cli2
```

We can generalize that to cover any remote user on any remote machine. Commands for that scenario would look like this:

```
ssh remoteuser@remotemachine
```

Or this:

```
ssh -l remoteuser remotemachine
```

Another part of the SSH stack is a command called `SCP`. We use SCP to copy files from one machine to another machine by using SSH as a backend (secure copy). So, let's use an example. Let's say that we want to copy a file called `source.txt` from the `student` user's home directory on `cli1` to the `student` user's home directory on `cli2`. We would use the following command to do that:

```
scp /home/student/source.txt student@cli2:/home/student
```

Or, if we were already in the `/home/student` directory on the source machine, we would use this:

```
scp source.txt student@cli2:/home/student
```

Generally speaking, `SCP` has a simple syntax:

```
scp source destination
```

It's just that the source and destination can have a lot of letters that need to be typed. Let's explain that point by using another interesting use case for SCP. We can use it to download files from remote machines to local machines as well. The syntax is similar but can be a bit confusing when we're doing it the first couple of times. So, let's say that we want to copy a file called `source.txt` from the home directory of the `student` user on `cli2` to the `/tmp` directory on `cli1`, logged in as the `student` user on `cli1`. We would use the following command to do that:

```
scp cli2:/home/student/source.txt /tmp
```

The syntax follows the same rule (`scp source destination`), it's just that the source is now a remote file, and the destination is a local directory. It makes sense, when we think about it.

The next step in our process is going to be installing secure shell keys. This means that – in our example – we will enable `passwordless login` from one server to the other. We can avoid that, but let's forget about that for the time being; we are going to cover it in a second as we are not discussing security implications here. We are only trying to get the environment ready for SSH and SCP from a local user (let's say, `student`) to a remote user (let's say, `student`). So, let's do that:

```
student@cli1:~$ ssh-keygen
Generating public/private rsa key pair.
Enter file in which to save the key (/home/student/.ssh/id_rsa):
Enter passphrase (empty for no passphrase):
Enter same passphrase again:
Your identification has been saved in /home/student/.ssh/id_rsa
Your public key has been saved in /home/student/.ssh/id_rsa.pub
The key fingerprint is:
SHA256:oLd8LyvLzhn6xUGQI1qB45uE4aB1YCfwsxpZrZ5G5n4 student@cli1
The key's randomart image is:
+---[RSA 3072]----+
|+=.o  ..         |
|*.B.. ..         |
|*B+.... .        |
|+* + . o         |
|+ B . . S        |
| X . o o .       |
|. =   + +        |
| o  E+.=..       |
|  ...oBo.o.      |
+----[SHA256]-----+
```

Figure 7.1 – Creating an SSH key with an empty private key

Let's now copy this key to the remote machine (cli2) and test if the SSH key copying process worked by trying to log in as that user. For the first part, we are going to use the command called ssh-copy-id (to copy the key to the remote machine), and then use SSH to try to log in to test if the SSH key was properly copied:

```
student@cli1:~$ ssh-copy-id student@cli2
/usr/bin/ssh-copy-id: INFO: Source of key(s) to be installed: "/home/student/.ssh/id_rsa.pub"
The authenticity of host 'cli2 (192.168.0.19)' can't be established.
ECDSA key fingerprint is SHA256:6+8ShLxNU1VOTA+zOBDrZr1OPpMIPn6SsdGxUvklakQ.
Are you sure you want to continue connecting (yes/no/[fingerprint])? yes
/usr/bin/ssh-copy-id: INFO: attempting to log in with the new key(s), to filter out any that are alr
eady installed
/usr/bin/ssh-copy-id: INFO: 1 key(s) remain to be installed -- if you are prompted now it is to inst
all the new keys
student@cli2's password:

Number of key(s) added: 1

Now try logging into the machine, with:   "ssh 'student@cli2'"
and check to make sure that only the key(s) you wanted were added.

student@cli1:~$ ssh cli2
Activate the web console with: systemctl enable --now cockpit.socket

Last login: Sun Sep 12 14:06:31 2021
[student@cli2 ~]$ _
```

Figure 7.2 – Copying the SSH key to the remote machine and testing if it works

As we can see, everything works from cli1 to cli2. Let's now repeat the same process in the opposite direction, because we are going to need that a bit later for another part of this recipe. First, let's create an SSH key:

```
[student@cli2 ~]$ ssh-keygen
Generating public/private rsa key pair.
Enter file in which to save the key (/home/student/.ssh/id_rsa):
Enter passphrase (empty for no passphrase):
Enter same passphrase again:
Your identification has been saved in /home/student/.ssh/id_rsa.
Your public key has been saved in /home/student/.ssh/id_rsa.pub.
The key fingerprint is:
SHA256:Ykag+jrPE+KWn8Ej3F2ZsEyoEYcPU36wXkimKPpz6nM student@cli2
The key's randomart image is:
+---[RSA 3072]----+
|  .o*.           |
|.+B.=.           |
|oo== =.          |
|o.+.=.o o        |
|o. . o++S        |
|oo+ .o..         |
|.+=*..           |
|.*++E            |
|o+**             |
+----[SHA256]-----+
```

Figure 7.3 – Creating an SSH key for student@cli2

Then, let's copy it to the remote server:

```
[student@cli2 ~]$ ssh-copy-id student@cli1
/usr/bin/ssh-copy-id: INFO: Source of key(s) to be installed: "/home/student/.ssh/id_rsa.pub"
The authenticity of host 'cli1 (127.0.1.1)' can't be established.
ECDSA key fingerprint is SHA256:6+8ShLxNU1VOTA+z0BDrZr10PpMIPn6SsdGxUvklakQ.
Are you sure you want to continue connecting (yes/no/[fingerprint])? yes
/usr/bin/ssh-copy-id: INFO: attempting to log in with the new key(s), to filter out any that are already i
nstalled
/usr/bin/ssh-copy-id: INFO: 1 key(s) remain to be installed -- if you are prompted now it is to install th
e new keys
student@cli1's password:

Number of key(s) added: 1

Now try logging into the machine, with:   "ssh 'student@cli1'"
and check to make sure that only the key(s) you wanted were added.

[student@cli2 ~]$ ssh student@cli1
Activate the web console with: systemctl enable --now cockpit.socket

Last login: Sun Sep 12 14:11:37 2021 from 192.168.0.16
[student@cli2 ~]$
```

Figure 7.4 – Copying the SSH key from cli2 to cli1 and testing if it works

We can see that in both of these examples, the remote server that we're connecting to doesn't ask us for a password. The reason for that is simple: when we were creating an SSH key, `ssh-keygen` gave us two very important things to input:

```
Enter passphrase (empty for no passphrase) :
Enter same passphrase again:
```

If we pressed the *Enter* key on the first question and confirmed it by pressing *Enter* again on the second one, that means that we created an SSH key that has an empty private key. And that's exactly what we did in our example. We didn't select any specific passphrase, therefore leaving it empty. If we wanted to use a custom private key, we just needed to type it in those two steps.

How it works...

As a protocol, SSH is an encrypted answer to the non-existent security of Telnet, `rlogin`, and FTP. These three plain-text protocols were easy to hack, especially in the good old days before we started using network switches (while we were still mostly using network hubs). Its first implementation goes way back to 1995. It can also be used as a tunneling protocol, and it was heavily used for that back in the day – for example, for proxying FTP and HTTP traffic. Nowadays, it's used more for tunneling for remote X applications (XDMCP) or even connections to SSH to servers behind an SSH-based tunneling host.

In simple terms, SSH works like this:

1. The SSH client connects to the SSH server, therefore starting the connection.
2. The server responds and gives the client its public key.
3. The server and client then try to negotiate the necessary encryption parameters, followed by a secure channel being opened between the server and client.
4. The application or user logs in the server.

For those of us familiar with SSL/TLS, it's kind of similar to both of these protocols as all of these protocols are TCP-based; they have a negotiating mechanism and are generally used for security purposes. Yes, they go about it in a slightly different way and their use cases are a bit different, but that still doesn't mean that they're vastly different in terms of the general principle.

The next stop on our journey is `rsync`, and we are going to explicitly use SSH as a backend to `rsync`. That's the reason why we made our SSH keys, especially the ones without an additional private key (passphrase). Let's now learn how to work with `rsync`.

There's more...

If you need more information about networking in CentOS and Ubuntu, make sure that you check out the following:

- How does SSH work: `https://www.hostinger.com/tutorials/ssh-tutorial-how-does-ssh-work`

- What is the **Secure Shell** (**SSH**) protocol: `https://www.sdxcentral.com/security/definitions/what-is-the-secure-shell-ssh-protocol/`

Learning how to use rsync

In our previous recipe, we worked with SSH from the client standpoint. We used SSH and `SCP` to both log in and copy files from source to destination. We discussed how to use a username/password combination to log in to a remote system, as well as how to use SSH key-based authentication. If we focus on SCP for a second, there's one thing that we didn't discuss, and that is how to *synchronize* the local source to the local destination, or, even better, how to create a scenario in which we synchronize the local source to a remote destination and vice versa between two Linux servers in place. This is where it's best to use `rsync`, a tool that's meant to do just that. Let's get cracking.

Getting ready

We will continue using our `cli1` and `cli2` machines, running Ubuntu and CentOS. Let's get ready by making sure that the necessary packages are installed. We need to use this command for Ubuntu:

```
apt -y install rsync
```

We use the following command for CentOS:

```
dnf -y install rsync
```

After that, we are ready to start.

How to do it...

We are going to talk about a couple of scenarios:

- Synchronization between local source and local destination
- Synchronization between local source and remote destination, or vice versa

There could be a number of other sub-scenarios, such as dealing with one-way sync and deleting files on source, rsync is just one subdirectory, and so on. We are just going to deal with these two in detail, and then add a couple of bits and pieces from these sub-scenarios.

Let's deal with the simple scenario first: how to synchronize a folder that's placed locally to another locally placed folder. Let's say that we want to synchronize (basically, create a backup of) the /etc folder, and that we want to synchronize it to the /root/etc folder. We can do that by using the following commands as root (using the cli1 machine as an example):

```
rsync -av /etc /root
```

The two options used, a and v, are there to use archiving mode (preserve permissions and ownerships) and verbose mode so that we can see the output of every copy operation. We don't need to create the /etc folder in the /root directory up front or put /root/etc as the destination folder because a folder named etc is going to be created automatically in /root upon command execution.

If we wanted to exclude some files from copying (for example, all files that have the .conf extension), we can do it like this:

```
rsync -av --exclude="*.conf" /etc /root
```

There are other cool options available in rsync that could make certain scenarios possible. Let's say that we want to copy files that are a maximum of 5 MB in size, or a minimum of 3 MB in size. We could do that by using the following syntax:

```
rsync -av --max-size=5M source destination
rsync -av --min-size=3M source destination
```

For example, if the source directory has a lot of large files in the second example (minimum size), we might want to add a --progress option to the rsync command so that we can have interactive output telling us about the progress being made.

Now let's work on one-way sync from a remote to a local destination. The opposite direction is almost the same, we just need to change the source and destination fields in `rsync`. So, let's say that we have a source directory on `cli2` called `/home/student/source`. That directory has files and subfolders; it has a hierarchy of files and folders. We want to synchronize that content to `cli1`, specifically, to the `/tmp` directory. Here's the content of our source directory:

```
[student@cli2 source]$ ls -alR
.:
total 0
drwxrwxr-x. 5 student student  42 Sep 12 17:35 .
drwx------. 7 student student 184 Sep 12 17:35 ..
-rw-rw-r--. 1 student student   0 Sep 12 17:35 1
drwxrwxr-x. 2 student student  24 Sep 12 17:35 2
drwxrwxr-x. 2 student student  24 Sep 12 17:35 3
drwxrwxr-x. 2 student student  24 Sep 12 17:35 4

./2:
total 4
drwxrwxr-x. 2 student student 24 Sep 12 17:35 .
drwxrwxr-x. 5 student student 42 Sep 12 17:35 ..
-rw-rw-r--. 1 student student 26 Sep 12 17:35 source.txt

./3:
total 4
drwxrwxr-x. 2 student student 24 Sep 12 17:35 .
drwxrwxr-x. 5 student student 42 Sep 12 17:35 ..
-rw-rw-r--. 1 student student 26 Sep 12 17:35 source.txt

./4:
total 4
drwxrwxr-x. 2 student student 24 Sep 12 17:35 .
drwxrwxr-x. 5 student student 42 Sep 12 17:35 ..
-rw-rw-r--. 1 student student 26 Sep 12 17:35 source.txt
[student@cli2 source]$
```

Figure 7.5 – Source directory on cli2, located at /home/student/source

This is what we should do, provided that we have the source material ready:

```
student@cli1:~$ rsync -rt student@cli2:/home/student/source /tmp
student@cli1:~$ cd /tmp/source
student@cli1:/tmp/source$ ls -alR
.:
total 20
drwxrwxr-x  5 student student 4096 Sep 12 15:35 .
drwxrwxrwt 13 root    root    4096 Sep 12 15:39 ..
-rw-rw-r--  1 student student    0 Sep 12 15:35 1
drwxrwxr-x  2 student student 4096 Sep 12 15:35 2
drwxrwxr-x  2 student student 4096 Sep 12 15:35 3
drwxrwxr-x  2 student student 4096 Sep 12 15:35 4

./2:
total 12
drwxrwxr-x 2 student student 4096 Sep 12 15:35 .
drwxrwxr-x 5 student student 4096 Sep 12 15:35 ..
-rw-rw-r-- 1 student student   26 Sep 12 15:35 source.txt

./3:
total 12
drwxrwxr-x 2 student student 4096 Sep 12 15:35 .
drwxrwxr-x 5 student student 4096 Sep 12 15:35 ..
-rw-rw-r-- 1 student student   26 Sep 12 15:35 source.txt

./4:
total 12
drwxrwxr-x 2 student student 4096 Sep 12 15:35 .
drwxrwxr-x 5 student student 4096 Sep 12 15:35 ..
-rw-rw-r-- 1 student student   26 Sep 12 15:35 source.txt
student@cli1:/tmp/source$ _
```

Figure 7.6 – rsync from the remote source directory

So, we just used one simple command, rsync -rt (-r means recursive, -t is to preserve times), with the source and destination as parameters, and the source directory was successfully transferred to our local directory. This is because we copied the SSH keys in the previous recipe, so we didn't need to do any authentication, which makes the overall process very easy and straightforward.

The next scenario is going to be about syncing the source and destination and then deleting source files. Specifically, we're syncing *files, not folders*, as there are different options for those scenarios. Let's see how that's done:

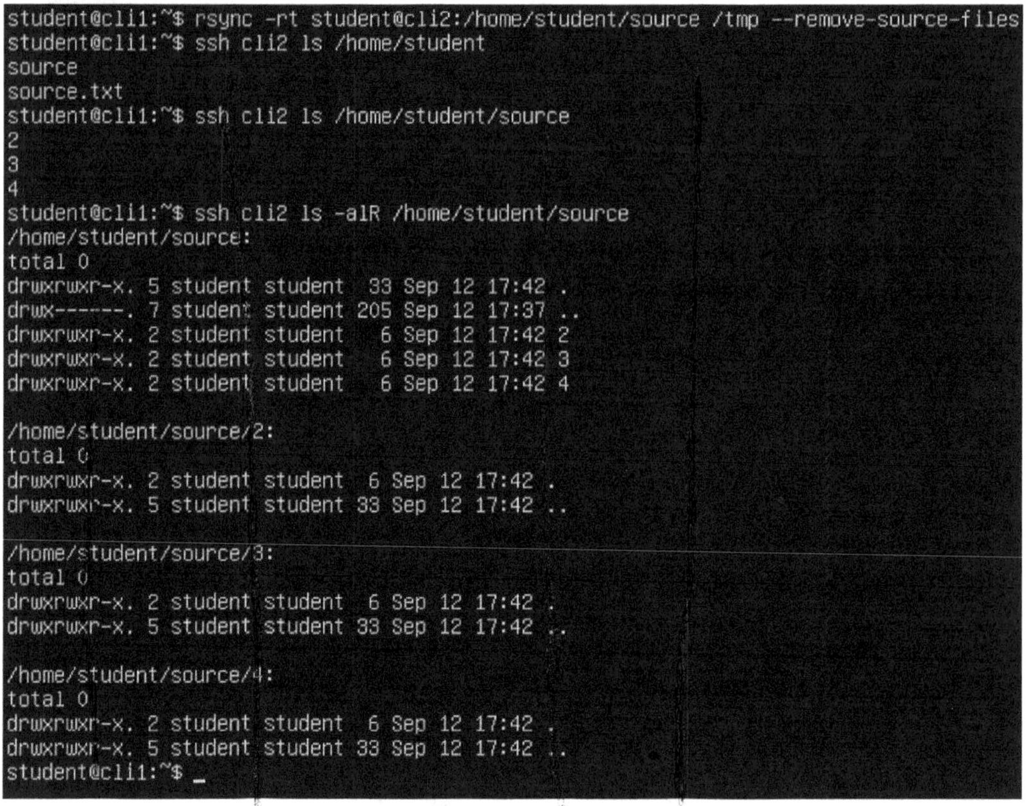

```
student@cli1:~$ rsync -rt student@cli2:/home/student/source /tmp --remove-source-files
student@cli1:~$ ssh cli2 ls /home/student
source
source.txt
student@cli1:~$ ssh cli2 ls /home/student/source
2
3
4
student@cli1:~$ ssh cli2 ls -alR /home/student/source
/home/student/source:
total 0
drwxrwxr-x. 5 student student  33 Sep 12 17:42 .
drwx------. 7 student student 205 Sep 12 17:37 ..
drwxrwxr-x. 2 student student   6 Sep 12 17:42 2
drwxrwxr-x. 2 student student   6 Sep 12 17:42 3
drwxrwxr-x. 2 student student   6 Sep 12 17:42 4

/home/student/source/2:
total 0
drwxrwxr-x. 2 student student  6 Sep 12 17:42 .
drwxrwxr-x. 5 student student 33 Sep 12 17:42 ..

/home/student/source/3:
total 0
drwxrwxr-x. 2 student student  6 Sep 12 17:42 .
drwxrwxr-x. 5 student student 33 Sep 12 17:42 ..

/home/student/source/4:
total 0
drwxrwxr-x. 2 student student  6 Sep 12 17:42 .
drwxrwxr-x. 5 student student 33 Sep 12 17:42 ..
student@cli1:~$ _
```

Figure 7.7 – rsync from the remote server using SSH keys,
and deleting source files after the download is done

Now, if we wanted to run the same scenario but delete all of the files and folders from `cli2` after the transfer is done, we'd need to separate that into two commands. Here's how it works:

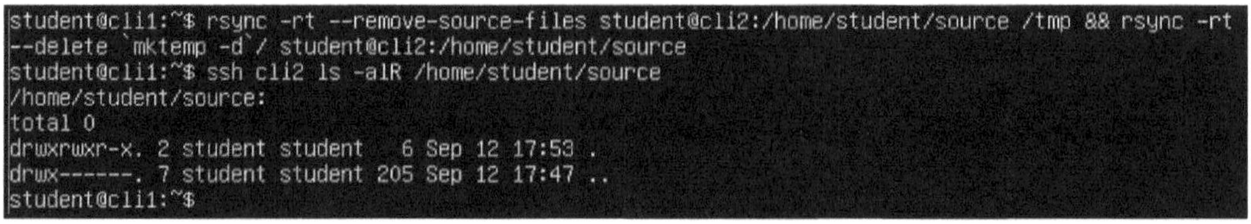

```
student@cli1:~$ rsync -rt --remove-source-files student@cli2:/home/student/source /tmp && rsync -rt
--delete `mktemp -d`/ student@cli2:/home/student/source
student@cli1:~$ ssh cli2 ls -alR /home/student/source
/home/student/source:
total 0
drwxrwxr-x. 2 student student   6 Sep 12 17:53 .
drwx------. 7 student student 205 Sep 12 17:47 ..
student@cli1:~$
```

Figure 7.8 – Removing source files from the remote source directory,
and then all subdirectories in the source directory

Now that we've shown this, we can also note a couple of other projects that will make it easier to do two-way sync. Projects such as Unison (`https://www.cis.upenn.edu/~bcpierce/unison/`) and bsync (`https://github.com/dooblem/bsync`) have implemented two-way sync methods that are very difficult to achieve by using `rsync`. Make sure that you check them out if you need two-way sync.

How it works...

`rsync` is a source-destination type of command, and that covers its syntax and mode of operation if we're using it interactively (no destination `rsync` service is involved). There can also be an `rsync` service involved, which usually changes the mode of operation significantly. It's important to point out that using `rsync` as a command (in combination with SSH) is most commonly done for backups. We've been using it in this fashion for 15 or more years in some of our environments, and it works perfectly.

`rsyncd` (the `rsync` service) is usually aimed at a completely different usage model – most commonly, software mirrors. If we want to create a local CentOS or Ubuntu mirror, the rule of the thumb is that we'll use `rsyncd`, as it allows us to do much more finely grained configuration in terms of what needs to be done as part of the `rsync` process. There might be other reasons to do it – for example, we can configure `rsyncd` to not use SSH and gain a bit of speed in doing so.

Now that we have discussed some of the key concepts of SSH, SCP, and `rsync`, it's time to move on to their – at least by default – much more insecure cousin, called `vsftpd`. We are going to make sure that we make it more secure, though, as there's absolutely no reason not to. So, let's get ready to configure `vsftpd`.

There's more...

If you need to learn more about `rsync`, we recommend the following links:

- How to set up an `rsync` daemon on your Linux server: `https://www.atlantic.net/vps-hosting/how-to-setup-rsync-daemon-linux-server/`

- 10 practical examples of the `rsync` command in Linux: `https://www.tecmint.com/rsync-local-remote-file-synchronization-commands/`

- 17 useful `rsync` (remote sync) command examples in Linux: `https://www.linuxtechi.com/rsync-command-examples-linux/`

Using vsftpd

The FTP service has been around for decades. Back in the mid-1990s, FTP was actually the vast majority of internet traffic. Yes, its importance in terms of traffic volume decreased over time, but it's not only that. FTP, all by itself, is a completely open, plain-text protocol. The latest revision that's been included in all major distributions is called vsftpd, and it's been there for more than a decade now. We are going to focus on three scenarios in this recipe: getting vsftpd to work, getting vsftpd to work with a user's home directories, and – last but not least – making vsftpd secure by implementing TLS and certificates. Let's start!

Getting ready

Keep the cli1 and cli2 virtual machines powered on and let's continue using our shell. Let's make sure that the necessary packages are installed by using our standard commands. So, for Ubuntu, use this command:

```
apt -y install vsftpd
```

For CentOS, let's use this command:

```
dnf -y install vsftpd
```

Then, let's enable them and start it. We're going to use the Ubuntu machine to show how vsftpd configuration should be done, but it's almost 100% the same on CentOS. So cli1 (Ubuntu) is going to act as a vsftpd server, and cli2 (CentOS) is going to act as an FTP client. So, let's run these commands on cli1:

```
systemctl start vsftpd
systemctl enable vsftpd
```

It would be prudent to configure firewalls to allow connections to necessary FTP ports (20, 21). So, on cli1, we need to do this:

```
ufw allow ftp
ufw allow ftp-data
```

On the client side (cli2), let's install lftp, a nice and simple-to-use ftp client, by using the following command:

```
dnf -y install lftp
```

Let's now configure vsftpd in accordance with the three scenarios that we mentioned.

How to do it...

Now that we have installed our packages, it's time to start configuring vsftpd on cli1. That means we need to go through some of the options in /etc/vsftpd.conf (usually, it's /etc/vsftpd/vsftpd.conf on CentOS).

Generally speaking, this configuration file is very well documented all by itself, so we should have no trouble configuring it to suit our needs. By default, it should let us use the FTP client to connect to it, but let's make a couple of changes from the very start. Let's allow anonymous FTP and not allow local users to log in. If we check the configuration file, that means that we need to configure the anon_root, anonymous_enable, and local_enable configuration options, so let's do that. Let's make sure that those two configuration lines look like this:

```
anonymous_enable=YES
local_enable=NO
anon_root=/var/ftp
```

We also need to create some directories for this configuration to work:

```
mkdir -p /var/ftp/pub
chown nobody:nogroup /var/ftp/pub
```

Restart the vsftpd service so that it works with the latest configuration:

```
systemctl restart vsftpd
```

On cli2, we have already installed lftp, and it is going to try to log in to the remote FTP server (cli1) anonymously by default. Let's see how that works:

```
[root@cli2 ~]# lftp cli1
lftp cli1:~> ls
drwxr-xr-x    2 65534      65534            4096 Sep 12 18:46 pub
```

Figure 7.9 – Testing the FTP connection by using lftp

We can see that we have no errors, but we also don't have any content in the directory that the anonymous FTP service uses. On Ubuntu, that directory is located at /srv/ftp, but we already changed the anonymous root directory to /var/ftp. Let's add a couple of files there and try to list the directory content in lftp:

```
[root@cli2 ~]# lftp cli1
lftp cli1:~> dir
drwxr-xr-x    2 65534        65534            4096 Sep 12 18:46 pub
lftp cli1:/> cd pub
lftp cli1:/pub> dir
-rwxr-xr-x    1 0            0                   0 Sep 12 18:36 ftptest1.txt
-rwxr-xr-x    1 0            0                   0 Sep 12 18:36 ftptest2.txt
-rwxr-xr-x    1 0            0                   0 Sep 12 18:36 ftptest3.txt
-rwxr-xr-x    1 0            0                   0 Sep 12 18:36 ftptest4.txt
lftp cli1:/pub> 
```

Figure 7.10 – Checking if we can see the files we created on cli1 by using the touch command

Let's now try to download these files. To do that, FTP has a command called get (similar to how HTTP has a get command). Let's now download these four files that we used the touch command on:

```
[root@cli2 ~]# lftp cli1
lftp cli1:~> ls
drwxr-xr-x    2 65534        65534            4096 Sep 12 18:46 pub
lftp cli1:/> cd pub
lftp cli1:/pub> ls
-rwxr-xr-x    1 0            0                   0 Sep 12 18:36 ftptest1.txt
-rwxr-xr-x    1 0            0                   0 Sep 12 18:36 ftptest2.txt
-rwxr-xr-x    1 0            0                   0 Sep 12 18:36 ftptest3.txt
-rwxr-xr-x    1 0            0                   0 Sep 12 18:36 ftptest4.txt
lftp cli1:/pub> get ftptest1.txt ftptest2.txt ftptest3.txt ftptest4.txt
Total 4 files transferred
lftp cli1:/pub> 
```

Figure 7.11 – Using FTP's get command to retrieve multiple files from the FTP server

If we wanted to upload files, we would need to use the put command but, of course, that wouldn't work as anonymous upload is forbidden by default (as it should be).

The next part of our scenario is to allow the user to log in to the user's home directory. That shouldn't be too hard, as we already mentioned the first option that we need to change, local_enable, and it needs to be set to YES. After that, we need to restart the vsftpd service. After we do that, we need to log in to the FTP server as a local user on the FTP server. Bearing in mind that we have a user called student there, let's log in to that one:

```
[root@cli2 ~]# lftp -u student cli1
Password:
lftp student@cli1:~> ls -al
drwxr-xr-x    6 1000        1000         4096 Sep 12 18:59 .
drwxr-xr-x    3 1000        1000         4096 Mar 14  2021 ..
-rw-------    1 1000        1000         3060 Sep 12 17:52 .bash_history
-rw-r--r--    1 1000        1000          220 Jun 18  2020 .bash_logout
-rw-r--r--    1 1000        1000         3771 Jul 27 09:39 .bashrc
drwx------    2 1000        1000         4096 Mar 14  2021 .cache
drwx------    3 1000        1000         4096 Jul 28 11:55 .config
drwx------    3 1000        1000         4096 Mar 16 22:38 .mozilla
-rw-r--r--    1 1000        1000          807 Jul 27 09:39 .profile
-rw-rw-r--    1 1000        1000           75 Jul 28 19:27 .selected_editor
drwx------    2 1000        1000         4096 Sep 12 12:11 .ssh
-rw-r--r--    1 1000        1000            0 Mar 14  2021 .sudo_as_admin_successful
-rw-------    1 1000        1000         7395 Jul 28 19:32 .viminfo
lftp student@cli1:~>
```

Figure 7.12 – Logging in as the student user via lftp (by using the -u option)

No problems so far. But all of these recipes were done on the premise that we're doing all this within the limits of our internal, secure network. What happens if our FTP server needs to be exposed to the internet? We don't want to use just regular, plain-text FTP as it would lead to disaster. So, the next step in our recipe is going to be to configure FTP with TLS.

We need to configure a couple of options in `vsftpd.conf`, and we can freely put these options at the end of that file:

```
rsa_cert_file=/etc/ssl/certs/ssl-cert-snakeoil.pem
rsa_private_key_file=/etc/ssl/private/ssl-cert-snakeoil.key
ssl_enable=Yes
ssl_tlsv1=YES
ssl_sslv2=NO
ssl_sslv3=NO
ssl_ciphers=HIGH
force_local_data_ssl=YES
force_local_logins_ssl=YES
ssl_request_cert=NO
allow_anon_ssl=YES
```

We need to configure these options in accordance with our security requirements. Most commonly, we want to enable TLS 1.2 or 1.3 (`ssl_ciphers=HIGH`, SSLv2, and v3=no). We can always not allow anonymous users to use SSL, and if we don't want to run client certificate-based authentication, we have to make sure to use the `ssl_request_cert=NO` option.

At the beginning of this configuration, we can see the `cert` file and the corresponding private key configuration options. We just used the built-in, self-signed certificates. Of course, we can create Let's Encrypt certificates or buy commercial ones instead and put them in the configuration here. It's all about the corporate security policy where we want to run this sort of configuration.

A quick note on FTP clients on Windows: a lot of people are using WinSCP to upload and download files and directories by using SCP, SFTP, FTP, WebDav, and Amazon S3 sources. If we use WinSCP, we have to use FTP configuration, TLS/SSL explicit encryption, and other relevant parameters accordingly. There are also other options available if we click on the **Advanced** button. For example, we can choose a minimum TLS level and similar options. As TLS v1.2 is the minimum that's recommended at this point in time, we could set those options to `1.2` for both the minimum and maximum versions. But if we've set up our `vsftpd.conf` as we recommended, there's no need to touch those options as **TLS v1.2** will be the only option available. We just wanted to mention these advanced options in case you need them.

That being said, here's a screenshot that will help in terms of basic options:

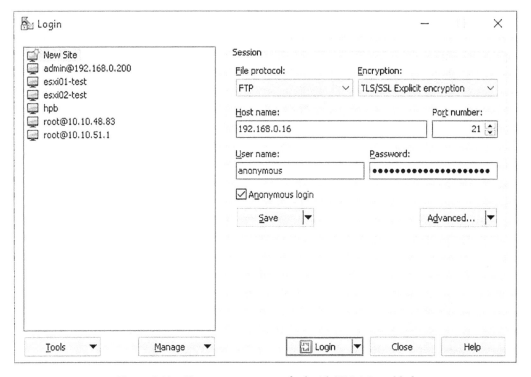

Figure 7.13 – How to connect to vsftpd with TLS 1.2 enabled

192.168.0.16 is the IP address of the cli1 machine. By using all of the options mentioned previously, we're able to log in anonymously to our vsftpd server and use it, just as we used it before we did the TLS configuration. But, bearing in mind that there were dozens and dozens of various types of attacks on the SSL protocol in the past couple of years (POODLE, BEAST, CRIME, BREACH, Heartbleed, SSL Stripping, using untrusted and fake certificate authorities, and so on), it's absolutely crucial that we pay close attention to every new attack and take all the necessary steps to mitigate those threats.

How it works...

vsftpd is an implementation of FTP, which means it's a TCP-based service that's used to upload and download files. Seeing as it's a TCP-based service, that means socket connections and reliable data transfer, which are essential to this service. Imagine if our file download or upload were to be unreliable; we definitely wouldn't like that. If we were to add an additional layer of security to it by using TLS, we'd still be using the same basic service, it's just that it'd be way more protected.

FTP uses ports 20 (ftp-data) and 21 (ftp). Both of these ports need to be allowed through the firewall for the FTP service to work. Port 21 is used as the *command* communication channel, while port 20 is used for data transfer, although there are implementations where port 21 can be used for both. There are some other options when using the FTP service (active FTP and passive FTP) but they are way beyond the scope of this book. Generally speaking, there's a reason why almost everybody uses SCP for file upload and download nowadays. Also, there's a reason why most of the distribution repositories and mirrors switched to using HTTPS-based delivery methods instead of FTP-based methods. There are exceptions, but they are more the *exception to the rule* types of situations, definitely not the standard.

FTP uses put and get commands to do two of its basic functions: upload (put) and download (get). These are two basic commands/methods that FTP uses, although we can create and delete content via FTP as well.

There's more

If you want to learn more about `vsftpd`, make sure that you check the following links:

- `vsftpd` home page: `https://security.appspot.com/vsftpd.html`

- `vsftpd.conf` man page: `https://security.appspot.com/vsftpd/vsftpd_conf.html`

- How does an FTP server work and what are its benefits: `https://www.ftptoday.com/blog/how-does-an-ftp-server-work-the-benefits`

- How to set up `vsftpd` for anonymous downloads on Ubuntu 16.04: `https://www.digitalocean.com/community/tutorials/how-to-set-up-vsftpd-for-anonymous-downloads-on-ubuntu-16-04`

8
Using the Command Line to Find, Extract, and Manipulate Text Content

Manipulating text is an everyday job for a full-time system administrator. It can happen for a variety of reasons – for example, you could just be trying to find a service option that you saw somewhere in some configuration file, without remembering what the name of the configuration file is. You know, those moments on Monday morning when you haven't had two cups of your favorite pick-me-up drink and your CPU hasn't booted properly yet? Or, maybe, when you're working with a text file that has a lot of content, but needs specific changes to be made, such as changing some configuration options from off to on, true to false, 0 to 1, and so on. This chapter is going to act as a prequel to one of the later chapters discussing shell scripting examples.

In this chapter, we are going to learn about the following:

- Using text commands to merge file content
- Converting DOS text to Linux text and vice versa
- Using `cut`
- Using `egrep`
- Using `sed`

Technical requirements

For these recipes, we're going to use one Linux machine – we can use `client1` from our previous recipes. It doesn't really matter which virtual machine gets used as all the commands that we are going to discuss in these recipes work the same way on all Linux distributions.

Using text commands to merge file content

Let's start with something simple – which is merging file content. Of course, we are only discussing text content here as merging binary files would be pointless. Our goal is to learn how to use two commands – `paste` and `cat` – to do simple things, such as concatenation and merging line by line. Let's start!

Getting ready

We just need one Ubuntu and one CentOS machine for this recipe. Here, we are going to use `cli1` and `cli2` to master these commands.

How to do it...

Starting with the simplest command for this chapter – `cat` – let's see some examples of what it does. If we type in a command such as `cat filename.txt` – if a file named `filename.txt` exists – we are going to get the content of that file on display. Let's check an example of this:

```
root@cli1:~# cat /var/log/auth.log
Sep 27 19:13:07 cli1 login[932]: pam_unix(login:auth): Couldn't open /etc/securetty: No such file or
 directory
Sep 27 19:13:08 cli1 login[932]: pam_unix(login:auth): Couldn't open /etc/securetty: No such file or
 directory
Sep 27 19:13:08 cli1 login[932]: pam_unix(login:session): session opened for user student by LOGIN(u
id=0)
Sep 27 19:13:08 cli1 systemd-logind[871]: New session 1 of user student.
Sep 27 19:13:08 cli1 systemd: pam_unix(systemd-user:session): session opened for user student by (ui
d=0)
Sep 27 19:13:12 cli1 sudo: pam_unix(sudo:auth): Couldn't open /etc/securetty: No such file or direct
ory
Sep 27 19:13:13 cli1 sudo: pam_unix(sudo:auth): Couldn't open /etc/securetty: No such file or direct
ory
Sep 27 19:13:13 cli1 sudo:   student : TTY=tty1 ; PWD=/home/student ; USER=root ; COMMAND=/usr/bin/su
 -
Sep 27 19:13:13 cli1 sudo: pam_unix(sudo:session): session opened for user root by student(uid=0)
Sep 27 19:13:13 cli1 su: (to root) student on tty1
Sep 27 19:13:13 cli1 su: pam_unix(su-1:session): session opened for user root by student(uid=0)
Sep 27 19:17:01 cli1 CRON[1421]: pam_unix(cron:session): session opened for user root by (uid=0)
Sep 27 19:17:01 cli1 CRON[1421]: pam_unix(cron:session): session closed for user root
```

Figure 8.1 – Using the cat command on a text file

So, we used the cat command to show the content of an auth.log file located in the /var/log directory. If we have been using this machine for a while, there will be other files with auth.log as a prefix, then a number, and the gz extension. Let's check:

```
root@cli1:~# ls -al /var/log/auth.log*
-rw-r----- 1 syslog adm  1245 Sep 27 19:17 /var/log/auth.log
-rw-r----- 1 syslog adm 67773 Sep 27 19:12 /var/log/auth.log.1
-rw-r----- 1 syslog adm  5386 Sep 18 23:39 /var/log/auth.log.2.gz
-rw-r----- 1 syslog adm  3781 Sep 11 23:30 /var/log/auth.log.3.gz
-rw-r----- 1 syslog adm  3358 Sep  4 23:30 /var/log/auth.log.4.gz
root@cli1:~# _
```

Figure 8.2 – Finding content that we are going to use

So, for the purpose of this recipe, let's use the `auth.log` and `auth.log.1` files. What happens if we want to have one file that contains both `auth.log` and `auth.log.1` content? We'd either open a text editor and do a bit of copy-pasting, or we can use `cat` to do that for us. The `cat` command can be used with multiple files at the same time, such as `cat auth.log auth.log.1`, which would show us the content of the first file followed by the content of the second file. The only thing that we need to do is to redirect the text output from that command to a new file, which we can easily do by using the `>` sign. Let's say that we want to save the output of this command to a file in the `/root` directory called `auth-full.log`. Here's how we'd do that:

```
root@cli1:~# ls -al /var/log/auth.log*
-rw-r----- 1 syslog adm  1245 Sep 27 19:17 /var/log/auth.log
-rw-r----- 1 syslog adm 67773 Sep 27 19:12 /var/log/auth.log.1
-rw-r----- 1 syslog adm  5386 Sep 18 23:39 /var/log/auth.log.2.gz
-rw-r----- 1 syslog adm  3781 Sep 11 23:30 /var/log/auth.log.3.gz
-rw-r----- 1 syslog adm  3358 Sep  4 23:30 /var/log/auth.log.4.gz
root@cli1:~# cat /var/log/auth.log /var/log/auth.log.1 > /root/auth-full.log
root@cli1:~# ls -al auth-full.log
-rw-r--r-- 1 root root 69201 Sep 27 19:31 auth-full.log
```

Figure 8.3 – Using the cat command to concatenate files

`cat` actually displays text files line by line, which is a property that we will heavily use in our chapters relating to shell script examples.

If for some reason we wanted to merge files line by line, we could've used the `paste` command. Let's see how that would work. Seeing that these files are just way too big, we are going to create two files. Let's say that the first file (named `first.txt`) will have the following content:

```
1 today
2 tomorrow
3 someday
```

The second file (named `second.txt`) will have this content:

```
may be good
may be even better
will be excellent
```

Now, let's use the `paste` command and check the result:

```
root@cli1:~# cat first.txt second.txt
1 today
2 tomorrow
3 someday
may be good
may be even better
will be excellent
root@cli1:~# paste first.txt second.txt
1 today    may be good
2 tomorrow      may be even better
3 someday       will be excellent
root@cli1:~# paste first.txt second.txt > third.txt
```

Figure 8.4 – Using the paste command to combine text files line by line

As we can see, the `paste` command combines two files line by line, by putting them one next to the other.

How it works...

These two commands are rather simple in operation:

- By default, `cat` displays the complete content of a file or set of files, line by line.

- By default, `paste` combines files line by line, side by side.

These are two very different approaches to text manipulation, both with real-life use cases.

Our next recipe is a simple one as well – how to deal with a situation when we transfer text files from Microsoft OSs to Linux in terms of making them usable in Linux. As we are going to see, there are some fundamental differences with `.txt` formats between Microsoft OSs and Linux, which makes the next recipe a necessity. Stay tuned for it!

There's more...

If you need more information about using `cat` or `paste`, make sure that you check out the following:

- Basic `cat` command examples in Linux: `https://www.tecmint.com/13-basic-cat-command-examples-in-linux/`

- The `paste` command in Linux (merge lines): `https://linuxize.com/post/paste-command-in-linux/`

Converting DOS text to Linux text and vice versa

This is a strange idea – you might have thought a `.txt` file is a `.txt` file, right? Wrong.

There are subtle differences between `.txt` file formats in DOS/Windows and Linux. Sometimes, those differences can make you mad in a matter of seconds. We've had our fair share of experiences of that – scripts not working as input files were prepared on Windows, not on Linux; different treatment of CSV files in Excel *by design*... sometimes it's just too funny when, after hours of deliberation, you realize that something as simple as a `.txt` file created on another OS can make such a mess. Let's explain what the problem is and work through it.

Getting ready

We just need one Ubuntu machine for this recipe. Let's say we are going to continue using `cli1` to master these commands. Furthermore, we need to install one package, called `dos2unix`. So, if we are using `cli1` (Ubuntu), we need to type in the following command:

```
apt-get -y install dos2unix
```

After this package is installed, we are ready to do our recipe.

How to do it...

Let's say that we created a `.txt` file called `txtsample.txt` in Notepad on Windows, which has the following content:

```
My first line in a file
My second line in a file
My third line in a file
My fourth line in a file
```

Then, we upload this file to our `cli1` machine and open it in vi or vim to check its content. This is what it looks like:

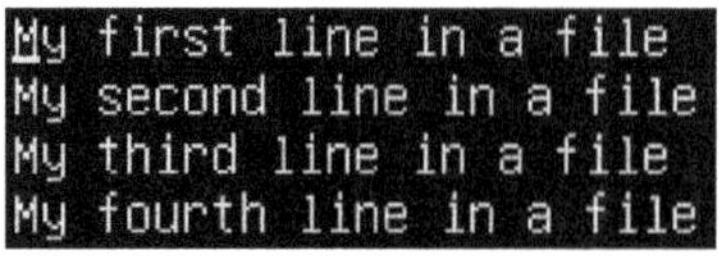

Figure 8.5 – What our file seems to look like

Everything seems fine, right? Now, let's do the same thing all over again, but this time, start vi or vim with the -b option. For example, use the vi -b txtsample.txt command and check the file content now. It should look like this:

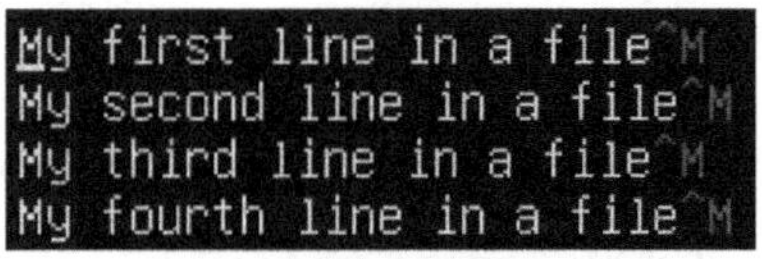

Figure 8.6 – What our file actually looks like

We can see **carriage returns** (**CRs**, those ^M characters) in the vi/vim editor now. This is one of those subtle differences between the way Notepad and Linux text editors treat .txt files. Linux shell commands aren't necessarily going to treat this type of text in a friendly manner, and sometimes scripts will not work properly because of these *extra* characters that Linux commands don't need.

The solution to this problem is a simple package and command called dos2unix that we installed in the *Getting ready* step of this recipe. After that, it's a simple procedure of typing in the following command:

```
dos2unix txtsample.txt
```

Let's open this file in vi with the same -b option now and check the file content:

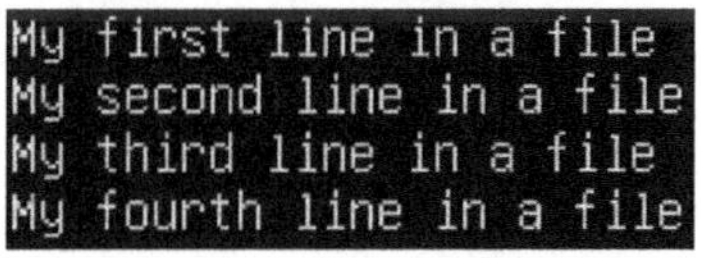

Figure 8.7 – End result – a file that's stripped of CRs

Now that's much better.

There are other examples of this approach – end-of-file characters, *invisible* characters that sometimes appear out of nowhere in Excel-exported CSV files. So, we have to make sure that we are aware of this problem and its simple solution.

We could also use tools such as tr, awk, and perl to do the same thing. Let's use tr as an example:

```
tr -d '\15\32' < input_dos_file.txt > output_linux_file.txt
```

Let's now explain how this works and why it's such a problem.

How it works...

A CR is a character that has been used through the years as a control mechanism to set the end of a line, and, as a result, start a new line of text. For those of us old enough to remember the old typewriter machines, the CR on a mechanical typewriter machine would be that funny lever that we had to pull to get to a new line. By extension, this is a part of ASCII code that helps with cursor positioning (beginning of the next line).

If we don't clear our `.txt` files of these characters (and others), we might have problems with scripting. In our last two chapters of this book with shell script examples, we're going to have multiple example scripts that use the `cat` command to input something from a `.txt` file into a loop. These characters might mess that procedure up, and we don't want that.

`dos2unix` and the mentioned `tr` command strip the input file of CRs. We might debate which method is better, but at the end of the day, it's about results, and both methods work. We prefer the `dos2unix` method; but, of course, you might prefer the `tr` way.

There's more...

If you need more information about converting DOS `.txt` files to Linux, refer to the following links:

- `dos2unix` man page: `https://linux.die.net/man/1/dos2unix`
- `yr` man page: `https://linux.die.net/man/1/tr`

Using cut

There are tools in IT that get elevated to greatness by the simple fact that they are great tools. The next three tools that we are going to use are tools that fit the description of some of the greatest CLI tools ever invented. For us, `cut` is the second greatest CLI command of all time; if you want to find out which command takes the coveted *#1* spot, stay tuned for the next recipes.

`cut` is a tool that can make our lives a lot easier if we're working with preformatted input. For example, it will easily work with CSV, as that's a formatted type of content that can be easily digested by `cut`. Let's learn about `cut` by doing some examples next.

Getting ready

We just need one Ubuntu machine for this recipe, so let's keep using `cli1`. The cut command is a standard part of any Linux distribution and that's how it should be, as it's more important than other commands, such as `ls`, `mkdir`, and `ps`.

How to do it...

Let's first create a sample CSV file. For example, we are going to create a CSV file with user data, and use `cut` on top of that file. Here's what we used for this recipe (CSV file content):

```
John,Fritz,johnny,Johnny the sales guru
Andrea,Flemming,andie,Andie the marketing guru
Alan,Buster,alan,Alan the IT man
Beth,Custer,Betty,Beth the big boss
```

Figure 8.8 – Sample input CSV file

We are now going to check what we can do with this file and the `cut` command. Let's start with some simple things. For example, first we are going to extract just names from this file, which translates to the first field (before the first comma sign):

```
root@cli1:~# cut -f1 -d, cut.csv
John
Andrea
Alan
Beth
```

Figure 8.9 – Extracting the first field from a standard-format file

By using the `cut` command and two switches, we were able to easily extract names from the CSV file. Now, let's add a bit more to the process. Let's extract the name and login (first and third fields):

```
root@cli1:~# cut -f1,3 -d, cut.csv
John,johnny
Andrea,andie
Alan,alan
Beth,Betty
```

Figure 8.10 – Extracting the first and third fields from our sample file

Furthermore, let's now extract the first three fields – name, surname, and login:

```
root@cli1:~# cut -f1-3 -d, cut.csv
John,Fritz,johnny
Andrea,Flemming,andie
Alan,Buster,alan
Beth,Custer,Betty
```

Figure 8.11 – Extracting a range of fields from our sample file

We could also choose to sort that output alphabetically:

```
root@cli1:~# cut -f1-3 -d, cut.csv | sort

Alan,Buster,alan
Andrea,Flemming,andie
Beth,Custer,Betty
John,Fritz,johnny
```

Figure 8.12 – Sorting output from a cut file

One other classic example is using `cut` to output fields from one specific field onward – for example, from the second field to the last field:

```
root@cli1:~# cut -f2- -d, cut.csv
Fritz,johnny,Johnny the sales guru
Flemming,andie,Andie the marketing guru
Buster,alan,Alan the IT man
Custer,Betty,Beth the big boss
```

Figure 8.13 – Extracting fields from the second field onward

If you're heavily into Microsoft PowerShell, this will kind of remind you of the `Import-Csv` PowerShell cmdlet, although the similarities end there, seeing as PowerShell is an object-based shell language.

As you can see from these examples, the `cut` command is very useful for situations where we have an input file that's in some sort of standard format that uses some character as a delimiter between fields. We can easily use it to extract content from our text files and to prepare text-based input for other actions that might follow using `cut` in a shell script.

How it works...

`cut` is a straightforward command that has only one prerequisite – we need to have an input file that's formatted in some sort of standard format. That translates to having a file with fields delimited by the same character. If that criterion is met, we can easily use `cut` to do wonders in one-liner commands and shell scripts.

Most commonly, we use two parameters with the cut command:

- The -f parameter is used to select which field(s) or field range we are going to use to be extracted by using the cut command.

- The -d parameter is used to select a delimiter, a character that separates our text strings.

We can also use it in conjunction with other commands, such as sort, tr, and uniq, to do further manipulation of the output extracted by using the cut command. We can even use its parameter called --output-delimiter to change the input delimiter to some other output delimiter.

This was a warmup exercise before the big star of the show – the egrep command – as its significance can't be overstated. Let's talk about egrep next.

There's more...

If you need more information about networking in CentOS and Ubuntu, make sure that you check out the following:

- The cut man page: https://man7.org/linux/man-pages/man1/cut.1.html

- Bash cut command examples: https://linuxhint.com/bash-cut-command-examples/

Using egrep

Using egrep, and regular expressions in general, is something like page one, chapter one stuff from the never-written *How to be both cool and incredibly useful* IT manual. It is, without a shadow of a doubt, the most useful command that was ever invented in the UNIX/Linux world for system administration. It doesn't really matter whether we're looking for a specific string in a file or set of directories, whether we're trying to find a line in a big text file where a specific string is located, or whether we're trying to find where a specific string isn't, egrep can do all of that for us. We are focusing on egrep specifically, as it supports both concepts that are behind this command – regular expressions and extended regular expressions. That's where we are going to start – first, by explaining the merits of using regular expressions, and then moving on from that to explain why egrep is such an important command. So, buckle up and get ready to go!

Getting ready

We are going to use two virtual machines for this recipe – the Ubuntu-based `cli1` and CentOS-based `cli2`. That's going to enable us to have more examples as logging configuration on Ubuntu is a bit different from CentOS, and CentOS's logging configuration makes it easier to drive some points home. So, start both of these virtual machines and let's get going.

How to do it...

Using regular expressions comes naturally to everyone if we take some time to get to know how to use them. By extension, our first *frontend* to regular expressions is using the `grep` or `egrep` command. These commands enable us to find a text string inside a text file or text input.

Let's use a simple example. In *Chapter 4, Using Shell to Configure and Troubleshoot Network*, we used the `ps` command to display running processes. Let's say that we want to do this now, but by using regular expressions. For example, we need to list all the processes on our Linux server that were started by user `student`.

If we start with the `ps` command first – for example, if we use the `ps auwwx` command – we are going to get an output like this:

```
root         1300  0.0  0.2  10372  5152 tty1     S    Sep27   0:00 sudo su -
root         1301  0.0  0.2   9368  4764 tty1     S    Sep27   0:00 su -
root         1302  0.0  0.2   7588  5344 tty1     S    Sep27   0:00 -bash
root        12821  0.0  0.4 245868  8576 ?        Ssl  Sep28   0:00 /usr/libexec/upowerd
www-data   278099  0.0  0.3 194584  6640 ?        S    00:00   0:00 /usr/sbin/apache2 -k start
www-data   278100  0.0  0.3 194584  6640 ?        S    00:00   0:00 /usr/sbin/apache2 -k start
www-data   278101  0.0  0.3 194584  6640 ?        S    00:00   0:00 /usr/sbin/apache2 -k start
www-data   278102  0.0  0.3 194584  6640 ?        S    00:00   0:00 /usr/sbin/apache2 -k start
www-data   278103  0.0  0.3 194584  6728 ?        S    00:00   0:00 /usr/sbin/apache2 -k start
root       281410  0.0  0.0      0     0 ?        S<   02:03   0:00 [loop7]
root       308549  0.0  0.0      0     0 ?        I    18:33   0:06 [kworker/1:0-cgroup_destroy]
root       311243  0.0  0.0      0     0 ?        I    20:09   0:01 [kworker/0:0-events]
root       312714  0.0  0.0      0     0 ?        I    21:05   0:00 [kworker/u256:1-ext4-rsv-conversi
on]
root       312807  0.0  0.0      0     0 ?        I    21:09   0:00 [kworker/1:1-events]
root       312855  0.0  0.0      0     0 ?        I    21:09   0:00 [kworker/0:1-events]
root       312894  0.0  0.0      0     0 ?        I    21:10   0:00 [kworker/u256:0-events_power_effi
cient]
root       313265  0.0  0.0      0     0 ?        I    21:25   0:00 [kworker/u256:2-events_power_effi
cient]
root       313335  0.0  0.1   8636  3236 tty1     R+   21:28   0:00 ps auwwx
```

Figure 8.14 – ps auwwx input, cut short for formatting reasons

Now, let's discuss the fact that this is a text output for just a second – it has a lot of letters and numbers. Also, let's focus on the fact that all processes running on our system are represented by a line in the `ps` command output – a line that, as seen in the figure, starts with the `root,` `www-data` string, or some other string that represents the user that owns the process. How about if we figure out a way of using that as a property to filter the `ps` command output as text, by using the idea of *line starts with* `student` for it?

If we wanted to do that, we would use a simple command: `ps auwwx | grep ^student`. As we discussed previously, the | sign means *we want to send the output of the first command to the second command*. Furthermore, `grep` means *we want to filter something out*. And this `^student` thing? That's what we call a regular expression pattern, with the `^` character being a regular expression symbol. Specifically, it's an anchor, which when used with `grep` or some other regular expression-aware command means *line starts with* the character, or a set of characters that follow. Let's put that theory into practice:

```
root@cli1:~# ps auwwx | grep -i ^student
student    1279  0.0  0.4  19432  9764 ?        Ss   Sep27   0:00 /lib/systemd/systemd --user
student    1280  0.0  0.1 108456  3692 ?        S    Sep27   0:00 (sd-pam)
student    1286  0.0  0.2   7220  4964 tty1     S    Sep27   0:00 -bash
```

Figure 8.15 – Using our first regular expression

So, here we go, we filtered our output by using the `student` string as a filter. We can also see that each appearance of that string is marked in red. This comes from the fact that `grep` commands use the `--color` option by default – an option that highlights the string that we were using as a condition for filtering.

Let's say that now we want to find all lines in our `ps auwwx` output that end with the `bash` string. We can easily use regular expressions for that. Here's how we'd do that:

```
root@cli1:~# ps auwwx | grep bash$
student    1286  0.0  0.2   7220  4964 tty1     S    Sep27   0:00 -bash
root       1302  0.0  0.2   7588  5368 tty1     S    Sep27   0:00 -bash
```

Figure 8.16 – Using a regular expression to find a string at the end of a line

So far, we've used the `ps` command as it's convenient and familiar. Let's now move on to other examples that are going to use text files as a basis. The first one that comes to mind is `/usr/share/dict/words`, a dictionary text file that contains more than 100,000 words. The format of that file is *one word per line*, so more than 100,000 words equals more than 100,000 lines. Let's try to find all the lines that have the `parrot` string in them. Here's the command and result:

```
root@cli1:/usr/share/dict# grep parrot /usr/share/dict/words
parrot
parrot's
parroted
parroting
parrots
```

Figure 8.17 – Using grep directly on a text file

So, the `grep` command can also be directly used on a text file, which is very useful when dealing with scripts, input files into scripts, and so on.

So far, so good. Let's make things a bit more complicated. Let's say that we need to find all lines in the same file (`/usr/share/dict/words`) that contain a string conforming to these rules:

- The line needs to start with p.
- The letter p needs to be *followed by a vowel*.
- A vowel needs to be followed by the `ta` string.

All of a sudden, things get much more complicated. Imagine having to find these words using a regular text editor. Writing down all these combinations would lead us to the following words that we're looking for:

- `pata`, followed by anything
- `peta`, followed by anything
- `pita`, followed by anything
- `pota`, followed by anything
- `puta`, followed by anything

In a regular text editor, it would take us five different sequential finds to find all the words, and even then, we'd have to press *next-next-next* for all the next appearances in case one of these string samples can be matched multiple times.

This is where regular expressions can be of great help. We could do this:

```
root@cli1:/usr/share/dict# grep ^p[aeiou]ta /usr/share/dict/words
petal
petal's
petals
petard
petard's
petards
pita
pita's
potable
potable's
potables
potash
potash's
potassium
potassium's
potato
potato's
potatoes
putative
```

Figure 8.18 – Finding a more complex string sample

By using the `^p[aeiou]ta` regular expression, we were able to find all of the words matching that criterion easily. When using these square brackets to input a regular expression sample, we are basically saying to the regular expression-aware command to search for *either a, or e, or i, or o, or u* as a character inside the regular expression string.

As we can see, getting to know regular expressions can be quite useful. Let's add some more of them with a short explanation:

Regular expression symbol	Meaning
`^`	Beginning of line
`$`	End of line
`[]`	Search for any character in the group
`[^]`	Inverse search for any character in the group
`[a-z]`	All lowercase letters
`[A-Z]`	All capital letters
`[a-zA-Z]` or `[a-Z]`	All uppercase and lowercase letters
`[0-9]`	Numbers from 0 to 9
`\w`	Word
`[:space:]`	Blank characters
`[:alnum:]`	Letters and digits
`\t`	Tab
`{3}`	Word length equals 3
`{3,}`	Word length is 3+ characters
`{3,8}`	Word length is from 3 to 8 characters
`*`	0 or more
`?`	0 or 1

Table 8.1 – Commonly used regular expression symbols and their meaning

There are many more regular expressions that we can use, but let's just start with these and then move on to more complex examples – for example, extended regular expressions.

How about matching numbers? Matching numbers with regular expressions is, let's say, *fun*. Especially if we need to find a number range – things get really complicated real soon if that's the case. Let's discuss this by using three examples – one for a single-digit number range, one for a simple double-digit number range, and one for a more complex double-digit number range. We can extrapolate how this logic would work on larger numbers. Let's start with a simpler example – for example, a number range from 0 to 5 – and just work with regular expressions, forgetting the `grep` command for a second. In regular expression terms, that would be `[0-5]`.

For our next example, let's use a simple double-digit number range – let's say, 14-19. In regular expression terms, we'd write that as 1[4-9].

This means 1 as a leading number (tens), and then the number range from 4 to 9, which equals 14-19.

So far, we specifically chose to use the grep command to do these first examples as it works with basic regular expressions. We need to add a -E switch to the grep command or start using the egrep command if we want to move on to extended regular expressions. So far, we've covered some basics, so it's time to make things a bit more complicated. Everything that we've discussed so far is what we call **Basic Regular Expressions (BREs)**. When talking about **Extended Regular Expressions (ERE)**, we need to keep in mind that, by using them, we are trying to match a specific set of strings, not just one specific string. Also, there are some differences in syntax, as they make a couple of things easier – we don't have to escape some characters. For example, BRE constructs such as \? and \| get replaced in EREs with ? and |. That makes the syntax cleaner and easier on the eye, so to speak. Let's work on some examples, first by continuing our *using regular expressions to match numbers* discussion.

What happens if we need to find a number range, for example, 37-94? Regular expressions can't work with multiple-digit numbers in ranges – we need to slice that into ones, tens, hundreds, and so on. And then, we need to use a very well-known concept called the union of a set (of ranges) to combine all of the ranges into one range that fits the regular expression that we wanted to find. Keeping in mind that we are going to need a set – a set of numbers, in this example – we are going to do this by using EREs. Let's see how that works with minimization in mind – we want to have a regular expression that's as short as possible. Knowing the fact that we need to use the union of a set, the simplest way to do this number range would be as follows:

- 37-39
- 40-89
- 90-94

In regular expression terms, those sets would be as follows:

- 3[7-9] for 37-39
- [4-8][0-9] for 40-89
- 9[0-4] for 90-94

As a regular expression, we'd write it like this:

3[7-9]|[4-8][0-9]|9[0-4]

Here, the | sign basically stands for *or*. This is a way of implementing a union set when using regular expressions.

Keeping in mind that Ubuntu's version of the `/usr/share/dict/words` file doesn't have a single number in it, we added a couple of numbers to the top of this file just so that we have something to test with. For example, we added the following to the top of this file:

- 41

- 58

- 36

- 95

- 94

We deliberately chose these numbers as they contain both numbers that are conformant to the regular expression that we made (41, 58, and 94 will be a match) and numbers that aren't (36 and 95). If we use `grep` with our regular expression on this file with added numbers, we will get the following output:

```
root@cli1:/usr/share/dict# egrep "3[7-9]|[4-8][0-9]|9[0-4]" /usr/share/dict/words
41
58
94
```

Figure 8.19 – Using a union set works and we found our matches

As we grow the number of digits that we're looking for, regular expressions get more and more complex. We should always minimize as much as we can, but there will be situations where we must make a lot of union sets to find something that we are looking for. That's just the name of the game, nothing to be scoffed at.

Our next example is going to find words (non-numbered only) that are 19 to 20 characters long, made of letters only. We're still using the same file as before, so let's see how we do that:

```
root@cli1:/usr/share/dict# egrep "^[A-Za-z]{19,20}$" /usr/share/dict/words
Andrianampoinimerina
chlorofluorocarbons
counterintelligence
counterrevolutionary
electrocardiographs
electroencephalogram
interdenominational
nonrepresentational
oversimplifications
uncharacteristically
```

Figure 8.20 – Finding words of specific length

Easy, right? We matched all the lowercase and uppercase characters and then said *words need to have 19 to 20 of those.*

Things tend to get even more complicated when you're trying to create a regular expression for some words that are common, yet different, especially if word length is variable, and even more so if number ranges are involved. But all of that doesn't bring us closer to explaining what the practical point, the real value of regular expressions, is. Everything that we did so far seems like general hocus pocus, *trying to find some text – why should we care about that?* Generally speaking, we were only describing the principle, as using generic examples helps. Realistically, there's absolutely no way to learn regular expressions by reading 10-15 pages of text. But we are now going to make an educated effort to move this story along to real-life, practical uses of regular expressions.

There are some common services that use `grep` and/or regular expressions constantly. We have to keep in mind that this recipe is about `grep` as a command, not about regular expressions only.

For example, regular expressions are heavily used for mail filtering. Checking the body of an email – basically text content – is easy if you have a regular expression-enabled mail filter. From an everyday system administrator standpoint, regular expressions are constantly used for parsing through log files and finding valuable information in them. Let's make that point now by using regular expressions on log files, as that's one of the most commonly used practices that's been happening for decades now.

Let's now switch to `cli2`, our CentOS-based system, and use `/var/log/messages` for the next examples. This file contains the main system log on CentOS, so it's perfectly suitable to use regular expressions to find something in it. Let's start with the simple stuff. For example, let's say that we need to find all log entries in `/var/log/messages` that were made on October 6, and even more specifically, at the ninth hour, by a service called `PackageKit`. Let's first check the structure of this log file – it looks like this:

```
Oct  6 09:47:27 cli2 PackageKit[1595]: uid 0 is trying to obtain org.freedesktop
.packagekit.system-sources-refresh auth (only trusted:0)
Oct  6 09:47:27 cli2 PackageKit[1595]: uid 0 obtained auth for org.freedesktop.p
ackagekit.system-sources-refresh
Oct  6 09:47:28 cli2 journal[1595]: Skipping refresh of media: Cannot update rea
d-only repo
Oct  6 09:47:28 cli2 journal[1595]: Skipping refresh of media: Cannot update rea
d-only repo
Oct  6 09:47:33 cli2 journal[1595]: Skipping refresh of media: Cannot update rea
d-only repo
Oct  6 09:47:33 cli2 journal[2161]: Only 0 apps for recent list, hiding
Oct  6 09:47:33 cli2 journal[2161]: hiding category audio-video featured applica
tions: found only 4 to show, need at least 9
Oct  6 09:47:33 cli2 journal[2161]: State change on system/package/appstream/des
ktop/firefox.desktop/* from updatable to installed is not OK
Oct  6 09:47:33 cli2 journal[1595]: Skipping refresh of media: Cannot update rea
d-only repo
Oct  6 09:47:34 cli2 journal[2161]: Only 0 apps for popular list, hiding
Oct  6 09:47:34 cli2 journal[2161]: hiding category graphics featured applicatio
ns: found only 2 to show, need at least 9
Oct  6 09:47:45 cli2 journal[1595]: Skipping refresh of media: Cannot update rea
d-only repo
Oct  6 09:48:03 cli2 systemd[1]: libvirtd.service: Succeeded.
```

Figure 8.21 – Format of a file that we're going to use grep on

As we were discussing in our previous recipe about the cut command, we can see that this file *kind of* has a standard format. There's a timestamp at the beginning, the hostname after that, the service and PID after that, and then some kind of text message. Also, notice the fact that the timestamp part has a very cool addition – its format is not Oct 6 with a single space between; it has two spaces. This is very important as it keeps the date format a fixed length when we get to double-digit dates, such as Oct 15. It just makes everything formatted nicely, which is cool.

So, the simple fact is that we can easily output this by using grep. Let's do it:

```
[root@cli2 log]# grep "Oct  6 09" messages | grep PackageKit
Oct  6 09:46:05 cli2 dbus-daemon[909]: [system] Activating via systemd: service
name='org.freedesktop.PackageKit' unit='packagekit.service' requested by ':1.61'
 (uid=42 pid=1382 comm="/usr/bin/gnome-shell " label="system_u:system_r:xdm_t:s0
-s0:c0.c1023")
Oct  6 09:46:05 cli2 systemd[1]: Starting PackageKit Daemon...
Oct  6 09:46:05 cli2 dbus-daemon[909]: [system] Successfully activated service '
org.freedesktop.PackageKit'
Oct  6 09:46:05 cli2 systemd[1]: Started PackageKit Daemon.
Oct  6 09:46:27 cli2 PackageKit[1595]: uid 0 is trying to obtain org.freedesktop
.packagekit.system-sources-refresh auth (only trusted:0)
Oct  6 09:46:27 cli2 PackageKit[1595]: uid 0 obtained auth for org.freedesktop.p
ackagekit.system-sources-refresh
Oct  6 09:47:27 cli2 PackageKit[1595]: uid 0 is trying to obtain org.freedesktop
.packagekit.system-sources-refresh auth (only trusted:0)
Oct  6 09:47:27 cli2 PackageKit[1595]: uid 0 obtained auth for org.freedesktop.p
ackagekit.system-sources-refresh
```

Figure 8.22 – Filtering out data that we wanted to filter out

The first part of the command greps out all messages from Oct 6 that starts with 09 in the hour part of the timestamp; then, we piped that out to another grep command that searches for the PackageKit string.

The next example – and this one becomes more common as we get older – is, let's say that we can't remember the name of a file where the firewall string is used. We remember that it was somewhere in the /etc/sysconfig folder, but we can't seem to remember what the filename was – one of those *it's early Monday morning, I haven't had time to wake up* moments. This is what we could do by using grep:

```
[root@cli2 sysconfig]# grep -r -i firewall *
ebtables-config:# Save current firewall rules on stop.
ebtables-config:# Saves all firewall rules if firewall gets stopped
firewalld:# firewalld command line args
firewalld:FIREWALLD_ARGS=
ip6tables-config:# are loaded after the firewall rules are applied. Options for
the helpers are
ip6tables-config:# Save current firewall rules on stop.
ip6tables-config:# Saves all firewall rules to /etc/sysconfig/ip6tables if firew
all gets stopped
ip6tables-config:# Save current firewall rules on restart.
ip6tables-config:# Saves all firewall rules to /etc/sysconfig/ip6tables if firew
all gets
iptables-config:# are loaded after the firewall rules are applied. Options for t
he helpers are
iptables-config:# Save current firewall rules on stop.
iptables-config:# Saves all firewall rules to /etc/sysconfig/iptables if firewal
l gets stopped
iptables-config:# Save current firewall rules on restart.
iptables-config:# Saves all firewall rules to /etc/sysconfig/iptables if firewal
l gets
```

Figure 8.23 – Using grep on a stack of files all at once

The grep option -r means *recursive* and the -i option means *case insensitive*. Also, the -r option ignores any symbolic links as it recursively moves through subdirectories. If we want that behavior to change, we can use the -r option, which will take symbolic links into account. As we have the capability to use recursive search through file contents, that means grep is going to dive into subdirectories and go through all files. That means we must add one caveat to our discussion – we really, *really shouldn't* use this on binary files, for obvious reasons. It would make a big mess on our terminal output, if nothing else.

Let's end this recipe by using egrep for a much more complex scenario involving EREs and different text patterns on text input, the caveat being that we want to see a bit of *context* around our text pattern matches. Let's say that we're trying to go through our dmesg boot log, and we are searching for all the hard disks – all the /dev/hd* and / dev/sd devices. We could use the following command for that purpose:

```
dmesg | egrep -C2 '(h|s)d[a-z]'
```

Let's check what the output would look like:

```
[root@cli2 sysconfig]# dmesg | egrep '(s|h)d[a-z]' -C2
[    0.000000] Linux version 4.18.0-305.17.1.el8_4.x86_64
(mockbuild@kbuilder.bsys.centos.org) (gcc version 8.4.1
20200928 (Red Hat 8.4.1-1) (GCC)) #1 SMP Wed Sep 8 14:00:07 UTC
2021
[    0.000000] Command line: BOOT_IMAGE=(hd0,msdos1)/vmlinuz-
4.18.0-305.17.1.el8_4.x86_64 root=/dev/mapper/cl-root ro
resume=/dev/mapper/cl-swap rd.lvm.lv=cl/root rd.lvm.lv=cl/swap
rhgb quiet
[    0.000000] Disabled fast string operations
[    0.000000] x86/fpu: Supporting XSAVE feature 0x001: 'x87
floating point registers'
--
[    0.000000] Built 1 zonelists, mobility grouping on.  Total
pages: 1032024
[    0.000000] Policy zone: Normal
[    0.000000] Kernel command line: BOOT_IMAGE=(hd0,msdos1)/
vmlinuz-4.18.0-305.17.1.el8_4.x86_64 root=/dev/mapper/cl-root
ro resume=/dev/mapper/cl-swap rd.lvm.lv=cl/root rd.lvm.lv=cl/
swap rhgb quiet
[    0.000000] Specific versions of hardware are certified with
Red Hat Enterprise Linux 8. Please see the list of hardware
certified with Red Hat Enterprise Linux 8 at https://catalog.
redhat.com.
[    0.000000] Memory: 3113700K/4193716K available (12293K
kernel code, 2225K rwdata, 7712K rodata, 2476K init, 14048K
bss, 256384K reserved, 0K cma-reserved)
--
[    0.023000] ... event mask:                  000000000000000f
[    0.023000] rcu: Hierarchical SRCU implementation.
[    0.023375] NMI watchdog: Perf NMI watchdog permanently
disabled
[    0.037028] smp: Bringing up secondary CPUs ...
[    0.037823] x86: Booting SMP configuration:
--
[    2.384538] scsi 2:0:0:0: Attached scsi generic sg0 type 0
[    2.384599] scsi 1:0:0:0: Attached scsi generic sg1 type 5
[    2.391073] sd 2:0:0:0: [sda] 41943040 512-byte logical
```

```
blocks: (21.5 GB/20.0 GiB)
[    2.391149] sd 2:0:0:0: [sda] Write Protect is off
[    2.391150] sd 2:0:0:0: [sda] Mode Sense: 61 00 00 00
[    2.391304] sd 2:0:0:0: [sda] Cache data unavailable
[    2.391305] sd 2:0:0:0: [sda] Assuming drive cache: write
through
[    2.401493]  sda: sda1 sda2
[    2.404998] sr 1:0:0:0: [sr0] scsi3-mmc drive: 1x/1x writer
dvd-ram cd/rw xa/form2 cdda tray
[    2.405000] cdrom: Uniform CD-ROM driver Revision: 3.20
[    2.406103] sd 2:0:0:0: [sda] Attached SCSI disk
[    2.413125] usb 2-2.1: new full-speed USB device number 4
using uhci_hcd
[    2.417859] sr 1:0:0:0: Attached scsi CD-ROM sr0
--
[    4.345611] usbcore: registered new interface driver btusb
[    4.396658] intel_pmc_core intel_pmc_core.0:  initialized
[    4.421252] XFS (sda1): Mounting V5 Filesystem
[    4.596411] XFS (sda1): Ending clean mount
[    4.706702] RPC: Registered named UNIX socket transport
module.
[    4.706703] RPC: Registered udp transport module.
```

If we take a look at the end of this output, we can see that the last two lines don't match our regular expression used to filter data. That's because we used the -C2 parameter with egrep, and that option enables egrep to show two lines preceding and two lines following our pattern match. We can divide that option into options -A and -B (after and before the match, respectively) if we want to specify a custom number of lines to appear after and before our pattern match. There are many more grep options available, but these are enough to get us started. We will add some more examples of using regular expressions and some topics we didn't cover here in the next recipe, about the sed command.

How it works...

grep is a pattern-matching command that can work in a variety of different ways – either as a standalone command that takes text files as input or as something that we pipe input to in a command set. Its purpose is clear – find specific text patterns in a large set of text. A couple of years ago, its default output changed a little bit as it used to show the lines matching our search pattern, while now it does that and it does it in color, by marking the found search pattern in red.

`grep` works by implementing the idea of text pattern search into a programmable command that's a regular part of shell scripts, as we'll explain in the last two chapters with practical examples of scripting. As such, it's an irreplaceable part of a system administrator's toolkit as it enables us to find important text data from one or many files, therefore bringing a sort of chaos into order.

The most commonly used options are as follows:

- `-E` (or `egrep`): By default, `grep` recognizes BREs only. If we use the `-E` parameter, it works with EREs.

- `-i`: Case-insensitive search.

- `-v`: Invert match, find the opposite of our search pattern: A, `-B`, `-C` – options providing *context* to our output, by showing A number of lines after, B number of lines before, or C number of lines before and after our pattern match.

- `-n`: Show the line number where the pattern match appears.

Let's continue our quest to use CLI-based utilities that can do important things on text content by moving on to look at the `sed` command in our next recipe. That will also give us some more scope to further our knowledge about regular expressions, as `sed` can use them to be even more useful than its usual, vanilla self.

There's more...

If you need more information about `grep` in Linux, you can check out the following links:

- grep man page: `https://man7.org/linux/man-pages/man1/grep.1p.html`

- 20 useful `grep` command examples in Linux: `https://www.linuxbuzz.com/grep-command-examples-linux/`

- Regular expressions in `grep`: `https://linuxize.com/post/regular-expressions-in-grep/`

- Stanford's regular expression cheat sheet: `http://stanford.edu/~wpmarble/webscraping_tutorial/regex_cheatsheet.pdf`

Using sed

In addition to our discussion about manipulating text, we absolutely must discuss `sed`. It's a go-to tool to solve so many problems where quick solutions are needed and a lot of text is involved. I can list more than a few examples just from the past couple of years where it saved my skin. For example, I had a couple of projects that required migrating WordPress sites from one domain to another. As it was something that needed to be done in a flash, testing migration modules wasn't an option. The simpler way was to just export the MySQL database, change *domain1* to *domain2*, and check whether it worked. Later, I had a couple more projects like that where it wasn't just a domain name change; subdomain name changes were needed too, and so on. Keeping in mind that it would take me weeks to do this manually on a database that was gigabytes in size – yes, `sed` really helped me out in those jams. Let's discuss the merits of using `sed` and learn by working on a couple of examples. In the last two chapters of this book, we're also going to go through some of these WordPress-based examples so that we can see how that's done in a jiffy.

Getting ready

Let's continue using the `cli2` machine from the previous recipe. If it's shut down, turn it back on so that we can learn about using `sed`.

How to do it...

When we're using `cut`, we are working with standard-formatted input that needs to be transformed in some way. When we're using `grep`, we are looking to find a text sample just for the purpose of finding that text sample. What happens if we need to find some text sample and change it to something else?

For most users, the answer is, *I'm going to open my favorite editor and do a search and replace.*

Precisely. And that's what `sed` is mostly about, especially if we have a file that's really big, gigabytes and more in size. It's about having the capability to do a search and replace, based on text patterns and regular expressions, without having to open an editor.

Have you ever tried to open a 1 GB text file in Notepad or Wordpad to do search and replace, and if so, how did that work out for you? Let alone having to read multiple large files on multiple systems, especially non-Windows systems?

Let's do some simple, a bit complicated, and more complicated scenarios with `sed`. The first scenario is going to be related to inserting and appending a bit of text:

```
echo "This should be first line" | sed 'i\THIS will be the
first line '
```

Let's check the result of this command:

```
[root@cli2 sysconfig]# echo "This should be first line" | sed 'i\THIS will be th
e first line '
THIS will be the first line
This should be first line
[root@cli2 sysconfig]#
```

Figure 8.24 – First example of using the sed command

The -i switch in this command does inline replacement. Keeping in mind that we used sed without any additional options, it didn't replace anything with anything, just inserted the line before the echo command, although logic suggests that the echo command's output should've been first.

Generally speaking, sed has the following syntax:

```
sed -parameter 'option/sourcetext/destinationtext/
anotheroption' sourcetextfile
```

As we can see, in our previous example, we didn't use any options at the beginning or end of the sed quotes; we only used source text. sed can also be used to extract specific lines from a file. For example, let's say that we want to extract lines 5-10 from the /var/log/messages file in our cli2 CentOS machine:

```
[root@cli2 ~]# sed -n '5,10p' /var/log/messages
Oct  6 11:32:56 cli2 systemd[1]: dnf-makecache.service: Succeeded.
Oct  6 11:32:56 cli2 systemd[1]: Started dnf makecache.
Oct  6 11:41:15 cli2 org.gnome.Shell.desktop[1868]: Window manager warning: last
 user time (6918770) is greater than comparison timestamp (6918769).  This most
likely represents a buggy client sending inaccurate timestamps in messages such
as  NET ACTIVE WINDOW.  Trying to work around...
Oct  6 11:41:15 cli2 org.gnome.Shell.desktop[1868]: Window manager warning: W3 a
ppears to be one of the offending windows with a timestamp of 6918770.  Working
around...
Oct  6 11:43:05 cli2 cupsd[968]: REQUEST localhost - - "POST / HTTP/1.1" 200 183
 Renew-Subscription successful-ok
Oct  6 11:43:05 cli2 org.gnome.Shell.desktop[1868]: Window manager warning: last
 user time (7028812) is greater than comparison timestamp (7028811).  This most
likely represents a buggy client sending inaccurate timestamps in messages such
as  NET ACTIVE WINDOW.  Trying to work around...
```

Figure 8.25 – Using sed to extract a specific part of a text file

The default way of operation for sed is to print every line, and if there is a substitution being made by our source/destination text configuration, it will print the substituted text instead of printing the original one. That's why we have -n, as we don't want to print any new lines as we're not doing any kind of substitution. '5,10p' means print from lines 5 to 10.

We can also do the opposite of that – let's say, we want to print all lines from a file, and delete lines 5 to 10. We can use the following command to do that:

```
# sed '5,10d' somefile.txt
```

We can also use sed to display some lines that are not successive in a text file, for example, display lines 20-25 and 40-100 in somefile.txt (-e used per expression):

```
sed -n -e '20,25p' -e '40,100p' somefile.txt
```

But that's all just using sed for some very pedestrian stuff. Let's now start using it for what we'd mostly use it for, which is string replacement.

Let's say that we have a text file called sample.txt with the following content:

This camera produces some weird sounds. Sometimes it buzzes, sometimes it squeals, and always manages to somehow produce un-camera-like high-pitched squeal that should only be audible to dolphins and whales. As a camera, it's good. As a camera with ability to record sound, it's useless. So, we need a camera.

Our first sed-based task is going to be to replace all instances of camera with microphone. For that, we need to use the following command:

```
sed 's/camera/microphone/g' sample.txt
```

As expected, the word camera gets replaced with the word microphone in the console, and the end result looks like this:

```
[root@cli2 ~]# sed 's/camera/microphone/g' sample.txt
This microphone produces some weird sounds. Sometimes it buzzes, sometimes it sq
ueals, and always manages to somehow produce un-microphone-like high-pitched squ
eal that should only be audible to dolphins and whales. As a microphone, it's re
ally good. As a microphone with ability to record sound, it's useless. So, we ne
ed a microphone.
```

Figure 8.26 – Using sed to replace words without saving these changes to our original file

If we want to replace camera with microphone and for these changes to be saved to the original file, we need to use the following command:

```
sed -i 's/camera/microphone/g' sample.txt
```

As expected, this is the end result:

```
[root@cli2 ~]# sed -i 's/camera/microphone/g' sample.txt
[root@cli2 ~]# cat sample.txt
This microphone produces some weird sounds. Sometimes it buzzes, sometimes it sq
ueals, and always manages to somehow produce un-microphone-like high-pitched squ
eal that should only be audible to dolphins and whales. As a microphone, it's re
ally good. As a microphone with ability to record sound, it's useless. So, we ne
ed a microphone.
```

Figure 8.27 – Using sed to replace words and saving those changes to the original file

The sed options s and g are for searching for a word or regular expression and then replacing it globally. By using the -i switch, we made that replacement operation in place, which means saving replacement changes to the original file.

Some more examples are as follows:

a) Insert blank lines for each non-empty line in the original file:

```
sed G somefile.txt
```

b) Keep the original file content by creating a backup file and do an inline replacement in the original file:

```
sed -i'.backup' 's/somethingtochange/somethingtobechangedto/g'
somefile.txt
```

c) Replace the word somethingtochange with somethingtobechangedto when the practice string appears in the line:

```
sed '/practice/s/somethingtochange/somethingtochangeto/g'
somefile.txt
```

d) Negation of the previous statement: Replace somethingtochange with somethingtochangeto and only replace it if the practice string does not appear in the line:

```
sed '/practice/!s/somethingtochange/somethingtochangeto/g'
somefile.txt
```

e) We can delete the line that matches some pattern:

```
sed '/somepattern/d'
```

f) Let's search for a number inline and append a currency symbol before the number (regular expression for finding numbers used, as well as a backslash for proper quoting):

```
sed 's/\([0-9]\)/EUR\1/g' somefile.txt
```

g) Let's replace /bin/bash with /bin/tcsh in /etc/password (a regular expression isn't needed here, but we have to use the \ character for correct interpretation of the / character as the / character is a special character):

```
sed 's/\/bin\/bash/\/bin\/tcsh/' /etc/passwd
```

As we can see, sed is a very powerful tool that can be used to do a lot of changes *on the fly*. We are going to show some more examples of using sed in scripts later in this book, specifically in the last two chapters with shell script examples. That will give us further insight into sed and its usefulness in everyday work.

How it works...

As a command-line text replacement utility, sed requires us to *explain* what we want to do to it. That's the reason why the structure of sed commands seems a bit *descriptive* – that's just because it is. It also has a lot of options and switches, which add to the overall usability and possible usage scenarios.

The basic command structure is usually something like this:

```
sed (-i) 's/something/tosomething/g' filename.txt
```

Or, it might be like this:

```
sed -someoption filename.txt
```

Of course, sed is often used in scripts, either standalone or as a part of a serial pipeline, something like this:

```
command1 |( command2 |) sed …..
```

Whichever way we use sed, it's essential to learn at least some of its switches and settings, starting with the most commonly used ones – s and g inside sed expressions and -i as a command-line parameter.

Say we have a command such as the following:

```
sed (-i) 's/something/tosomething/g' filename.txt
```

Obviously, this has multiple important options. The `-i` option, as we mentioned, is all about interactive change that's going to implement our search-and-replace criteria in the original file. Without it, we are going to get results from our sed command to the screen, basically results written to the console. Options inside the quotes – s and g – are the most-used sed options, and they mean search and globally replace, that is, in the whole file.

We could do the same thing without the `-i` option, by doing this:

```
sed 's/something/tosomething/g' inputfile.txt > outputfile.txt
```

But, as you might imagine, this requires more typing and is generally more complicated.

The sed command-line option `-n` can be used to suppress output to the terminal, and that's the reason why it's used often. If we have a large text file that we're modifying and we aren't using the `-i` option, this might be the go-to option to use if we don't want our console filled with text data.

One more very useful option is the `-f` option, as it allows us to use an input sed file with replacement definitions. Say we run the following command:

```
sed -f seddefinitionfile.sed inputfile.txt > outputfile.txt
```

We create a seddefinitionfile.sed file that contains this:

```
#!/usr/bin/sed -f
s/something/tosomething/g
s/somethingelse/tosomethingelse/g
```

We can use these options to do multiple sed transformations in one command. We just need to create sed definitions in the file and use it.

The next chapter in this book is going to introduce us to the world of shell scripting – and the whole second half of the book is about shell scripting. We will get to use all the tools that we discussed up to now there, and combine them to create shell scripts, some of the most used programming-based administration tools ever. Take a short break and get ready to shell script!

There's more...

If you need more information about networking in CentOS and Ubuntu, make sure that you check out the following:

- sed manual page: `https://man7.org/linux/man-pages/man1/sed.1p.html`

- 50 sed command examples: `https://linuxhint.com/50_sed_command_examples/`

- sed quick reference guide: `https://kwiki.kryptsec.com/books/sed-editor/page/sed-quick-reference-guide`

9
An Introduction to Shell Scripting

We have come to the part that defines one of the things that Unix (or Linux) is known for – its scripting. When it comes to the so-called *Unix philosophy*, being able not only to use tools that the command line offers to you but also being able to create your own is an amazing ability, using shell tools that do one thing really well.

Scripting is exactly that – the ability to create simple (and complex) tools that, at their core, are a set of commands performing a certain task. We need to clear one thing up before everything else – there is a distinction that some people make between programming and scripting. Strictly speaking, all scripting is programming, but not all programming is scripting. We are talking about disciplines that follow the exact same premises, logic, and ways of thinking, but at the same time, there are major differences between the two. When we talk about scripting, we are in reality creating files that are going to get *interpreted* when running, and that means that the shell (or some other *interpreter*) is going to read the file line by line and then run the commands. There is another option, and that is to create text files that are *compiled* before being run. Usually, this is faster than interpreting them, but at the same time, it both requires a few extra steps and is not as flexible as scripting.

We are not going to waste our time on anything connected to compiled applications; in this book, we'll be strictly dealing with scripts.

We will cover the following recipes in this chapter:

- Writing your first Bash shell script
- Serializing basic commands – from simple to complex
- Manipulating shell script input, output, and errors
- Shell script hygiene

Technical requirements

For this recipe, we're going to use a Linux machine. We are using the same setup as in other chapters:

- A virtual machine with Linux installed, any distribution (in our case, it's going to be *Ubuntu 20.04*)
- Bash – the default shell for every major distribution out there

Scripts in this chapter, and in all the other chapters covering scripting, will probably run in any distribution using Bash. The power of scripting is exactly in this compatibility; if a machine runs Linux, it will run almost any script, and the only problems are going to come from what the script itself expects on the server.

Writing your first Bash shell script

Before we do a simple `Hello World!` shell script, let's quickly talk about the shell itself and what does it do on a normal Linux machine. The simplest way of describing it is that the shell is the connection between the user (us) and the kernel (the part of the operating system in charge of everything). We have already talked about that before, but we need to make some points here to make it easier to explain some concepts.

The shell is an application that usually displays a prompt and finds and runs whatever command we give it. This is called the **interactive shell** and is the most-used way of working in Linux. This is what all the **Command-Line Interface** (**CLI**) business is about – having an interface that enables us to execute whatever we need:

Figure 9.1 – A typical root shell

There is, however, another mode of operation for a shell called **non-interactive mode**. This covers all the instances of the shell when it is not acting based on our commands from the command line but instead by reading a file (our script) line by line and executing the commands. Obviously, we cannot interact with the commands directly, so the mode is aptly called non-interactive.

Bear in mind that we can interact with the script during execution if we wish (and plan) to; the name refers only to the lack of direct interaction with the shell or having no CLI available. At the same time, this interaction limit means that we get to see our script run as fast as possible, at any time that we need it. Pair that with a myriad of tools at our disposal in a normal Linux system, and we have an extremely powerful feature available to finish our tasks.

Getting ready

Let's quickly run a few commands and find out about our current shell:

```
demo@ubuntu:~$ ps -p $$
    PID TTY          TIME CMD
   5329 pts/0    00:00:00 bash
demo@ubuntu:~$ echo $SHELL
/bin/bash
```

What happened here? The first command we used was ps, and it gave us information about which shell is currently running or, to be precise, which shell is responsible for the current execution of commands that we issue. Using $$ as the process number, we are asking the ps command to give us the number of the process assigned to our current shell. We performed a small trick here – $$ is a Bash internal variable that gives us the running PID of a process.

The other command that we used is echo, and we used its $SHELL variable that automatically resolves to whatever the user's current shell is.

However, there is a big difference between these two commands. Depending on the circumstances, they can give us completely different results, since they are referring to completely different things. Let's explain – each user has their *assigned* shell, which is going to be executed when the user logs in. The result of the echo command is going to give you that, and the shell itself is defined in the /etc/passwd file, in the line that describes a particular user. So basically, the output of the command is going to provide the name of your default shell.

At the same time, every user can run any shell available on the system as a command, and automatically get this shell as their *current* shell. By that, we mean that this shell is going to process whatever the user is typing into the command line. This is important, since your script can be run from the command line using a different shell than the one you are supposed to be using, based on information in the /etc/passwd file.

Your shell doesn't have to be bash. You can also choose any shell that is available on your system, or you can even install shells that are not currently available on the system you are using but are available as packages.

With that in mind, when it comes to scripting, bash is the shell of choice, even for users running other shells, since bash will run on most, if not all, Linux machines.

Now, let's talk a little bit about the editors used for scripting.

For chapters dealing with scripting in this book, we are going to use vim or vi; however, the script examples are going to be displayed as text without any color. We already covered a lot of editors in a separate chapter. Since the topic of text editors tends to be pretty divisive, and we are a bit pragmatic on this, our suggestion is to use whatever works for you.

Vim, JOE, nano, vi, Emacs, gedit, Sublime Text, Atom, Notepadqq, Visual Studio Code, and so on are what's available, but it's up to you to choose. For simple scripts, any editor will work, and more often than not, you will choose something that already exists on the system you are working on, just because you need to make a small change in a script.

When developing scripts on your own machine, you are probably going to go with something more complicated, simply because it makes working much easier. Vim is a good example, since it provides syntax highlighting and formatting for bash. Advanced editors are going to provide you with more functions, but our opinion is that you shouldn't rely on too many bells and whistles, since this will make you dependent on a certain application that may or may not be available to you at all times:

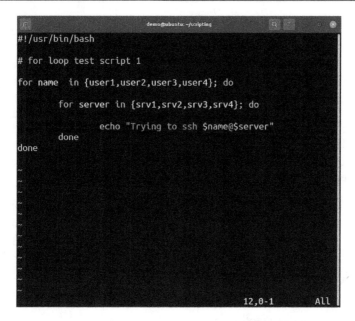

Figure 9.2 – A script opened in Vim – note the color highlights and indentation

In the end, you will end up using two editors, one for developing your scripts and one on servers you are deploying your scripts to. Remember that you will inevitably have to do some debugging on the systems you are deploying your scripts to, so be prepared that some of your normal tools will be missing. Don't become dependent on them.

How to do it...

Let's create our first script and see what it is about the shell that we are so concerned with:

```
#!/bin/bash
# Hello world script, V1.0 2021/8/1
echo "Hello World!
```

First, what does the script actually do? It simply writes Hello World! to the standard output and exits. Although we have three lines in the script itself, only one of them does something, while the other two have different purposes.

What we are going to do is first unpack the meaning behind the first two lines and then pay attention to the echo command.

Note that we are not counting empty lines in the script, although your editor might. In scripting, empty lines are meaningless to the interpreter, so we are using them to make the script more human-readable and ignoring them when talking about the script – Bash does the same thing when executing the script.

As a rule, if we use the # character anywhere in the script, an interpreter is going to treat everything that comes after that character *in the same* line as a comment. Our first two lines are comments, but the first line in the script is special – it defines the shell that is going to run commands in the scripts while also being a comment. This sequence is called a *shebang*. We need to explain that.

Under Linux, scripting is not confined to using bash only or even using any other Bash-compatible shell. In your script, you can actually use whatever scripting language you want. Other than being Bash, it can be Python or Perl – you can use whatever language is available on the system and that you know how to write scripts in.

Normally a script is run by an interpreter. The interpreter is basically an application that is able to understand what's inside the file and then is able to run the commands that are inside the file, one by one. All the interpreters that we mentioned (Python, Bash, and Perl) use simple plaintext files as inputs, so there needs to be a way of telling what kind of script is in a particular file to let the system know how to run it.

This can be done in two different ways – one way is to actually call the script directly using the right interpreter, such as the following:

```
demo@ubuntu:~/scripting$ bash helloworld.sh
Hello World!
```

This simply makes sure we are using the right interpreter for the script; it doesn't make our script any more readable to the system or another user.

Now, consider another way of doing this. Let's make the script executable, and then run it directly:

```
demo@ubuntu:~/scripting$ chmod u+x helloworld.sh
demo@ubuntu:~/scripting$ ./helloworld.sh
Hello World!
```

The difference between these two is subtle but important, although the end result is going to be the same, since we are running the same script.

In the first example, we are explicitly telling the system to use a particular interpreter and run our script. In the second example, we are telling the system to run the script using whatever interpreter it needs, and this is when the first line of the script plays its crucial role. What the current shell is going to do is take the first line (the shebang) and try and find the interpreter that this line points to. If it finds it, the system is going to use this interpreter to run whatever is in the rest of the file. The end result is simple – if we follow the convention and put our interpreter as the first-line comment, the system is going to be able to run our script even if we don't mention it explicitly.

If the first line is something other than the name of the interpreter, our script will work only if we call it by explicitly using the name of the interpreter – if we try to run it directly, the system is going to throw an error.

How it works...

Linux does not use extensions as a way of identifying files, so a script can have any name that is possible on the filesystem; the extension does not have to be .sh. All this means is that in order to have our script work universally, we need to think about the right formatting for the first line.

The next line in the script is our comment, which identifies the name of the script and the version. Of course, any script will function without comments such as these or, in general, without any comments at all, but comments are extremely important in scripting. We will pay much more attention to the comments later in this chapter.

The third line is the one actually doing the work, and it simply displays the string to the standard output that is assigned to the script. Standard input, output, and error handling are things that we will also deal with a little bit later.

This is our first script. We explained a lot in this part of the chapter, and we focused on everything other than the actual command that performs the script task, but we had to deal with a lot of other things.

There's more...

We are going to be dealing with scripts a lot more in this chapter and the next few, but we have links to get you started:

- The Bash manual – containing everything about the Bash shell: `https://www.gnu.org/savannah-checkouts/gnu/bash/manual/bash.html`
- Bash scripting cheat sheets: `https://devhints.io/bash`
- How to get a shell script to actually work: `https://linuxcommand.org/lc3_wss0010.php`

Serializing basic commands – from simple to complex

Scripts are nothing more than a list of commands that are executed in a particular order. In its most basic structure, the order is completely linear without any decision making, loops, or conditional branching.

Commands are executed from first to last, from top to bottom, and from the start of the line to its end. Even if it sounds simple and not very useful, this way of creating scripts can have its uses, since it enables us to quickly run a predefined set of commands instead of repeating them from the command line. In other words, there are problems that require more than a single one-line command but are not complicated enough to require complex logic. This is not to devalue complex Bash scripting logic, as there's a lot of automation in IT that can be implemented by using Bash scripting.

Let's now imagine a simple task like that, something that we will be using as a recurring example. We are going to create a simple backup script. Our task is as follows:

1. Create a directory under /opt/backup that has today's date in its name.

2. Copy all the files from the /root folder to this directory.

3. Send an empty email to the root user that will simply say that a backup was done.

4. Add a line to a file named donebackups.lst in the /root folder that will have today's date in it.

Getting ready

Before we start – a disclaimer. This is a simple script that does not make too much sense for a number of reasons. The most important is that it is oblivious to the context it is running in. We quickly need to talk about this and the other problems first, and then we'll compose a script that solves this task.

What do we mean by *context*? No matter which way we choose to run them, scripts are run by a user, in so-called user space, and they have some things that define their environment that we usually call context.

Context is the entire environment that is running the script and offers the following questions:

*   Which user is running the script?

*   What permissions does this script have?

*   Is the script run from the command line as a tool or as a background task?

*   What directory is the script running from?

Other than these, there are usually several other things that can be relevant for the running script, and we will talk about those as we progress through this book.

Right now, we need to make clear that context is extremely important and that our scripts should never in any way or form take for granted any element of it. If we want our scripts to run correctly, we should presume nothing, and instead, we should check everything that we expect to be in a certain state.

A script that is able to check and decide what the possible problems are in the environment that started it requires two things that we haven't yet talked about – controlling the flow of the script and interacting with the system. Right now, it is obvious that we still don't know how to do that.

Not being able to test things in a script means that, after all, we are going to presume a lot when creating this particular script. Be warned – this is usually the first thing that will lead us into problems.

Not thinking things through before we even type a single letter is usually the source of all problems; scripts are rarely so simple that they can be created without planning in advance.

The main reason we are talking about this now is to try and put you in the right mindset when creating your scripts.

How to do it...

So, how do you create a script? Before you even begin, you should do the following:

- Define your tasks.
- Research commands that you are going to use.
- Check permissions and things required for the individual commands to succeed.
- Try individual commands before using them in the script.

Think about things that you are presuming:

- If you are reading from or writing to any files, do you expect the files to be there, or do you need to create them?
- If you are referencing some file or directory, does it even exist, and do you have the right permissions?
- Are you using some command that needs to be installed or configured beforehand?
- Are you referencing files by an absolute or relative path?

This is only the tip of the iceberg that is usually called **sanity checks**, and the term *sanity*, in this case, refers to the state your script operates in. **Sane state** is what we say when we mean that everything is okay. Any deviation from this, or any error or problem that makes your script behave in an unexpected way, is a problem. This is why we need to think in advance. And that's also the reason why sanity-checked code can take much more effort than regular code that just does the basics.

But believe us – these kinds of checks will help keep not only the sanity of the environment but also your own sanity intact when dealing with complex tasks.

Right now, we are bravely ignoring all this and dealing with the basics. For our backup script, we are going to presume that both the /root and /opt directories exist and are accessible to whatever user is running the script. In our particular case, this means that our script is going to work ONLY if run by a superuser, since this user is required to be able to access files under /root.

Also, we are going to presume that an email system of some kind exists and is running on the local computer. We will also presume that the log file mentioned in the last step is writeable to our script.

> **Important Note**
> When running a script, you are making a lot of presumptions like these, and if any of them is not correct, your script is going to fail in some way. Your main job as a script creator is to prevent this.

What commands are we going to use?

Our first task is going to be to create a directory using today's date in its name. Here, we are going to presume that this directory does not exist, and we are going to create it no matter what. This is a significant deviation from the logic that we are going to use later – a command like this would usually check at least if the directory existed, and if the command itself succeeded. A script should fail in some graceful way if any of these are not true, or it should create directories that it needs.

Before you say, *But wait – I can already do that; I know how to do testing in a single command line,* let's go back and talk again about how a script is run.

At this moment, we are trying to create something that has no logic to control the flow of the script so that commands are run one line at a time. We could try to cheat and do some checks in each individual line, but since we are not controlling the flow of the entire script, this can be even more dangerous than doing no checks at all.

An interpreter is going to run all the commands no matter what, and even if we are checking for problems and spot them, we will end up running all the commands in the script. The right thing to do if something is not right is to control how the *script* behaves, not how a *single command* behaves. No matter how simple or complex a scripting task is, you should always think about the aforementioned context and what your script is doing to it. If something fails, your script needs to decide – is the failure something that can be dealt with, or do you need to abort the execution of the entire script?

If you are aborting execution in the middle of the script, is there something you need to do before your script ends? Usually, when something that forces you to abort a task happens, you will have some way of notifying the system and the user that there was a problem. Sometimes, that will not be enough – your script will have to clean up after itself.

It could be a matter of deleting some files, but it could be that you need to revert a change or even hundreds of changes that you made to the system. Every failure should be evaluated not only on its severity but also on how it affects the system and the state that your script created on that system.

How it works...

After creating what is probably the world's largest disclaimer on why our script is so basic, let's get down to work.

Task by task, how are we going to solve this?

Creating a directory is simple; we are going to avoid using Bash shell expansion, and we will use the `date` system command. We are cheating a little bit here, since we are referencing the system environment, but the task is simply impossible to accomplish without it, and we are not relying on something that is inherent to Bash itself. Note here that this also demonstrates that there are usually multiple ways to do any one thing in scripting. The only difference is your creativity.

The first command will be something like the following:

```
root@ubuntu:/home/demo/# mkdir /opt/backup/backup$ (date \
+%m%d%Y)
```

Please note that we are running this as the `root` user. Let's quickly check what happened:

```
root@ubuntu:/home/demo/scripting# ls /opt/backup/
backup08202021
```

We can see that our directory has been created. Now, let's deal with the copy operation:

```
root@ubuntu:/home/demo/scripting# ls /opt/backup/
backup08202021/
root@ubuntu:/home/demo/scripting# touch /root/testfile
root@ubuntu:/home/demo/scripting# cp /root/* /opt/backup/
backup'date +%m%d%Y'
cp: -r not specified; omitting directory '/root/snap'
root@ubuntu:/home/demo/scripting# ls /root
snap   testfile
root@ubuntu:/home/demo/scripting# ls /opt/backup/
backup08202021/
testfile
```

We were able to create a test file and copy it to our directory. Pay attention to the way we are referencing the directory that we are copying to – since we don't know *when* this script is going to run, there is no way for us to know what the current directory we need to copy to is.

To avoid the hassle of having to read and parse directories, we are simply recreating the directory name the same way we did when we created the directory itself. There is a possible bug here – if, by some freak chance, your script is run exactly at midnight, it is possible that the part where the directory is created is run before midnight and the part we are using to copy the files is run after midnight. This can create an error, since the names will not match. The chance of this happening is slim, and we are not going to plan for it.

In a large script, things like this can and will create big problems if not addressed correctly.

Now, let's do mail functionality:

```
root@ubuntu:/home/demo# mail -s "Backup done!" root@localhost
Command 'mail' not found, but can be installed with:
apt install mailutils
root@ubuntu:/home/demo/scripting# apt install mailutils
Reading package lists... Done
Building dependency tree............................
```

The error here is important. We did it to show the importance of testing commands. In this case, we tried sending mail, and it made us realize that the command we are expecting to use is not installed by default.

In order to run this script, we actually need to have the `mail` command installed. It is going to be there on servers that have mail service configured but is going to be absent on normal workstations. Since our backup script should work on any server, we need to solve that.

Blindly installing a package using a package manager is usually safe; the system will either install the package or update it if it was already installed.

Now, we are going to try that command again, and once again, it is going to fail:

```
root@ubuntu:/# mail -s "Backup was done!" root@localhost
Cc:
Null message body; hope that's ok
root@ubuntu:/# man mail
You have mail in /var/mail/root
root@ubuntu:/# mail -s "Backup was done!" root@localhost < /
dev/null
mail: Null message body; hope that's ok
```

We didn't actually fail, but when we invoked the first command, it required us to input some data. It was asking us for the `Cc` address, and we had to press *Ctrl + D* to finish the body of the email.

This is another reason to test commands before using them in a script.

After realizing that we need to do something to make this command run unattended and reading the manual, we can see it is a matter of simply providing no inputs by redirecting `/dev/null` into the command.

Now, for the last command that we need to do, we implement the actual reporting of the backup:

```
root@ubuntu:/home/demo# date +%m%d%Y >> /root/donebackups.lst
```

Remember, we need to append to a file. Additionally, what we want to do is reference the file directly using an absolute path; after all, we have no idea where we will be when running this script.

Okay, we have tried and tested all the commands. How does our script actually look? It's not complicated:

```
#!/bin/bash
mkdir /opt/backup/backup'date +%m%d%Y'
cp /root/* /opt/backup/backup'date +%m%d%Y'
```

```
mail -s "Backup was done!" root@localhost < /dev/null
date +%m%d%Y >> /root/donebackups.lst
```

Now, let's run it.

```
root@ubuntu:/home/demo/scripting# bash backupexample.sh
cp: -r not specified; omitting directory '/root/snap'
mail: Null message body; hope that's ok
```

There are a few things that need our attention. First, note that we have some output from the script that we didn't expect. This is normal and is a direct result of what we saw when we tested our command – some commands threw errors. The reason we are seeing this error is going to be explained in the next part of this chapter. Another thing that we need to note is that other than the errors, we have no other output. The only way we have to tell whether our script was successful or not will be by having the script itself report it – we need to check the mail and the file mentioned in the script to check whether everything is correct. That points us to another thing we are going to need – the logging. We are going to deal with logs and debugging in later chapters.

Now, let's elaborate a little on how your script communicates with the environment.

There's more...

- Different operators for chaining in Bash: `https://www.thegeekdiary.com/6-bash-shell-command-line-chaining-operators-in-linux/`
- Formatting dates in Bash: `https://www.cyberciti.biz/faq/linux-unix-formatting-dates-for-display/`

Manipulating shell script input, output, and errors

There are only a few things that are as pragmatic as the idea behind the concept of standard input and standard output on Linux.

Since the start of Unix, the idea of interoperability between different applications and tools installed on a system was one of the primary prerequisites that every script, tool, and application had to follow.

Simply put, if you wrote any tool on a system, you could count on three separate channels of communication to your surroundings. Based on the concept of ANSI C input/output streams called **standard output** and **standard input**, everything that runs in a shell can communicate in three ways – it can receive inputs from standard input, it can output results and information to standard output, and it can report errors to a separate output that is marked just for this task as **error output**.

Pair this idea with the concept that every tool should output text-only information with minimal formatting, and should be ready to accept text input if it is required, and you have a framework that is simple but amazingly robust and portable.

Getting ready

When we create scripts, we are going to use these concepts a lot, in a myriad of different ways. Before we can do that, we need to make sure that we understand what inputs and outputs actually exist and are available to us, and what is the usual way of using them when scripting. After that, we will deal with some recommendations and how to conform to certain well-established best practices when it comes to user interaction.

Even before that, we need to define some things. Standard input, output, and error are just special cases of something called **file descriptors**. To simplify things a bit, we will not be spending too much time on what a file descriptor actually is; for the sake of this chapter, let's just say that it is a way of referencing an open file.

Since in Linux everything is considered to be a file, we are basically just assigning a number to something we can write to, read from, or both read and write to, depending on the context. Obviously, our options on reading and writing depend on what the actual device referenced is.

By default, your script will *open* communication with three *files* that it can use. Standard input is going to be connected to a keyboard; your script is going to accept information from the keyboard unless you change it to something else, such as some other file or output of another script or application.

Standard output is, by default, set to the *console* or the screen you are running your script from. In some circumstances, we will also call a screen connected physically to your server a **console**, but that is not part of what we are dealing with right now. The reason we are mentioning this is to avoid unnecessary confusion.

We cannot read from a screen or write to a keyboard, and this is why we are usually referring to them as *consoles*, which is a common name that more or less describes both the keyboard and the screen. There is a lot more to learn here, but for now, we are going to leave it at this.

How to do it...

To better explain these two things, you can do something simple – run a command called `cat` without any arguments. When executed like this, any command including `cat` will accept standard input and then output the result to the standard output. In this particular case, `cat` is going to do it one line at a time, since it waits for a line separator before it outputs information.

In reality, this means that `cat` is going to echo whatever you type in, line by line, until you use *Ctrl + D* to signal a special character called **End of Transmission** (**EOT**), which tells the system that you decided to end typing.

This will end the execution of the application. In the screenshot, it looks like we typed each line twice; in fact, one is our input, and the other one is the output from the command:

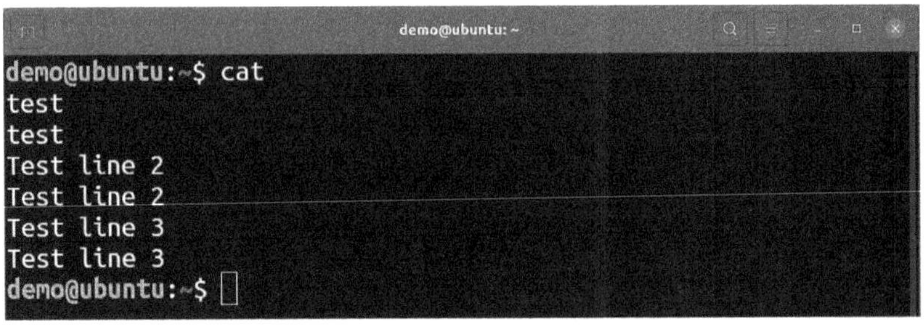

Figure 9.3 – cat – the simplest command to demonstrate standard input and output

There is also standard error, which also defaults to the screen but is a separate stream of data; if we output something to standard error, it will be displayed in a way that looks exactly the same as standard output, but that can be redirected if we need to.

The reason why there are two separate streams handling output is simple – we usually want to have some data as the result of our script, but we do not want errors to be part of it. In this case, we can redirect data into some file and redirect errors to the screen, or even to another file, and then deal with them later.

Now, remember when we mentioned that standard input, output, and error are special instances of a file descriptor? Bash can actually have nine file descriptors open at the same time, so there are many more things we can do when writing out something in our scripts. This is, however, rarely done, since almost everything can be accomplished by using only the default ones. For now, just remember the following:

- Standard input is file descriptor number 0.
- Standard output is file descriptor number 1.
- Standard error is file descriptor number 2.

Why do these numbers matter? By using some special characters in the command line and in our scripts, we can do a lot if we know only these three numbers. First, how do we stop a script displaying something on the screen, and how do we output it to a file instead? By simple use of the > character.

Sometimes, you will see a command line containing 1> instead of just >. This is exactly the same as using a single > character, but it is sometimes written like this to make sure you understand that you are redirecting standard output.

You are probably familiar with this form of redirection, since this is one of the first things you learn when dealing with the command line. An important thing to note is that we can redirect to a file in two different ways, depending on what do we want to do with the file if it already exists.

By using > `filename`, we are going to redirect whatever the script outputted to the standard output to the file named `filename`. If the file does not exist, it will be created, and if the file exists, it will be *overwritten*.

By using just one more bracket, the >> `filename` redirection will be different in the way it treats files that already exist. If we redirect using this symbol, we are going to *append* data into an already existing file; data is going to go to the end of the file.

Having mentioned 1>, we need to deal with the way more popular 2> symbol that represents standard error. When something wrong happens in our script, it is going to output it as an error. Usually, you will notice it if you just redirect the script output to a file; if you fail to mention 2>, you will see that only errors are going to appear on the screen, while everything else is going to end up in the file.

If we actually want to output the result of errors in a particular file, we can do that by using 2> `errorfilename`, and the script will write its errors into a file called `errorfilename`.

There is also the possibility that we want to output everything into a single file, and there are two ways to do this. One is to do both redirections in one command line separately, using the same filename for both redirects. This has its own advantage of being easy to read when we are trying to understand where the outputs are going.

The main disadvantage is that this redirection is probably the most used one when it comes to dealing with scripts, especially when we run them unattended, and this makes it *harder* to read in most environments. Of course, there is a simple solution to this – instead of using two separate redirects, we can use a single one by using the `&> filename`. In the Bash environment, this means that we want to redirect both standard error and output to the same file:

```
demo@ubuntu:~/scripting$ bash helloworld.sh 1> outputfile \
2>errorfile
demo@ubuntu:~/scripting$ bash helloworld.sh &>outputfile
```

Please note that this trick works only if redirecting both the output and errors to one file; if the output files are different, we need to specify them explicitly one by one.

When we started discussing outputs, we said that there can be more than just the three predefined ones, and the way to handle them is logical. If we decide to redirect something outputted to file descriptor number 5, the way to handle it in the command line would be to just redirect `5> filename`. This is something you will not see every day, but it can be extremely useful if you need to create more than one log file or need to create different outputs to different destinations from the same script. This approach is seldom used, since it is much easier to handle redirection directly from the script, and by using variables in the script, anyone debugging your scripts is going to have a much easier job.

Up to this point, we were dealing with redirection from *outside* of our scripts. It is time to move on to how to use this in our everyday work.

The main thing that we are going to do using redirection is to log things. There are a couple of approaches here. One is to simply use the `echo` command in the script and then do the redirection for the whole script – for example, we can create a simple script that just prints four lines of text:

```
#!/usr/bin/bash
echo "First line of text!"
echo "Second line of text!"
echo "Third line of text!"
echo "Fourth line of text!"
```

Let's name it `simpleecho.sh` and run it using Bash:

```
demo@ubuntu:~/scripting$ bash simpleecho.sh
First line of text!
Second line of text!
```

```
Third line of text!
Fourth line of text!
demo@ubuntu:~/scripting$
```

Now, we are going to redirect it to a file:

```
demo@ubuntu:~/scripting$ bash simpleecho.sh > testfile
demo@ubuntu:~/scripting$ cat testfile
First line of text!
Second line of text!
Third line of text!
Fourth line of text!
```

Okay, we can see that our file now contains output for the echo commands. For the sake of showing how errors work, we are going to insert an intentional error into our script:

```
#!/usr/bin/bash
echo "First line of text!"
echo "Second line of text!"
echo "Third line of text!"
bad_command
echo "Fourth line of text!"
```

Now, we are going to do the same procedure again, first starting the script and then redirecting it to see what happened:

```
demo@ubuntu:~/scripting$ bash simpleecho.sh
First line of text!
Second line of text!
Third line of text!
simpleecho.sh: line 5: bad_command: command not found
Fourth line of text!
demo@ubuntu:~/scripting$ bash simpleecho.sh > testfile
simpleecho.sh: line 5: bad_command: command not found
demo@ubuntu:~/scripting$ bash simpleecho.sh &> testfile
demo@ubuntu:~/scripting$ cat testfile
First line of text!
Second line of text!
Third line of text!
```

```
simpleecho.sh: line 5: bad_command: command not found
Fourth line of text!
```

The main thing to take away here is that error output is always separate from standard output, so we are not going to see errors in our file unless we specifically redirect them.

Everything up to this point was simple, since our script was using just standard output. Often, communicating with users is not so simple because we want to have our script to be able to provide some information on screen and to a special log file. Things are similar when working with unattended scripts; having the ability to redirect script output to a certain file is nice, but more often, we are going to make our script use a particular log file by itself, without the need for the user or the administrator to do any redirection when executing the script.

The process required to do this is remarkably simple – we can use the redirection at the command level to redirect our output to a file. There is only one thing you have to remember here. Redirection into a file is limited to a single command; if you redirect anything, the file is going to be closed as soon as the command finishes. This is important primarily because you will always need to append to a file; if you forget to do so, that file is going to get rewritten with new data, making it useless as a log. Since a log is usually used to track multiple executions of a script or service, you will almost universally append to files.

How it works...

Let's now expand our initial script, adding a little bit of logging. What we are going to do is write separate logs that will contain information about the actions that the script took while running. We are going to write this information to a log file located in the directory that the script was invoked from. This means that our script will at any one time be able to use three separate channels for output; in addition to standard output and standard error, we are also using our log file. The big difference between log files and standard output is that our log is hardcoded, and there is no way of redirecting it to another file. Of course, a solution for this problem exists, but we are not going to spend too much time on it; we already said it is possible to use one of the other file descriptors and map output to it, forwarding the output to whatever stream we need later. This is seldom used, since it requires additional attention when running a script:

```
#!/usr/bin/bash
echo "We are adding four lines of text!" >> simplelog.txt
echo "First line of text!"
echo "Second line of text!"
echo "Third line of text!"
```

```
echo "Fourth line of text!"
echo "Exiting, end of script!" >> simplelog.txt
```

This approach gives us additional flexibility, since we do not have to forward standard output in order to have logs; our script already does that. This means that we can start the script either from the command line or as an unattended task and get the same results in the logs. Of course, we can always use redirection and make sure that every output is written and saved.

There's more...

- Standard input and output (in C): https://www.technologyuk.net/ computing/software-development/computer-programming/c-programming/basic-io.shtml

- Standard input and output (in Bash, with examples): https://tldp.org/LDP/ abs/html/io-redirection.html

Shell script hygiene

Commenting is not just something you can do; it's an art by itself. In this part of the chapter, we are going to deal with comments in order to make your life easier when writing scripts, but the advice and best practices that are given here are easily used in any programming language that we can think of. Really understanding how to comment in a useful way is something that you're going to need to learn, since it will help anybody going to use your scripts after you are done writing them.

So, what are comments? Possibly the easiest way to describe them is to say that they are documentation on what the script is intended to do, how the script works, and who has created the script, and they provide more information on technical details of the script, such as when it was created.

Commenting is something that you should automatically want to do. Nobody is perfect and nobody has a perfect memory. Comments are there to help you remember what you did inside some script and to provide anybody else with guidance on how the script works and what the different things are that they need to know if they need to change anything in the script.

One more important point is that commenting is not the same as providing documentation. Sometimes, people are going to say that they don't need documentation because they already have comments in their code, but this is completely wrong. Unless you're talking about a script that has only 10 lines of code or so, comments are going to help you to understand what the script is doing without looking up the whole documentation, which saves you a lot of time.

Getting ready

So, let's now talk about different types of comments. When writing code, there is always a part that involves commenting on individual procedures or parts of the script, expected input and output, data types, and data in general.

In Bash, comments are universally started by the # sign. Bash does not recognize multiline comments, unlike some other programming languages. What this means is that we need to pay attention so that every line that contains a comment starts with #. The only exception of sorts is the first line in the script, which contains the interpreter that is going to run the script, but the interpreter continues working after that line, so we can say that every line that starts with # is actually a comment. The shell is going to ignore everything inside the comment or, to be more precise, it is going to completely ignore the lines that contain the comment. So, understand that the comments are written for you and for other people that are going to be dealing with your scripts. Try to make them easy to understand, precise, and avoid repeating what can actually be deduced from the command itself. For example, if you have a command that is echoing something, try not saying, *Okay, this command is going to echo...* whatever text you're trying to output to a user, but try to explain why. This is especially useful when commenting on cryptic output that contains a lot of variables.

You can and should write comments in front of every block of code in your script, but you should also comment at the start and the end of the script.

How to do it...

Let's start with the beginning of the script. What should be the first comment? The first line should, of course, be the name of the interpreter, and after that, we usually give information about the script itself. Generally, the script should start with a comment that gives information on who wrote it, when, and if it is part of the project that is responsible for the script itself.

This part should also state technical things such as licensing distribution, limitation on warranties, and who is and is not allowed to use the script.

Having done the header, we should also deal with the arguments and information on how a script should be run and what to expect in terms of input. If there is something special about the input, such as expected types, the number of arguments, or some prerequisites that need to exist or run before the script is running, they should be stated somewhere at the beginning of the script.

Now, we come to the functions. We are going to deal with the concept of functions later, but we do need to talk about how to comment on them, since this also applies to any other block of code. This is because functions are, by themselves, modular and written as a separate block of code.

Sharing something separated inside a function gives us the opportunity to comment. We should use this part of the comments to describe what the function or the module does, what variables we are going to change and need, the arguments that your function is going to take, what the function is going to do, and what the output to the function is going to be. If we are dealing with some sort of nonstandard output – for example, if we are dealing with logging to a separate file – we should state that in the function header. We should also note all the return codes that the function outputs if it changes the exit status of the script.

There are useful ways of using comments to create reminders for yourself and for others to inform what still needs to be done in a script, which are called **to do comments**. They are usually written in capital letters – **TODO**.

We should also note that there is something called a `heredoc` notation that is sometimes used when we need to create large blocks of comments. This notation uses shell redirection in a very specific way so that it can provide the header and the footer for the comment block without using common signs. We're going to provide you with an example of this notation, since you will run into it when you analyze other people's scripts, but we're not going to use it in our scripts. The main reason for that is that it tends to make scripts less readable.

For example, this is a perfectly valid way of creating a comment:

```
#!/bin/bash
echo "Comment block starts after this!"
<<COMMENTBLOCK
    Comment line
    Another comment line
    Third one
COMMENTBLOCK
echo "This is going to get executed"
```

Now, what do we actually comment?

Let's start with some general things:

- State clearly who wrote the script and when was it created.

- Version your script – if there are any changes, update the version so that you can track which script you are using on different computers.

- Explain any complicated part of the code – things such as regular expressions, calling outside sources, and general references to anything outside of your script should be commented.

- Comment on individual blocks of code.

- Clearly note all the old parts of code that you commented out and left in your script.

We'll talk a bit about all these points.

Clearly marking the script author and creation date is crucial. Your scripts are probably going to end up being maintained by people other than you. The worst thing that can happen to you when you open a script with a couple of hundred lines of code is not knowing who to talk to when something goes wrong. Some people think that they are going to avoid being constantly pestered by other admins by not signing the script, but this is simply wrong. You wrote that script; be proud of it.

After you mention the author, always note when the script was created. This helps people prioritize possible changes, especially to some outside resources that you may be using in the script itself. Also, write when the last change was made, since it is relevant information for everybody maintaining the script, including you.

After mentioning changes, learn to version. Versioning is a way of keeping track of different changes that you make in a script and making sure you know which version you are using at any given moment. Versioning itself is a simple concept of using a scheme that enables you to track how your script progresses and what has changed.

There are a couple of ways this can be done, since there is no official standard on how to write down versions, although a lot of people tend to use semantic versioning (`https://semver.org/`). Usually, versions more or less strictly follow either changes in the source code or the time when a particular version was created. Both schemes have their merits, but when writing scripts, we think that tracking changes is a much better idea, since we can deduce very little from versions using dates as a reference.

Before we commit to any versioning scheme, we will quickly go over some examples. The way we handle versions in different software is directly related to the type of software we are dealing with and the number of changes between versions.

General-purpose applications usually stick to a *normal* versioning scheme that has a structure using two numbers representing major and minor versions of the application. For example, we can have App v1.0, then App v1.1, then App v2.0, and so on. The first number represents major changes made to the applications; the second number usually represents minor changes or bug fixes. This is practically the norm for large applications on the market today.

In our scripts, we are going to use the same scheme, but we are going to implement semantic versioning, so a version will be `1.0.0` or `3.2.4`. The third number represents small changes and makes sense when the number of changes is small, but changes are significant. Note that some applications take this approach to the extreme, so you will inevitably run into things such as Version `2.1.2.1-33.PL2`. When dealing with scripts, this will just complicate your work, so don't do it.

Another way of dealing with versions is by referencing time, as most operating systems do now. So, for example, there is Ubuntu 20.04 and 20.10, representing releases that came out in April and October of 2020 respectively. The reason for this is the enormous number of changes. Releasing a new version of an entire operating system each time something changes is simply impossible; you would need to release a new version practically every few hours.

There is also a sequential numbering scheme that is usually paired with one of the two approaches we mentioned. Microsoft uses this versioning style, having major releases with names such as *Windows 10*, update releases named something such as 20.04 or 21H1 that represent the time of the release, and then using build versions to denote minor changes in the operating systems.

All of these schemes have their good and bad sides, but whatever you choose, we have only one recommendation – stick to it. Don't mix different versioning schemes, since it will confuse people.

While talking about versioning, we should also talk about change tracking. When creating *a new version* of a script, most of the time, you will make many changes to the script itself. It can be that you will fix bugs, or make your code quicker or more reliable. Some of these changes will have to be documented in some way other than by increasing the version number. This is important in order to remember what you did to the script. There are a couple of ways to do that. One is to keep track of all the changes in a separate file (usually, we use a `ChangeLog` file for that purpose). This makes your comments and the script itself much more legible, but now you have another file whose updates you need to care about. It also makes it easier for everyone else to read the code, as it gets developed and changed with each new version. Another way is to keep a list of all the changes in the script itself. A benefit of that is that you can quickly check what has changed, but your script now has extra text that you need to skip through. There is also a version that keeps the changes in line with the place where they were created, usually before the line of code that contains the changes.

Let's see how all this looks in practice:

```bash
#!/bin/bash
# V1.2 by Author, under GPLV2 licence
# V1.0 - Hello world script, V1.0 1/8/2021
# V1.1 - Added changes to comments on 2/8/2021
# V1.2 - Added more changes to comments 3/8/2021
echo "Hello World!"
```

We are going to stop here, since we will cover this and a lot more in further chapters, learning as we go along.

There's more...

- Commenting in Bash – examples: `https://git.savannah.gnu.org/cgit/bash.git/tree/examples`
- Identifying files in Linux: `https://man7.org/linux/man-pages/man4/magic.4.html`

10
Using Loops

In the previous chapter, we started dealing with scripting, and we did a lot of learning about how scripts work and how they are structured. However, we missed a huge topic in scripting – influencing the order in which commands are executed when a script is run. There are a couple of things that we need to cover here, since there are multiple ways we can influence what is going to be the next command executed in a script.

We are going to start with a concept we call **iterators** or, more commonly, **loops**. There are a lot of things in everyday tasks that need to be done repeatedly, usually changing just one small thing in every iteration. This is where looping comes in.

We will cover the following recipes in this chapter:

- The `for` loop
- `break` and `continue`
- The `while` loop
- The `test-if` loop
- The `case` loop
- Logical looping with `and`, `or`, and `not`

The for loop

When we talk about loops, we usually make a distinction based on the place in the execution where the variable we are using changes its value. `for` loops, in that respect, belong to the group where a variable is set before each iteration and keeps its value until the next iteration is run. The most common task that we are going to perform by using a `for` loop is going to be using the loop to iterate through sets of things, usually either numbers or names.

Getting ready

Before we begin introducing different ways of using a `for` loop, we need to address its abstract form:

```
for item in [LIST]
Do
   [COMMANDS]
done
```

What do we have here? The first thing to note is that we have some reserved keywords that make Bash understand that we want to use a `for` loop. In this particular example, `item` is actually the name of the variable that will hold one value from the list in each loop iteration. The word `in` is a keyword that further helps us understand that we are going to use a set of values that we currently call a list, although it can be different things.

After the list, there is a block that defines what commands we intend to run each time a loop is performed. Currently, we are going to work with this block as a single entity, containing commands that will be executed one after another without interruptions. Later in this chapter, we are going to introduce some conditional branching that will enable us to cover more possible workflow solutions, but for now, a block is uninterruptable.

What will probably surprise you is that a `for` loop is often used directly from the command line, even more often than in scripts. The reason is simple – there are a lot of tasks we can accomplish by using a simple `for` loop, and complicating them by creating scripts is to be avoided. A `for` loop in this form looks a little different than in the one we showed as our first example, the main difference being the semicolon that separates the keywords when we are using a single line to write out the loop.

How to do it...

Let's start with a simple example. We are going to run through a list of servers and echo one in each iteration of the loop. Note that the shell takes our command from the prompt and repeats it before executing it:

```
root@cli1:~# for name in srv1 srv2 srv3 ;do echo $name; \
done;
Srv1
Srv2
srv3
```

Using `echo` as a placeholder for commands when testing a loop is common. We are going to use this style of debugging a lot in our examples. `echo` as a command is probably the most useful one in this context since it changes nothing and, at the same time, enables us to see what the actual output is going to be.

When creating a list of objects, we don't have to use any special characters to separate individual entries; bash is going to treat space as a delimiter, and as long as we separate our values by a space, `bash` is going to understand our intentions. A little later, we are going to show you how to change the character that separates a value in the list, but a space will work in almost all the circumstances.

The list we are using in the iterations can be explicitly defined, but more often than not, we are going to need to create it when we run our loop, in the command line or the script.

A typical example of this is running a loop on a set of files in a directory. The way to do this is to use shell expansion. This means letting the shell run a command and then using its output as the list for the `for` loop. We can do this either by specifying a command in backticks (`` ` ``) or using `bash` notation of $(command). Both ways have the same result – a command is run and then piped to a list.

Our example is going to be a loop that iterates through the current directory and runs the `file` command on each individual file, giving us information on what this particular file actually is. We are still on the command line:

```
root@cli1:~# for name in `ls`; do file $name; done;
donebackups.lst: ASCII text
snap: directory
testfile: empty
```

Now, let's deal with something more interesting. Often, we need to use numbers in our loops, either to count something or to create other objects. Almost all programming languages have some sort of loop that enables this. Bash is a sort of an exception to this rule, since it can do it in a couple of different ways. One is to use the echo command and a little bit of shell expansion to accomplish this task.

If you are unfamiliar with this, giving echo an argument that consists of a number formatted in curly brackets will make it output all the numbers in the interval that you specified:

```
demo@cli1:~/scripting$ echo {0..9}
0 1 2 3 4 5 6 7 8 9
```

To use this in a loop, we simply do the same trick as we did in the previous example:

```
root@cli1:~# for number in `echo {0..9}`; do echo $number; \
done;
for number in `echo {0..9}`; do echo $number; done;
0
1
2
3
4
5
6
7
8
9
```

We are not limited to using a fixed step in the interval; if we simply mention an interval followed by a number, this number will be considered as a step value. A step value is basically a number that your variable is going to be incremented with in each loop iteration.

We are going to try a simple loop using multipliers of 20:

```
root@cli1:~# for number in `echo {0..100..20}`; do echo \
$number; done;
for number in `echo {0..100..20}`; do echo $number; done;
0
20
40
```

```
60

80

100
```

We can combine shell expansion the same way that we normally do in the command line and create different values for our loop. For example, in order to create server names for three groups of servers, each containing six servers, we can use a simple one-line loop:

```
root@cli1:~# for name  in srv{1,w,m}-{1..6}; do echo $name; \
done;
srv1-1
srv1-2
srv1-3
srv1-4
srv1-5
srv1-6
srvw-1
srvw-2
srvw-3
srvw-4
srvw-5
srvw-6
srvm-1
srvm-2
srvm-3
srvm-4
srvm-5
srvm-6
```

Of course, loops can be embedded within each other by simply placing the inner loop into the do-done block of the outer loop. In this particular example, we are using shell expansion to loop through a list of values in both loops:

```
root@cli1:~# for name  in {user1,user2,user3,user4}; do \
for server in {srv1,srv2,srv3,srv4}; do echo "Trying to ssh \
$name@$server"; done;done;
Trying to ssh user1@srv1
Trying to ssh user1@srv2
Trying to ssh user1@srv3
```

```
Trying to ssh user1@srv4
Trying to ssh user2@srv1
Trying to ssh user2@srv2
Trying to ssh user2@srv3
Trying to ssh user2@srv4
Trying to ssh user3@srv1
Trying to ssh user3@srv2
Trying to ssh user3@srv3
Trying to ssh user3@srv4
Trying to ssh user4@srv1
Trying to ssh user4@srv2
Trying to ssh user4@srv3
Trying to ssh user4@srv4
```

How it works...

Now is the time to slowly switch from the command line to how we can use these loops in scripts. The biggest difference here is that `for` loops are way easier to read when formatted in a script.

For our first example, we are going to mention another way of creating a set of numbers in a loop, a so-called *C-style loop*. As the name suggests, this loop takes its syntax from the C language. Each loop has three separate values. Of these, the first two are compulsory; the third is not. The first value is called either the *initialization value* or the *start value*. It gives us the value of the variable in the first loop iteration. One thing to note here is that we need to assign the initial value explicitly, which significantly differs from the usual style used in the *normal* `for` loop.

The second value in this loop variation is the *test condition*, occasionally known as the *boundary condition*. This represents the last valid value that our loop iterator will have before we finish the loop or, to put it more simply, the largest number if we count incrementally.

The third value can be omitted; it will default to 1. If we use it, this is going to be the default step or increment that our loop is going to use.

Theoretically, this C-style `for` loop will look like this:

```
for ((INITIALIZATION; TEST; STEP))
Do
   [COMMANDS]
done
```

In reality, it has a more complex syntax, but it will look very familiar to all of you with experience in programming in C, as the name suggests:

```
for ((i = 0 ; i <= 100 ; i=i+20)); do
   echo "Counter: $i"
done
```

Before we go on, let's look at an example of a loop that we have already used but formatted as it would be in a script:

```
#!/usr/bin/bash
# for loop test script 1
for name  in {user1,user2,user3,user4}; do
        for server in {srv1,srv2,srv3,srv4}; do
               echo "Trying to ssh $name@$server"
        Done
done
```

As we can see, the only real difference here is the formatting and the omission of semicolons that directly stems from not having to parse the entire script in one line.

See also

In order to understand looping, you will probably need quite a few examples. Start with these links:

- `https://linuxhint.com/30_bash_loop_examples/`
- `https://tldp.org/HOWTO/Bash-Prog-Intro-HOWTO-7.html`

break and continue

Up until now, we haven't really done any conditional branching in our scripts. Everything we did was linear, even loops. Our script was able to execute commands line by line, starting from the first one, and if we had a loop, it was running until our conditions that we stated at the loop start were met. This means that our loops have a fixed, predetermined number of iterations. Sometimes, or to be more precise often, we need to do something that breaks this idea.

Getting ready

Imagine this example – you have a loop that has to iterate a number of times *unless* a condition is met. We said that our loops have the number of iterations fixed at the start of the loop, so we obviously need a way to end the loop prematurely.

This is why we have a command called break. As the name suggests, this command breaks the loop by escaping from the command block it is included in and finishing the loop, regardless of the conditions that were used in the definition of the loop. The main reason why this is important is to establish control over the loop and handle any possible state that requires you to not finish the job you started in your loop. One more thing to note is that the break command is not limited only to for loops; it can be used in any other block of code, a thing that will become more useful later when we learn other ways to structure our scripts into blocks.

How to do it...

It is always easy to start with an example, but in this particular case, we are going to start with an overall view of how this command works. We are going to use abstract commands instead of actual ones to help you understand the structure of this loop. After this, we are going to create some real-world examples:

```
for I in 1 2 3 4 5
do
#main part of the loop, will execute each time loop is started
   command1
   command2
#condition to meet if we need to break the loop
   if (break-condition)
   then
#Leave the loop
      break
```

```
    fi
#This command will execute if the condition is not met
    statements3
done
command4
```

What is going on here? The `for` loop by itself is a normal loop that gets executed using 1, 2, 3, 4, and 5 as values. `command1` and `command2` are going to get executed the way we expect them to be at least once, since they are the first thing after the start of the loop.

The `if` statement is where things get interesting. We will talk a lot more about `if` statements, but we need to mention them here in their most basic form. Here, we have something called a break condition. This can be anything that can be resolved to a logical value. That means the result of our condition has to be either `true` or `false`. If the result is false, our condition is not met, and the loop continues by executing `command3`, looping back to the beginning of the loop, and assigning the next value to our variable.

We are more interested in what happens if the break condition evaluates to true. This means that we have met our condition and need to run the block of code that follows. A simple `break` statement is here, and it has no arguments. What will happen next is that the script will immediately exit the loop and go to execute `command4` and whatever is after it. The important thing is that `command3` will not be run in this case, and the loop will not repeat, regardless of the value of the loop variable.

There is another statement called `continue` that can also be useful, although it is not used as much as `break`. Continue also breaks the loop in a way, but not permanently. Once you use `continue` in the loop, the program flow is going to immediately go to the start of the loop block without executing the remaining statements.

How it works...

Having talked about the abstract structure, it is time to create an example.

Imagine we are counting using a `for` loop, but we want to break out of it as soon as we hit the number 4 as the value of our variable. Of course, we could do this by simply specifying number 5 as the upper value that we are counting, but we need to show how the loop works, so we are going to break out of it using the `break` statement:

```
#!/usr/bin/bash
# testing the break command
for number  in 1 2 3 4 5
do
```

```
echo running command1, number is $number
echo running command2, number is $number
if [ $number -eq 4 ]
        Then
                echo breaking out of loop, number is $number
                Break
fi
echo running command3, number is $number
done
```

It's time to break down our script, but we are going to run it before we do that:

```
demo@cli1:~/scripting$ bash forbreak.sh
running command1, number is 1
running command2, number is 1
running command3, number is 1
running command1, number is 2
running command2, number is 2
running command3, number is 2
running command1, number is 3
running command2, number is 3
running command3, number is 3
running command1, number is 4
running command2, number is 4
breaking out of loop, number is 4
```

Our sample script looks a lot like our *abstract* example, but we used actual the echo command to emulate something that should happen. The most important part that we need to talk about is the if command; everything else is as we said in the first part of this recipe.

We mentioned that we need to have a condition for our break statement to make any sense. In this particular case, we are using if with a test condition; basically, we are telling bash to compare two values and let us know whether they are the same or not. In bash, there are two ways to do this – one is to use the = operator we are used to, and another is to use the -eq or equals operator. The difference between these two is that = compares strings, while -eq compares integers. We will go into much more detail in later recipes, since they are important in scripting.

Now, let's see how the `continue` command works. We are going to slightly modify our script so that it skips over the third command once it hits 3 as the value of the variable:

```
#!/usr/bin/bash
# testing the continue command
for number  in 1 2 3 4 5
do
echo running command1, number is $number
echo running command2, number is $number
if [ $number -eq 3 ]
        Then
                echo skipping over a statement, number is \
$number
                Continue
fi
echo running command3, number is $number
done
```

What we did is a simple change in the `if` statement; we changed the condition so that it checks whether the variable value is equal to 3, and then we created a command block that skips over the rest of the loop when our condition is satisfied. Running it is simple:

```
demo@cli1:~/scripting$ bash forcontinue.sh
running command1, number is 1
running command2, number is 1
running command3, number is 1
running command1, number is 2
running command2, number is 2
running command3, number is 2
running command1, number is 3
running command2, number is 3
skipping over a statement, number is 3
running command1, number is 4
running command2, number is 4
running command3, number is 4
running command1, number is 5
running command2, number is 5
running command3, number is 5
```

The only real point here is noting that we finished all our iterations; the only thing we skipped was one instance of running the third command in the script. Also, note that the `continue` command in the loop is going to skip over everything up to the end of the current loop and go back to the beginning, while a `break` statement is going to skip the entire loop and not repeat it.

See also

Interrupting command flow can be a problem at first. More information is available at these links:

- `https://tldp.org/LDP/abs/html/loopcontrol.html`
- `https://linuxize.com/post/bash-break-continue/`

The while loop

Up until now, we have dealt with loops that have a fixed number of iterations. The reason is simple – if you are using a `for` loop, you need to specify *for* what values your loop is going to run, or what values your variable is going to have while in the loop.

The problem with this approach to looping is that sometimes you don't know in advance how many iterations you are going to need to do something. This is where the `while` loop comes into play.

Getting ready

The most important thing you need to know about the `while` loop is that it does its testing at the start of the loop. This means that we need to structure our script to run *while* something is true. This also means that we can make a loop that will never get executed; if we create a `while` loop that has a condition that is not met, `bash` is not going to run it at all. This has a number of great advantages, since it gives us the flexibility to use our loop as many times as we need without thinking about boundaries, and we can still use `break` when we need to get out of the loop before our condition is met.

How to do it...

A `while` loop looks even simpler than a standard `for` loop; we have a condition that must be met, and a command block that is going to get executed. If the condition is not met, commands will not run and `bash` is going to skip over the block and continue running whatever is after the end of the `done` statement that terminates the block:

```
while [ condition ]; do commands; done
```

Condition, in this case, is the same logical condition that we mentioned earlier. There is also another way of using the `while` loop by having something called `control-command`, a command that runs and directly provides information for the loop to start. We are going to use this one a lot, since it enables us to, for example, read a file line by line, without specifying how many lines it has beforehand:

```
while control-command; do COMMANDS; done
```

How it works...

As usual, we are going to give a few examples. First, we are going to repeat the task we already accomplished using the `for` loops. The idea is to loop until our value reaches 4 and then finish the loop. Note that the value can be a string, not necessarily a number:

```
#!/bin/bash
x=0
while [ $x -le 4 ]
do
    echo number is $number
    x=$(( $x + 1 ))
done
```

There are a few little things we need to emphasize. The first one is the condition we used. In our `for` loop, we compared whether the value is 4 and then used `break` to get out of our loop. In this case, we cannot do that; if we check whether the value of our x variable is 4, the loop will never run, since the initial value is 1.

In a `while` loop, we need to check for the opposite – we want our loop to run until the value becomes 4, so the condition has to be true in all cases *except* when our variable is exactly 4.

Thankfully, the very same `while` keyword helps in creating the condition.

We mentioned that instead of a condition, we can have a command. A typical example that you are going to use often is reading a file. We can do this using a `for` loop, but it would be needlessly complicated. `for` loops need to know the number of iterations before we even start a loop. In order to solve this problem using a `for` loop, we would need to count the lines in a file before we can start looping, and this would be both complicated and slow, since it requires us to open the file twice – first to count the lines and then to read them in the loop.

A much simpler way is to use a `while` loop. We simply run the loop while our command gives us some output – in this case, while it reads from a file. As soon as the command fails, the loop is over:

```bash
#!/bin/bash
FILE=testfile.txt
# read testfile and display it one line at a time
while read line
do
        # just write out the line prefixed by >
        echo "> $line"
done < $FILE
```

You will notice that there are a few things we haven't yet seen in the scripts. The first one is the use of the variables. We sort of already did that when we were dealing with the `for` statements, but here you can see both how a variable is declared and how it is used. We'll talk a lot more about this later. Another thing is how we actually read the file. The `read` command has no arguments; it is intended to be used with standard input. Since we know how to redirect inputs and outputs, we are going to just redirect whatever is in the file as the input of the `read` command. This is why we used redirection in the last line of the script. It may look awkward, but it is the way to do it.

Sometimes, we have a reason to use a loop that never finishes, a so-called infinite loop. It looks counterintuitive, but this kind of loop is extremely common in scripts when we need to run the script over and over again and have no idea how many iterations we need. Sometimes, we may even want our script to run continuously and then use the `break` statement to stop it if something happens. An infinite `while` loop is simple; just use : as the condition:

```bash
#!/bin/bash
while :
```

```
do
        echo "infinite loops [ hit CTRL+C to stop]"
done
```

See also

- `https://linuxize.com/post/bash-while-loop/`
- `https://tldp.org/LDP/Bash-Beginners-Guide/html/sect_09_02.html`
- `https://www.redhat.com/sysadmin/bash-scripting-while-loops`

The test-if loop

When strictly talking about loops, we usually divide them into `for` and `while` loops. There are some other structures that we sometimes call loops, even though they are more structured like a block of commands. Other names for these could be `decision` loops or `decision` blocks, but for legacy reasons, they are usually referred to as `test-if` loops, `case` loops, or `logical` loops.

The primary idea behind this is that any decision-making part of the code actually branches the code into different paths containing blocks of commands. Since branching and decision-making is probably the most important thing you will do in your scripts, we are going to show you some of the most commonly used structures that will find their way, more or less, into any script you make.

Getting ready

For this recipe, the most important thing is to understand that for any conditional branching, or for that matter, any conditions that you put in your code, you will use logical expressions. Logical expressions are, simply put, statements that can be either true or false.

Take, for example, statements such as the following:

- The `something.txt` file exists.
- The number 2 is greater than the number 0.
- The `somedir` directory exists and is readable by the user Joe.
- The `unreadable.txt` file is not readable by any user.

Every statement here is something that can be either true or false. The most important thing here is that there are no other logical states that we can define about any of the statements. Another thing is that every statement here refers to a particular object, a file, a directory, or a number, and gives us some attribute or state of that object.

Having this in mind, we are going to introduce shell testing as a concept and then use it to help us work on our scripts.

How to do it...

We already introduced the concept of the `if` statement using `condition` to branch to one of the *evaluated* blocks of code. This condition has to be met, which means it needs to be resolved into a `true` or `false` statement. The `if` command is then going to decide which part of the code is going to run.

This evaluation is also called *testing*, and there are two ways of doing it in shell. The `bash` shell has a command called `test` that is sometimes used in scripts. This command takes an *expression* and evaluates it to see whether the result is true or false. The result of the command is not printed in the output, but instead, the command assigns its *exit status* to the appropriate value.

Exit status is a value that each command will set after finishing, and we can check it from inside the command line, or from our script. This status is usually used to either see whether there were any errors executing a particular command or to pass some information, such as a logical value of a tested expression.

In order to test exit status, we can use a simple `echo` command. Let's do a few examples using a simple expression and the `test` command.

The first example uses the `echo` command to write out what the exit status was of the `test` command. In all the examples, `0` means `true` and `1` means `false`:

```
demo@cli1:~/$ test "1"="0" ; echo $?
0
```

So, how come we got a result that says that `1=0` is true? We made a syntax error (on purpose) to show you probably the most common mistake in scripting. All commands will usually use a very strict syntax, and *test* is not an exception. The problem with this particular command is that it will not show an error; instead, it is going to just treat our expression like it is *one single argument* and then decide it is `true`.

We can check this by using a completely nonsensical argument, such as a single word:

```
demo@clil:~/$ test whatever ; echo $?
0
```

As you can see, the result is logically true, even if it does not make any real sense. In reality, `test` requires spaces to understand which part of the expression is the operator and what are the operands. The right way to write our previous example is as follows:

```
demo@clil:~/$ test "1" = "0" ; echo $?
1
```

This is the result we expected. To check, we are going to try evaluating another expression:

```
demo@clil:~/$ test "0" = "0" ; echo $?
0
```

So, this one is true. This is completely what we expected. The reason we are using quotation marks here is that we are not actually evaluating numbers; we are comparing *strings*. What if we remove the quotation marks?

```
demo@clil:~/$ test 0 = 0 ; echo $?
0
```

This also works okay; just to check, we are going to retry with something that should be false:

```
demo@clil:~/$ test 0 = 1 ; echo $?
1
```

The result is also completely what we expected to see. Let's now try something else. We said there is a difference between comparing numbers and strings. A number has the same value regardless of the number of zeroes preceding it:

```
demo@clil:~/$ test 01 = 1 ; echo $?
1
```

Our command now states that these two are not equal. Why? Because the *strings* are not equal. Bash uses different operators to compare strings and numbers, and since we used the 1 for the strings, these values are not the same. The same goes for using them in quotation marks, just to show how quotes are handled:

```
demo@cli1:~/$ test "01" = "1" ; echo $?
1
```

The operator that we should have used for integer comparison is -eq; it will understand that we are comparing numbers and compare them accordingly:

```
demo@cli1:~/$ test "01" -eq "1" ; echo $?
0
```

Regardless of whether we are using quotes or not, the result should be the same:

```
demo@cli1:~/$ test 01 -eq 1 ; echo $?
0
```

For the last example, we are going to see what happens when we confuse the operators the other way around and try to compare strings using the integer comparison:

```
demo@cli1:~/scripting$ test 0a -eq  0a ; echo $?
bash: test: 0a: integer expression expected
2
```

What does this result mean? First, our test tried to evaluate the condition and realized there is an error in comparison, since it cannot compare a string and an integer or, to be more precise, that an integer cannot contain letters. We got the error in our output, so the command exited with the 2 status, which signifies an error. The result logically makes no sense, so the result is neither 0 nor 1.

The next thing we need to do is implement what we learned in actual scripting, but before that, we need to address one more thing. There are two ways to create our tests. One is by explicitly using the `test` command. Another is by using square brackets ([ ]). While we are going to use `test` a lot when we need to run something in the command line depending on some condition, when using the `if` statement, we are going to use square brackets most of the time, since they are easier to write and look better when glancing over the script. Just to make sure, here is one of the expressions we used, written in a different way. Please pay attention to the spaces inside the brackets; there needs to be a single space between the brackets and the expression we are using:

```
demo@cli1:~/$ [ 01 -eq 1 ] ; echo $?
0
```

How it works...

We are going to write a small script that is going to test whether a file exists in the directory the script was run from. For that, we need to talk a little about some other operators that we can use.

If you take a look at the `man` page for the `test` command or at a `bash` manual, you will see that there are many different tests we can do, depending on what we want to check; the most common ones we are going to use are probably the following (taken directly from the man pages for `test(1)`):

- The -d file: The file exists and is a directory.
- The -e file: The file exists.
- The -f file: The file exists and is a regular file.
- The -r file: The file exists and read permission is granted.
- The -s file: The file exists and has a size greater than zero.
- The -w file: The file exists and write permission is granted.
- The -x file: The file exists and execute (or search) permission is granted.

Let's create a script using this:

```bash
#!/usr/bin/bash
# testing if a file exists
if [ -f testfile.txt ]
        then
                echo testfile.txt exists in the current directory
        else
                echo File does not exist in the current directory!
    fi
```

Probably the most important thing to learn here is the structure and the use of the `else` statement. There are two blocks or *parts* of code we define in an `if` statement – one is called `then` and the other `else`. They do as their names suggest; if the condition we used in the statement evaluates as true, then the `then` code block is going to get executed. If the condition is not met, then the `else` block will be run. These blocks are mutually exclusive; only one of them is going to get run.

Now, we are going to deal with a topic that will sometimes confuse you. We already mentioned that a script has a context it is running in. Among other things, there are two things you need to know every time your script is running – where it was run from and which user ran the script.

These two pieces of information are crucial, since they define how we are going to reference the files we need and what permissions we will have from inside of the script.

Our next task is going to be to create a script that will show us how to deal with all of this. What we are going to do is test whether the script can read and write the `root` directory and whether the directory even exists. The reference we are going to make to this directory is going to be relative, so we are going to presume that our script is being run from the `/` directory, which is usually false. Then, we are going to try and run the script in different directories and under different users, comparing the results:

```bash
#!/usr/bin/bash
# testing permissions and paths
if [ -d root ]
        then
```

```
                echo root directory exists!
        else
                echo root directory does NOT exist!
fi
if [ -r root ]
        then
                        echo Script can read from the directory!
        else
                        echo Script can NOT read from the directory!
fi
if [ -w root ]
        then
                        echo Script can write to the directory!
        else
                        echo Script can not write to the directory!
fi
```

As you can see, we are basically testing for three different conditions. First, we are trying to see whether the directory exists at all and, after that, whether the script has read and write permissions.

First, we are going to try and run this as the current user in the directory that the script is created in. Then, we are going to go to the / directory and run it from there:

```
demo@cli1:~/scripting$ bash testif2.sh
root directory does NOT exists!
Script can NOT read from the directory!
Script can not write to the directory!
demo@cli1:~/scripting$ cd /
demo@cli1:/$ bash home/demo/scripting/testif2.sh
root directory exists!
Script can NOT read from the directory!
Script can not write to the directory!
```

What does all this tell us? Our first run was unable to find the directory since we were using a relative path in the script. This makes the directory that the script is run from important.

Another thing we learned is how our checks work. We can independently check whether a file or directory exists, and different permissions that the current user has on a particular file. We are going to show that by running the script under a root user using the sudo command:

```
demo@cli1:~/scripting$ cd /
demo@cli1:/$ sudo bash home/demo/scripting/testif2.sh
[sudo] password for demo:
root directory exists!
Script can read from the directory!
Script can write to the directory!
```

As soon as we change the context, we can see that the same script is not only able to see that the directory is there but also has full rights to use it.

Now, we are going to completely change our script to demonstrate how we can embed our checks into one another. Our script will once again test whether the root directory is in the current directory, but this time, the script is going to check whether it has read and write rights only if the directory exists. After all, it makes no sense to see whether you can read a directory that isn't there:

```
#!/usr/bin/bash
# testing permissions and paths
if [ -d root ]
        then
                echo root directory exists!
                if [ -r root ]
                    then
                        echo Script can read from the \
directory!
                    else
                        echo Script can NOT read from the \
directory!
                fi
                if [ -w root ]
                    then
                        echo Script can write to the directory!
                    else
```

```
                        echo Script can not write to the \
directory!
                 fi
         else
                 echo root directory does NOT exists!
 fi
```

Now, we are going to run it in two directories to see if our script works; the main difference should be the output. Also, when you have a nested structure such as this one, always try to keep your indentation consistent. This means that you always should try to keep commands in the same block indented in such a way that it is immediately obvious where each command belongs:

```
demo@cli1:~/scripting$ bash testif3.sh
root directory does NOT exists!
demo@cli1:~/scripting$ cd /
demo@cli1:/$ bash home/demo/scripting/testif3.sh
root directory exists!
Script can NOT read from the directory!
Script can not write to the directory!
```

We have now seen what can be done with different tests and conditions in bash. The next topic is similar to this one – the case statement or case loop.

See also

- https://www.thegeekdiary.com/bash-if-loop-examples-if-then-fi-if-then-elif-fi-if-then-else-fi/

- https://tldp.org/LDP/Bash-Beginners-Guide/html/sect_07_01.html

- https://ryanstutorials.net/bash-scripting-tutorial/bash-if-statements.php

The case loop

Up until now, we have dealt with basic commands that allow us to do things we need when trying to write a script, such as looping, branching, breaking, and continuing program flow. A case loop, the topic of this recipe, is not strictly necessary, since the logic behind it can be created using a multi-nested group of individual if commands. The reason we are even mentioning this is simply because case is something that we are going to use a lot in our scripts, and the alternative of using if statements is both difficult to write and read, and complicated to debug.

Getting ready

One could simply say that a case loop or case statement is just another way of writing multiple if then else tests. Case is not something that can be used in place of a normal if statement, but there is a common situation in which a case statement makes our lives a lot less complicated and our scripts much easier to debug and understand. But before we go into that, we need to understand a little bit about variables and branching. Once we start using if statements, we are quickly going to realize that they can be used, more or less, in two distinct ways. The first is the one everyone thinks about when thinking about an if statement – we have a variable and we compare it to another variable or a value. This is common and often done in a script. Something a little less common is when we have to compare a variable to a list of values. This happens most often when we need to sort things into groups or run a block of code depending on the user input.

User input is probably the most popular reason a case statement is used. In scripts, this is often used once we start using arguments. Our scripts will have to reconfigure things based on what arguments the user chose when running the script. We will take a look at that a little bit later when we start dealing with passing arguments to a script, which will exclusively use case statements to run appropriate commands.

User menus are another thing that is solved by using case statements; to generalize, each time a user has a multiple-choice answer to a question, this is going to get handled by a case statement.

How to do it...

The best way to explain a case statement is by creating an example. Let's say that a user starts a script, and they have four choices of what they want the script to do. Right now, we are not prepared to deal with how they will input their choice, so let's just presume that there is a variable called $1 that contains one of these values – copy, delete, move, and help. Our script will have to run the appropriate part of the code based on user input. In fact, this is the way arguments are handled, but we will talk about that later.

Our first version is going to use the if - then - elif loop:

```
#!/usr/bin/bash
# $1 contains either copy, delete, move or help


if [ $1 = "copy" ]
        then
                echo you chose to copy!
        elif [ $1 = "delete" ]
                then
                        echo you chose to delete!
        elif [ $1 = "move" ]
                then
                        echo you chose to move!
        elif [ $1 = "help" ]
                then
                        echo you chose help!
else
                echo please make a choice!
fi
```

This works, but it has two problems. One is that it throws errors if there are no arguments given, since this means we are comparing a value to a variable without a value. The other problem is that this is complicated to read, even if we pay extra attention to using the right indentation. We are going to redo this using a case statement:

```
#!/usr/bin/bash
# $1 contains either copy, delete, move or help
case $1 in
        copy)    echo you chose to copy! ;;
        delete) echo you chose to delete! ;;
        move)    echo you chose to move! ;;
        help)    echo you chose help!   ;;
        *)    echo please make a choice!
esac
```

The first thing you will notice is how simple and clean this looks. As well as being easier to write, the code is much easier to read and debug if we need to. There are just two simple things to pay attention to – the end of the statement block is defined as `esac`, which is `case` spelled backward, similar to how the `if` statement is terminated by `fi`. Another thing is that you have to use `;;` to terminate a line, since that's what's used to delimit choices in the `case` loop.

When matching values, you can also use limited regular expressions; this is the reason that the `*` `glob` is used to symbolize *zero or more characters*.

How it works...

Now that we know a lot more about scripting, we are going to do a simple script that searches for a string in a directory and lets us know what happened. We don't care about where the text is; we just want to know whether there is text that we used somewhere in the directory we ran our script in.

The things we need to know before we even start are as follows:

- `$1` is going to hold a string value that is going to be the text we are searching for.
- `$?` holds the `exit` value of a command that was just completed in the script.
- `grep` as a command returns either `0` if it found something, `1` if it didn't, or `2` if there was an error.
- There is a special device called `/dev/null` that can be used if we need to silence some output.

Thanks to the `case` statement, this is a trivial task:

```
#!/usr/bin/bash
# $1 contains string we are searching for

grep $1 * &> /dev/null
case $? In
        0)      echo Something was found! ;;
        1)        echo Nothing was found! ;;
        2)         echo grep reported an error! ;;
esac
```

For the last script, we are going to use `case` to combine another script from this chapter that was testing a directory and put it into a larger script. We are going to create a script that will be given a command and a filename as arguments. The command is going to be either `check`, `copy`, `delete`, or `help`. If we specify either `copy` or `delete`, the script will check whether it has the permissions to do the task and then the `echo` command that it would normally call.

If we specify `check`, the script is going to check for permissions on a given file:

```bash
#!/usr/bin/bash
# $1 contains either check, copy, delete or help
#script expects two arguments: a command and a file name
case $1 in
        copy)
          echo you chose to copy!
          if [ -r $2 ]
        then
          echo Script can read the file use cp $2 ~ to copy to \
your home Directory!
        else
        echo Script can NOT read the file!
        fi
            ;;
        delete)
          echo you chose to delete!
            if [ -w $2 ]
                then
                echo Script can write the file, use rm $2 to \
remove it!
                else
                echo Script can NOT read the file!
            fi
            ;;
        check)
```

```
        if [ -f $2 ]
            then
                echo File $2 exists!
                if [ -r $2 ]
                    then
                            echo Script can read $2!
                    else
                            echo Script can NOT read $2!
                fi
                if [ -w $2 ]
                    then
                            echo Script can write to $2!
                    else
                            echo Script can not write to
$2!
                fi
            else
                echo File $2 does NOT exist!
        fi  ;;
    help)
        echo you chose help, please choose from check, copy or \
delete!  ;;
        *)    echo please make a choice, available are copy \
check delete and help!
esac
```

What we have done here is combine everything that we have done so far into a script that actually does something. The only thing we haven't mentioned before is $2 as the second argument in the script. In this case, we use it to get the filename we need to run the commands. This is how it all looks when run from the command line:

```
demo@cli1:~/scripting$ bash testcas4.sh check testfile.txt
File testfile.txt exists!
Script can read testfile.txt!
Script can write to testfile.txt!
demo@cli1:~/scripting$ bash testcas4.sh check testfile.tx
File testfile.tx does NOT exist!
```

See also

When it comes to using `case` in your scripts, you will soon realize that a lot of examples are copied and pasted between sites. The following links are two good sources:

- `https://tldp.org/LDP/Bash-Beginners-Guide/html/sect_07_03.html`

- `https://www.shellhacks.com/case-statement-bash-example/`

Logical looping with and, or, and not

There is no way to escape logic when it comes to computers. We already dealt with some things you can do with evaluating conditions, but there is a lot more that can be done in `bash`.

In this recipe, we are going to deal with different logical operators that help us with scripting in general. First, we are going to deal with what can be done on the command line, and then we are going to use that in scripts.

Getting ready

First, let's quickly talk about logic operators. So far, we mentioned expressions that have a value of true and false. We also mentioned a lot of different expressions that are built into `bash`, since they provide functionality crucial for everyday work on the command line. Now is the time to talk about logical operators that help us combine expressions and create complex solutions. We are going to start with the usual operators:

- `&&` (the logical AND)

- `||` (the logical OR)

The interesting thing about these is that they can be used directly on the command line. The command line in `bash` basically has four ways of executing commands. One is to run them one by one on each line. This is the usual way we work in interactive mode.

How to do it...

Sometimes, we need (or want) to run multiple commands on one line. This is mostly done by using `;` to separate commands, such as the following:

```
demo@cli1:~/scripting$ pwd ; ls
/home/demo/scripting
```

```
backupexample.sh   errorfile   forbreak.sh   forcontinue.
sh   forloop1.sh   helloworld.sh   helloworldv1.sh   outputfile
readfile.sh   testcas1.sh
```

As we can see, it is exactly the same as if we ran each command by itself, but shell just executes them in a row. We already used this when we tested different expressions using the `test` command. We needed to check what the exit status was of that command, so we always used `echo` directly after running the test.

Sometimes, however, we can use a little logic to create shortcuts. This is where logical operators come into play. The remaining two ways to run multiple commands use them to not only run the command but also to run them conditionally.

Imagine we want to perform a command after we make some kind of test – for example, we want to open a file but only if the file actually exists. We could write an `if` statement here, but it would make absolutely no sense to complicate things like that. This is where we can use the logical AND:

```
demo@cli1:~/scripting$ [ -f outputfile ] && cat outputfile
Hello World!
demo@cli1:~/scripting$ [ -f idonotexist ] && cat outputfile
demo@cli1:~/scripting$
```

In general, using `&&` between commands tells `bash` to run the command on the right only if the command on the left succeeded. In our example, this means that we have a file named `output` in our directory. On the left, we are doing a quick test if this file exists. Once this is successful, we run `cat` to output the file contents.

In the second example, we intentionally used the wrong filename, and the `cat` command hasn't been run, since the file is not there.

Another logical operand we can use is the logical OR. The operator to use is `||` in the same way as before. This operator instructs `bash` to run the command on the right only if the command on the left failed:

```
demo@cli1:~/scripting$ [ -f idonotexist ] || cat outputfile
Hello World!
demo@cli1:~/scripting$ [ -f outputfile ] || cat outputfile
demo@cli1:~/scripting$
```

This is the exact opposite of the previous example. Our `cat` command ran only when the test failed. A structure such as this is sometimes used in scripts to create fail-safes or to quickly run things such as updates.

What is nice is that this enables us to immediately do something, depending on the test:

```
demo@cli1:~/$[ -f outputfile ] && echo exists || echo not \
exists
Exists
demo@cli1:~/$[ -f idonotexist ] && echo exists || echo exists \
not
exists not
```

These operators also exist in test expressions, allowing us to create different conditions that would otherwise require multiple if statements.

How it works...

Testing a condition is hopefully now completely familiar to you. We are going to try and combine a few of them to explain what different operators can do. For example, if we want to quickly check whether a file exists and is readable, we can do it by either testing whether it is readable or explicitly combining those two things into one statement:

```
demo@cli1:~/$ [ -f outputfile ] &&  [ -r outputfile ] ; echo \
$?
0
```

These tests are going to be most useful when dealing with strings and numbers. For instance, we can try and find whether a number is within an interval in a script, as follows:

```
#!/usr/bin/bash
# testing if a number is in an interval
if [ $1 -gt 1 ]
    then
        if [ $1 -lt 10 ]
            then
                echo Number is between 1 and 10
        else
            echo Number is not between 1 and 10
        fi
    else
        echo number is not between 1 and 10!
fi
```

We are going to run this script, but before we even do that, we can see that it looks complicated, more than it should be. It is not just the fact that we have to use two `if` statements to make sure that we handle both parts of the outside interval; this script demands a lot of explanations, even though it is only a couple of lines long. Does it work? Yes, as we can see here:

```
demo@cli1:~/scripting$ bash testmultiple.sh 42
Number is not between 1 and 10
demo@cli1:~/scripting$ bash testmultiple.sh 2
Number is between 1 and 10
demo@cli1:~/scripting$ bash testmultiple.sh -1
number is not between 1 and 10!
```

Now, we are going to use logical operators to optimize our script:

```
#!/usr/bin/bash
# testing if a number is in an interval
if [[ $1 -gt 1  &&  $1 -lt 10 ]]
        then
                        echo Number is between 1 and 10
        else
                        echo Number is not between 1 and 10
 fi
```

We are using double brackets here because we have to. There are multiple ways to achieve the same goal, and there are some older versions of the syntax, but it is best practice to use double brackets when dealing with multiple expressions.

See also

Dealing with logical operators is, in part, complicated because there are so many of them. You can find much more information here:

- `https://linuxhint.com/bash_operator_examples/#o23`
- `https://opensource.com/article/19/10/programming-bash-logical-operators-shell-expansions`

11
Working with Variables

Variables are one of the most important things in programming. Being able to store and then use values in our code is as important as being able to make decisions in our scripts using `if` statements.

We will cover the following recipes in this chapter:

- Using shell variables
- Using variables in shell scripting
- Quoting in the shell
- Performing operations on variables
- Variables via external commands

We are going to cover the most important things you need to know about variables, but as with almost everything else, this chapter will require you to practice.

Technical requirements

The machine you can use for these recipes is the same as in the previous chapters on scripting—basically, anything that can run bash is going to work. In our case, we are using a **virtual machine** (**VM**) with Linux and Ubuntu 20.10 installed.

So, start your VM, and let's get cracking!

Using shell variables

Variables are something that you probably understand, even if only conceptually. We are not talking about programming here; our everyday life is full of variables. Basically, a variable is something that holds a value and that can provide us with that value once we need it.

Getting ready

In everyday language, we could say an activity such as driving is full of variables. This means that the weather temperature, the amount of ambient light, the quality of the road surface, and many other things are going to change as you move along. Even though they are changing all the time, it is important that at any given point, we are able to see what the actual *value* of the weather is, what is the actual value of the temperature, how much light we have, and how the road behaves or how it is structured.

This is what we mean by variables and looking variables up.

As soon as we establish what the weather is actually like, it stops being a variable since it has an actual value. Variables work the same way when we're talking about programming. What we do is we give a name to a space that we are going to use to store some value. In our code, we refer to this space to store and read values from it. Depending on the language, this *space* can *hold* different things, but right now, we just refer to the variable as something that can hold a value.

In bash, variables are a lot simpler than in many other languages, and they can basically hold two different types of values. One is a string; it can be any sequence of numbers and letters, and it can include special characters.

Another one is a number, and the only reason that there is a difference between those two types of variables is that some operators and some operations are different when we are dealing with strings or dealing with numbers.

How to do it...

When you start to work with variables, there are two things that you need to learn.

First, you need to know how to assign a value to a variable. This is usually called *assigning* a variable or *instancing* a variable. A variable has a name and a value. In bash, when we want to create a variable, we are simply going to choose a name and assign a value to it. After that, our shell knows that this is a variable, and it keeps track of the value or values we assign to it. Before we assign a value, a variable simply does not exist, and any reference to it will be invalid.

So, how do you choose a name for a variable?

Every variable has its own name, which is used to reference a variable inside the script or inside your working environment in the shell. The choice of name is completely up to you. The name should be something that you can easily remember and something that you will not confuse with other variables. A good choice is usually either something that identifies what purpose the variable has or a completely abstract name that will hint at what the meaning is of the variable.

One thing that you should always avoid using when naming variables are keywords, especially those that already have a meaning in bash. For example, we cannot use continue as a variable name since this is the name of a command. This will inevitably generate an error since the shell is going to get confused about what to do with the variable itself.

We mentioned environment variables. In an interactive shell, there are quite a few variables that are used to store information about your environment. This information describes different things that are required by different applications—things such as the username, your shell, and so on.

Let's do a few quick examples. We assign a variable exactly as we mentioned, by giving a value to a name. In our case, we are going to assign a value string value to a variable called VAR1:

```
demo@cli1:~$ VAR1=value
```

That was easy. Now, let's read from the variable we just created:

```
demo@cli1:~$ echo $VAR1
value
```

As we can see, in order to read the variable, we need to prefix the variable name with the $ character. Also, we need to use the same case in the variable name that we used when creating the variable itself, as names are case-sensitive.

If we don't do that, we are not going to get any useful value out of our `echo` command, but be very aware that neither of these examples gave us any errors:

```
demo@cli1:~$ echo var1
Var1
demo@cli1:~$ echo $var1

demo@cli1:~$ echo VAR1
VAR1
```

We made these errors on purpose to make a few small points. When using an `echo` command, we tell it to display a string. If the string contains a variable name, it has to be prefixed; otherwise, the `echo` command is just going to output it directly as it was written, without the variable value.

As we said, names are case-sensitive, but if we make a mistake, there won't be any errors displayed—we will simply get an empty line. This can be changed, and we will deal with this behavior later when we start using variables in scripts.

Let's now do something else—we'll try to use our variable in a script. Remember that we assigned a variable in the shell, but now, we are going to reference it in a script.

The script is going to be the simplest possible—create a file, name it `referencing.sh`, and enter the following code:

```
#!/bin/bash
#referencing variable VAR1
echo $VAR1
```

What happens when we run it? Let's have a look:

```
demo@cli1:~$ bash referencing.sh
demo@cli1:~$ echo $VAR1
value
```

We see we have a problem. When we are reading the variable from the command line, everything is fine, but this variable does not exist inside our scripts. The reason is not as simple as it seems, though. We mentioned contexts and environment variables before. Each variable exists in the current environment and is not implicitly inherited by any command. When we start a script, we are actually creating a new environment and a new context that inherits all the variables that are marked as inheritable. Since we just assigned a value to our variable and didn't do anything else to it, this variable will remain visible only to our shell, and not to any commands or scripts that we run from it.

To fix this, we will need to *export* a variable. Exporting means flagging our variable to tell the environment that we want the value of the variable to be available to the commands and scripts that are running as its child processes. To do that, we need to use a command called `export`. The syntax couldn't be simpler:

```
demo@cli1:~$ export VAR1
demo@cli1:~$ bash referencing.sh
value
demo@cli1:~$
```

As we can see, our script now knows the value of our variable, and it got inherited from the `bash` shell.

If we just type in `export`, we will see a list of all the variables that are exported and available to our scripts:

```
demo@ubuntu:~$ export
declare -x COLORTERM="truecolor"
declare -x DBUS_SESSION_BUS_ADDRESS="unix:path=/run/user/1000/bus"
declare -x DESKTOP_SESSION="ubuntu"
declare -x DISPLAY=":0"
declare -x GDMSESSION="ubuntu"
declare -x GJS_DEBUG_OUTPUT="stderr"
declare -x GJS_DEBUG_TOPICS="JS ERROR;JS LOG"
declare -x GNOME_DESKTOP_SESSION_ID="this-is-deprecated"
declare -x GNOME_SHELL_SESSION_MODE="ubuntu"
declare -x GNOME_TERMINAL_SCREEN="/org/gnome/Terminal/screen/dca9c087_5f12_4f3f_
84a2_f7328904cfb5"
declare -x GNOME_TERMINAL_SERVICE=":1.82"
declare -x GPG_AGENT_INFO="/run/user/1000/gnupg/S.gpg-agent:0:1"
declare -x GTK_MODULES="gail:atk-bridge"
declare -x HOME="/home/demo"
declare -x IM_CONFIG_PHASE="1"
declare -x LANG="en_US.UTF-8"
declare -x LESSCLOSE="/usr/bin/lesspipe %s %s"
declare -x LESSOPEN="| /usr/bin/lesspipe %s"
declare -x LOGNAME="demo"
declare -x LS_COLORS="rs=0:di=01;34:ln=01;36:mh=00:pi=40;33:so=01;35:do=01;35:bd
=40;33;01:cd=40;33;01:or=40;31;01:mi=00:su=37;41:sg=30;43:ca=30;41:tw=30;42:ow=3
4;42:st=37;44:ex=01;32:*.tar=01;31:*.tgz=01;31:*.arc=01;31:*.arj=01;31:*.taz=01;
```

Figure 11.1 – Different exported variables exist for every user

Notice one important thing: every line starts with the `declare -x` command, followed by a variable name and value. This points us to another extremely useful command: `declare`.

When we are creating a variable and giving it a value, we are using only part of what can be done with variables in `bash`. Remember how we exported the variable? Variables have attributes that are additional information about how the variable should behave. Having a variable being exported is one of the attributes, but we can also make a variable read-only, change the variable name case, and even change the type of information that the variable holds. For all that, we use `declare`.

How it works...

The only thing left to do is to give you more information about environment variables.

The environment can be, depending on your system and its configuration, huge. It contains a lot of things, and it is different from system to system because variables in the environment and their values are dependent on different programs and options installed on your particular system. For example, if you use a shell other than `bash`, you may have different variables specific to that shell. If you use **GNU Network Model Object Environment** (**GNOME**) or **K Desktop Environment** (**KDE**) as your **graphical user interface** (**GUI**), there are different variables that each have a specific meaning. To see what your environment looks like, you can use either `declare -p` or `env`.

The difference between those two is very important. The `declare` statement is a `bash` built-in command. It will read every variable there is in the environment and show you all of them. `env`, on the other hand, is an application. It will run, create its own environment to run in, and then show you all the variables in that environment:

demo@ubuntu:~$ env
SHELL=/bin/bash
SESSION_MANAGER=local/ubuntu:@/tmp/.ICE-unix/2797,unix/ubuntu:/tmp/.ICE-unix/279
7
QT_ACCESSIBILITY=1
COLORTERM=truecolor
XDG_CONFIG_DIRS=/etc/xdg/xdg-ubuntu:/etc/xdg
XDG_MENU_PREFIX=gnome-
GNOME_DESKTOP_SESSION_ID=this-is-deprecated
GNOME_SHELL_SESSION_MODE=ubuntu
SSH_AUTH_SOCK=/run/user/1000/keyring/ssh
XMODIFIERS=@im=ibus
DESKTOP_SESSION=ubuntu
SSH_AGENT_PID=2761
GTK_MODULES=gail:atk-bridge
PWD=/home/demo
LOGNAME=demo
XDG_SESSION_DESKTOP=ubuntu
XDG_SESSION_TYPE=x11
GPG_AGENT_INFO=/run/user/1000/gnupg/S.gpg-agent:0:1
XAUTHORITY=/run/user/1000/gdm/Xauthority
GJS_DEBUG_TOPICS=JS ERROR;JS LOG
WINDOWPATH=2
HOME=/home/demo

Figure 11.2 – The environment can be checked at least two ways, but we usually use the env command

We are going to mention some of those most important:

- USER—Holds the username of the current user. This is extremely important if you need to check under which user the script is running. An alternative to this is to run the whoami command.

- PWD—Holds the absolute path to the current directory. This is also important to any script since it can help you find which running directory the script was called from. An alternative to this command is pwd.

- LOGNAME—Provides the same information as USER, specifically the username of the logged-on user, hence the name.

- SHELL—Contains the entire path to the current user's login shell. This is not the same as the running shell; we can run any shell and work from it, and this variable returns what our login shell is set to. This value comes from the /etc/passwd file.

- SHLVL—When you run your shell initially, you are one level into your environment. What this means is that there is nothing else running *above* your shell—or, to be more precise, your shell was started by your system directly. As you work, you can run other shells, scripts, and even shells inside shells. Each time you run a shell inside your shell, you increase your SHLVL. This is useful when trying to find out whether your script was run from another shell or directly by the system.

- PATH—PATH contains a list of directories that your shell is going to look in when trying to find any command that you try to execute. Since almost everything on Linux is a command, this piece of information is crucial—if a certain path is not in the PATH variable, it won't be searched, and commands from it can only be executed if you reference them directly. This is useful if you don't want to reference commands directly all the time, or you have some reasons to prefer a command in one directory over another.

Before we go on to the next recipe, there is another way to get variables listed, and that is by using set without any parameter:

```
demo@ubuntu:~$ set
BASH=/usr/bin/bash
BASHOPTS=checkwinsize:cmdhist:complete_fullquote:expand_aliases:extglob:extquote
:force_fignore:globasciiranges:histappend:interactive_comments:progcomp:promptva
rs:sourcepath
BASH_ALIASES=()
BASH_ARGC=([0]="0")
BASH_ARGV=()
BASH_CMDS=()
BASH_COMPLETION_VERSINFO=([0]="2" [1]="11")
BASH_LINENO=()
BASH_SOURCE=()
BASH_VERSINFO=([0]="5" [1]="0" [2]="17" [3]="1" [4]="release" [5]="x86_64-pc-lin
ux-gnu")
BASH_VERSION='5.0.17(1)-release'
COLORTERM=truecolor
COLUMNS=80
DBUS_SESSION_BUS_ADDRESS=unix:path=/run/user/1000/bus
DESKTOP_SESSION=ubuntu
DIRSTACK=()
DISPLAY=:0
EUID=1000
GDMSESSION=ubuntu
GJS_DEBUG_OUTPUT=stderr
```

Figure 11.3 – set not only shows you variables but is also capable of configuring the shell

Of course, since there is a lot of variables active at any given time, it is much better to use some sort of filtering:

```
demo@ubuntu:~$ set | grep SSH
SSH_AGENT_PID=2761
SSH_AUTH_SOCK=/run/user/1000/keyring/ssh
demo@ubuntu:~$ set | grep SHELL
GNOME_SHELL_SESSION_MODE=ubuntu
SHELL=/bin/bash
SHELLOPTS=braceexpand:emacs:hashall:histexpand:history:interactive-comments:moni
tor
demo@ubuntu:~$ set | grep DISPLAY
DISPLAY=:0
demo@ubuntu:~$ set | grep USER
USER=demo
USERNAME=demo
    local -a dirs=(${BASH_COMPLETION_USER_DIR:-${XDG_DATA_HOME:-$HOME/.local/sha
re}/bash-completion}/completions);
demo@ubuntu:~$
```

Figure 11.4 – The only way to quickly find things is to use grep

See also

We are going to give you just the place to start since this topic is massive:

- `https://tldp.org/HOWTO/Bash-Prog-Intro-HOWTO-5.html`
- `https://ryanstutorials.net/bash-scripting-tutorial/bash-variables.php`

Using variables in shell scripting

Variables sometimes look simple enough—they are there to enable you to put a changing value in your code. The problem is that in this simplicity, there are a couple of things you should know about where you actually place a variable—in something called a context. We are going to deal with that in this chapter.

Getting ready

When we're talking about scripting, things are a little different than they are when we are working in an interactive environment. Every environment variable that is available to you when you use the interactive shell is also available to you in the script. There is, however, one important thing you must always remember. As we said earlier, your script is running in a certain context. This context is defined by the user that has run the script. In a previous chapter, we wanted you to make sure that you have appropriate permissions to do tasks that you need in the script.

In this recipe, we are going to make sure you understand this also applies to variables. Unless we have explicitly set the variable in our script, we need to make sure that the one we got from the environment is something that we expect. Also, a lot of times, we will simply check if the variable is there in the first place since it may not be exported from the shell and will hence be invisible to us.

There is also a special class of variables that are set right at the moment the script is run and contain a certain amount of information very important to successfully running a script.

What we are going to do is to start with how the script interacts with the shell using variables.

How to do it...

We are, as always, going to start simple. First, we are going to do the most basic thing we can do—Hello World, but with variables:

```
#!/bin/bash
# define a variable
STRING="Hello World!"
# output the variable
echo $STRING
```

This is basically what we mentioned before but in a script. We have created a variable, assigned it a value, and then used that value to output text.

Now, let's try something more useful. When writing scripts, there are things that we need to calculate or prepare in some way so that we can use them in different parts of the script. Variables are a good way to do this clearly so that they can be reused in the code.

For example, we can create a string that will contain today's date. We can then use a variable instead of running the appropriate command every time to create a date in a given format over and over again:

```
#!/bin/bash
# we are using variable TodaysDate to store date
TodaysDate=$(date +%Y%m%d)
# now lets create an archive that will have todays date in \
the name.
tar cfz Backup-$TodaysDate.tgz .
```

After we run this one, the output is going to be interesting:

```
demo@clil:~/variables$ bash varinname.sh
tar: .: file changed as we read it
demo@clil:~/variables$ ls
Backup-20210920.tgz   varinname.sh
```

We can see that the file was created correctly and that our date looks OK. What we didn't expect was the error. The reason for the error is simple—`tar` starts creating files by first creating an output file and then reading the directory it must archive. If the archive file is created in the directory it is trying to archive, this means that the `tar` command will try to run on the archive itself, creating this error. This is normal in these circumstances but try to avoid doing this archive loop. The solution is to archive to a place outside of the directory we are archiving.

Now for the fun part—passing arguments to your scripts. Up to this point, we have made scripts that were completely unaware of their surroundings. We need to change that since we need to be able to both pass information to our script and make our script report back what has happened.

Any script, regardless of the way it was executed, can have arguments. This is so common that we usually don't even think about it. Arguments are basically strings that come after the script name when we execute the script.

This is precisely how arguments work in scripts—the shell takes whatever is in the command line that started the script and passes it along using a variable that has a number as the name. Here's an example:

```
#!/bin/bash
# we are going to read first three parameters
# and just echo them
echo $1 $2 $3
# we will also use $# to echo number of arguments
echo Number of arguments passed: $#
```

Now, here's how we can run it in a few different ways:

```
demo@clil:~/variables$ bash parameters.sh
Number of arguments passed: 0
```

If we don't give it any parameters, everything works as well as if we give it three parameters we expect:

```
demo@cli1:~/variables$ bash parameters.sh one two 3
one two 3
Number of arguments passed: 3
```

But let's try to use more than three:

```
demo@cli1:~/variables$ bash parameters.sh one two 3 four
one two 3
Number of arguments passed: 4
demo@cli1:~/variables$ bash parameters.sh one two 3 four five
one two 3
Number of arguments passed: 5
```

We see a problem here. Variables that hold the parameter value are *positional*, and it is up to us to correctly reference everything in the parameter line. The way to do it is to read the number of `arguments` variable, and then create a loop of some kind to read the arguments.

You may be wondering: *What about $0?* Programmers tend to count from zero, not from one, and this is no exception—there is a variable called `$0` and it contains the name of the script itself. This is extremely convenient for scripting. We are creating a script called `parameters1.sh` and running it:

```
#!/bin/bash
# reading the script name
# and just echo
echo $0
```

As we can see, this script could not be simpler. But in this simplicity is one neat trick:

```
demo@cli1:~/variables$ bash parameters1.sh
parameters1.sh
demo@cli1:~/variables$ cd ..
demo@cli1:~$ bash variables/parameters1.sh
Variables/parameters1.sh

demo@cli1:~$ bash /home/demo/variables/parameters1.sh
```

```
/home/demo/variables/parameters1.sh
demo@cli1:~$
```

The point we are trying to make here is that the variable holds a value that contains not only the name of the script but also the entire path that was used to run the script. This can be used to determine how the script was run if we are running from `crontab` or another script.

To continue, we need to learn about a new concept—the `shift` statement.

There are two ways to parse arguments to a script—one is by using a loop that is going to run for $# iterations, which means that we are going to run for each argument that the script has once. This is a completely valid way, but there is also another, rather more elegant way to deal with this problem. `shift` is an in-built statement that enables you to parse your arguments one at a time without knowing how many of them there are.

How it works...

The way shifting works is completely intuitive once you understand what it does. Let's quote from the `help` page:

```
demo@cli1:~/variables$ help shift
shift: shift [n]
    Shift positional parameters.
    Rename the positional parameters $N+1,$N+2 ... to $1,$2 ...
If N is
    not given, it is assumed to be 1.
        Exit Status:
    Returns success unless N is negative or greater than $#.
```

Basically, we only need to read the $1 parameter and then invoke `shift`. The command is going to delete this parameter and shift all of them to the left, making the next one $1, and so on.

This enables us to do these kinds of things:

```
#!/bin/bash
while [ "$1" != "" ]; do
    case $1 in
        -n | --name )
            shift
```

```
            echo Parameter is Name: $1
      ;;
      -s | --surname )
            shift
            echo Parameter is Surname: $1,
      ;;
      -h | --help )     echo usage is -n or -s followed by a \
 string
            exit
      ;;
      * )               echo usage is -n or -s followed by a \
 string
            exit 1
   esac
   shift
done
```

We need to explain a few things here. The reason we are using shift instead of a for loop is that we are parsing arguments that can be different options. Our script has three possible switches: -n that can be written down as —name, -s that can also be used as -surname, and -h or —help. After the first two arguments, our script expects to have some string. If none of the arguments is used or we choose -h, our script is going to write a small reminder on the usage parameters.

If you tried to do this in a for loop, you would have a problem—we would need to read the option, store it somewhere, read the option parameters in the next loop, and then loop again, trying to decide if what follows is an option or an argument.

By using shift, things are much simpler—we read an argument, and if we find any option we shift it; the parameters then become stored in $1 and we can print and use them.

If we don't find an option, we simply ignore what is inside the variable.

See also

The topic of using arguments is very complicated and is needed in almost every script. So, there are open source solutions for that, such as these:

- `https://dev.to/unfor19/parsing-command-line-arguments-in-bash-3b51`

- `https://www.baeldung.com/linux/use-command-line-arguments-in-bash-script`

Quoting in the shell

Quotes are something that we take for granted, not only in Linux but also in a lot of other applications. In this recipe, we are going to deal with how quotes work, which quotes to use, and how to make sure that your quoted part of the script behaves as you intended.

Getting ready

Using quotes is incredibly important in Linux, not only in shell scripts but also in any other application that uses text. In this context, quotes behave pretty much the same way as brackets do in mathematical expressions—they offer us the way to change how an expression is evaluated. Almost all command-line tools use a space as a delimiter that tells the tool where one string ends and another one begins. You probably ran into this when you tried to use a file or a directory that has a space in its name. Usually, we solve this problem by using an escape character (\), but it makes it much easier to read if we apply quotes.

This is not the only reason we use quotes, so we are going to pay much more attention to them right now.

First, we must define different quotation symbols that we can use and outline what they mean:

- Double quotation marks: " " " "

  Used to quote strings and stop a shell from using a space as a delimiter. This quotation style will use shell expansion characters such as $, `, \, and ! as expansion characters, not quoting them but instead replacing them in the normal way. You will use this quotation style all the time.

- Single quotation marks: '

  These behave almost exactly the same as double quotes, but with an important twist. Everything inside single quotation marks is treated *as is* and will not be changed in any way. Even if you use special characters, this will have no influence—they are going to be used as part of a string.

- Backticks: " ` "

  The backtick character is sometimes considered a quote and often mistaken for a single quote.Note that this is a completely separate character—on a standard **United States** (**US**) keyboard, you can find it in the upper row, on the key left of the number *1* key, furthest to the left. The difference is in the slope of the character, so the name *backtick* really means that it is oriented differently than the quote character. In the shell, it is used to run a command—or, to be more precise, to run a command and then use its output in its place.

Even though backticks are not strictly quotes, in most learning materials you may find them mentioned as such. This is either because they look like quotes, or because they are the most probable character to get changed automatically to a quote in any text editor.

How to do it...

To understand quotes, we are going to make a few script examples, starting with a simple `if` statement, just to remind you what it looks like. We are going to create a file called `quotes1.sh` using this code:

```
#!/bin/bash
directory="scripting"
# does the directory exist?
If [ -d $directory ]; then
        echo "Directory $directory exists!"
else
        echo "Directory $directory does not exist!"
fi
```

Once we run this, the results are as we expected:

```
demo@cli1:~/variables$ bash quotes1.sh
Directory scripting does not exist!
```

Now, let's just make one small change in `quotes1.sh` and save it as `quotes2.sh`:

```bash
#!/bin/bash
directory='scripting'
# does the directory exist?
if [ -d $directory ]; then
        echo 'Directory $directory exists!'
else
        echo 'Directory $directory does not exist!'
fi
```

In this case, when we run the command, the result is going to be quite different. Since we used single quotes, the shell is not displaying our variable, and instead, we are seeing our actual variable name with its prefix:

```
demo@cli1:~/variables$ bash quotes2.sh
Directory $directory does not exist!
```

There is also a special case that we need to mention, and that is when we use double quotes inside single quotes and the other way around. In the case of double quotes being outside, they will negate the single quotes, so we get the usual expansion of variables. This time, create a file called `undeterdouble.sh` and get this code typed into it:

```bash
#!/bin/bash
directory='scripting'
# does the directory exist?
echo "'Directory $directory is undetermined since we have no \
logic in this script'"
```

When we run it, we get this:

```
demo@cli1:~/variables$ bash undeterdouble.sh
'Directory 'scripting' is undetermined since we have no logic
in this script'
```

Notice that the shell inserted another set of quotes to separate the variable value and the rest of the string.

If we turn it the other way around, we are going to end up with everything being quoted, since the single quotes mean just that:

```
#!/bin/bash
directory='scripting'
# does the directory exist?
echo '"Directory $directory is undetermined since we have no \
logic in this script"'
```

Notice there are no additional quotes in the string:

```
demo@cli1:~/variables$ bash undetersingle.sh
"Directory $directory is undetermined since we have no logic in
this script"
```

How it works...

The shell needs to know when to expand variables and when not to do this. Spaces are also a big problem in scripting—more often than not, your script is going to completely miss some part of the string because it will cut it up into single words divided by spaces.

Both quotes have their uses, but you are going to be using double quotes most of the time. The reason is that you will usually have a string with spaces but also with different variables in it. By using double quotes, you will have your variables expanded while keeping the text.

See also

When it comes to single and double quotes, there are only a couple of resources since they are straightforward:

- https://bash.cyberciti.biz/guide/Quoting
- https://www.gnu.org/software/bash/manual/html_node/Quoting.html

Performing operations on variables

Variables are great since they can hold any value that we can think of. Often, we need more than just holding a value inside a variable. In this recipe, we are going to deal with a lot of different things that we can do to a variable, sometimes changing it and sometimes completely replacing it.

Getting ready

In order to be able to change variables, you will need to understand one simple concept. bash cannot change the variable itself; we are going to mention this a little later, but if you need to change something in a variable, you will have to reassign it.

How to do it...

There is a lot of things that can be done to a variable. Sometimes, we want to know more about what it contains; sometimes, we need to change something in order to use it later; or, we may simply want to know if the variable even has a value.

In this recipe, we are going to use the command line a lot since it makes explaining things much easier.

Before we begin, we are going to introduce one thing we haven't mentioned yet: arrays.

An array is a variable that holds separate strings divided by spaces. You could say it's a string itself, but for a lot of reasons to do with flexibility, bash is able to address different parts of the array individually, keeping the values in one variable.

We are going to define an array that will have four strings in it. The way to define a variable is by using brackets and enclosing strings inside them:

```
demo@clil:~/variables$ TestArray=(first second third fourth)
```

Now, we can see how many elements there are in our array. This is where things get a little strange. Remember when we said that counting in bash starts at zero?

```
demo@clil:~/variables$ echo ${#TestArray[@]}
4
```

We see that we got the right information—our array has exactly four elements. The way we got this was by using curly brackets together with some special characters. Our expression starts with $ {, which tells bash that we are going to do something with an array. Then comes the # sign, which means that we are expecting a count of something, either the length or number of elements. After that, we have our array name followed by square brackets and the @ sign inside brackets. In shell syntax, this tells bash that we want all elements in the array.

Translated into plain English, this command says: show me the count of how many elements there are in the TestArray array.

But beware—things are extremely sensitive when it comes to syntax. For example, if you omit the `[@]` part, this is a completely valid command, but it gives you also completely different information:

```
demo@cli1:~/variables$ echo ${#TestArray}
5
```

The number we get is actually the length of the first string in the array, not the array itself. This is because if we try to just use the array name, we are going to get only the first string as a result:

```
demo@cli1:~/variables$ echo ${TestArray}
first
```

To avoid this, we should always use square brackets and a number inside them. This is the right way of referencing the positions of strings in our array. Have in mind that the first string has an index of `0`:

```
demo@cli1:~/variables$ echo ${TestArray[2]}
third
demo@cli1:~/variables$ echo ${TestArray[0]}
first
demo@cli1:~/variables$ echo ${TestArray[1]}
second
demo@cli1:~/variables$ echo ${TestArray[@]}
first second third fourth
```

Now that we have seen how to reference arrays and their parts, let's see if a variable even exists and what is the way to check its length. We already know how to do that—we just need to use `${#variablename}` to have the shell output the length:

```
demo@cli1:~/variables$ TestVar="Very Long Variable Contains \
Lots Of Characters"
demo@cli1:~/variables$ echo $TestVar
Very Long Variable Contains Lots Of Characters
demo@cli1:~/variables$ echo ${#TestVar}
46
```

As we can see, since we put a string in the quotes, our variable contains all the spaces and characters in a single string. The length is then correctly calculated.

What about checking if a variable exists by looking at its length?

```
demo@cli1:~/variables$ echo $VariableThatDoesNotExist
demo@cli1:~/variables$ echo ${#VariableThatDoesNotExist}
0
```

The length is in this particular case 0. If you are not used to this kind of calculation, you will probably expect not to get a valid number but to have the shell report that the variable is not defined, but bash does it differently.

The next thing we can do is do substitutions of variables. An extremely useful thing is being able to check if a variable has a value, and if it doesn't have a value, just substitute another value in its place. In other words, before you use a variable, always make sure it has a value since bash is by default going to return an empty result if the variable is not defined. Here's an example:

```
demo@cli1:~/variables$ echo ${TEST:-empty}
empty
demo@cli1:~/variables$ echo $TEST
demo@cli1:~/variables$ TEST=full
demo@cli1:~/variables$ echo $TEST
full
demo@cli1:~/variables$ echo ${TEST:-empty}
full
```

What we are doing here is testing if the TEST variable has a value. If not, we are going to output empty as a string. As soon as our variable is set, the output is going to revert to the value of the variable.

How it works...

The things we have mentioned up to now were simple substitutions of a whole variable. What is much more common is having to change something inside a variable. This can be done using a special syntax. What we can do is extract strings from our variable. This is not going to change the variable itself; instead, we need to save this string into another variable if we need it for something later. The syntax we are going to use is shown here:

```
${VAR:OFFSET:LENGTH}
```

VAR is the variable name. OFFSET and LENGTH are self-explanatory—they basically mean *take this many characters starting from this exact position*. The easiest way to explain this functionality is to show you a couple of examples:

```
demo@cli1:~/variables$ echo $TestVar
Very Long Variable Contains Lots Of Characters
demo@cli1:~/variables$ echo ${TestVar:5:4}
Long
demo@cli1:~/variables$ echo ${TestVar:5:13}
Long Variable
demo@cli1:~/variables$ echo ${TestVar:5}
Long Variable Containg Lots Of Characters
demo@cli1:~/variables$ echo ${TestVar:5:}

demo@cli1:~/variables$ echo ${TestVar:5:-4}
Long Variable ContainsLots Of Charac
demo@cli1:~/variables$ echo ${TestVar:5:-10}
Long Variable Contains Lots Of
```

Notice that we can also use negative numbers. If we do that, we are going to get the part of the string from the given offset up to the last *X* characters, *X* being the negative number we used.

The last thing we wanted to show you is replacing patterns in variables. For that, we use this syntax:

${VAR/PATTERN/STRING}

The same things apply as when we talked about extracting parts of the variable—we are not changing the variable itself, we are just modifying the output:

```
demo@cli1:~/variables$ echo ${TestVar/Variable/String}
Very Long String Contains Lots Of Characters
demo@cli1:~/variables$ echo $TestVar
Very Long Variable Contains Lots Of Characters
```

See also

Variable operations contain a lot more possibilities. Check them out here:

- `https://tldp.org/LDP/Bash-Beginners-Guide/html/sect_10_03.html`

- `https://opensource.com/article/18/5/you-dont-know-bash-intro-bash-arrays`

Variables via external commands

Sometimes, while writing a script, you will have to run a certain command and then use its output to do something in your script. A complicated way to do that is by using redirection. We say *complicated* because once you have to use redirection, you are unable to use it for other things. You could redirect to different file descriptors, but that is going to complicate things even more.

Getting ready

You will soon notice that it is hard to separate different things related to shell commands and functions. The reason for this is that there are a few fundamental rules that then get repeated in a different way. We are going to mention some of them a few times through this book, not because we like redundancy but because you need to completely understand those rules to be able to write good scripts.

This is why shell expansion exists, and there are two ways to put it into action to accomplish our task.

How to do it...

There are two syntaxes we can use for this. One is by enclosing the command with all its parameters into backticks, like this: `command`. Another is by using `$(command)`. Both have the same result—whatever is the output of the command is going to get translated into a group of strings and used instead of the original command:

```
demo@cli1:~/variables$ ls
Backup-20210920.tgz    parameters.sh    quotes2.sh
undetersingle.sh
parameters1.sh         quotes1.sh       undeterdouble.sh
varinname.sh
demo@cli1:~/variables$ echo $(ls)
```

```
Backup-20210920.tgz parameters1.sh parameters.sh quotes1.sh
quotes2.sh undeterdouble.sh undetersingle.sh varinname.sh
demo@cli1:~/variables$ echo `ls`
Backup-20210920.tgz parameters1.sh parameters.sh quotes1.sh
quotes2.sh undeterdouble.sh undetersingle.sh varinname.sh
```

This was just to show you how this sort of expansion behaves. Using a single `echo` command makes no sense; we are going to try with something more complicated:

```
#!/usr/bin/bash
# testing extension on list of files
for name  in $(ls) ;              do
            for exten in .pdf .txt; do
                        echo "Trying $name$exten"
      done
done
```

What we are doing is getting a list of files from the current directory, and then using this list to try different extensions. This way of working with files is the most common thing you will use in your scripts. When iterating like this, there are going to be either files or lines in the file:

```
demo@cli1:~/variables$ bash forexpand.sh
Trying Backup-20210920.tgz.pdf
Trying Backup-20210920.tgz.txt
Trying forexpand.sh.pdf
Trying forexpand.sh.txt
Trying parameters1.sh.pdf
Trying parameters1.sh.txt
Trying parameters.sh.pdf
Trying parameters.sh.txt
Trying quotes1.sh.pdf
Trying quotes1.sh.txt
Trying quotes2.sh.pdf
Trying quotes2.sh.txt
Trying undeterdouble.sh.pdf
Trying undeterdouble.sh.txt
Trying undetersingle.sh.pdf
Trying undetersingle.sh.txt
```

```
Trying varinname.sh.pdf
Trying varinname.sh.txt
```

This shell capability is amazing but it has its own limitations, the main one being that the output of the command inside brackets has to be *clean*. By cleanliness, we mean that it has to contain only the information that can be directly used as parameters. Consider this minuscule change in our script:

```
demo@cli1:~/variables$ cat forexpand.sh
#!/usr/bin/bash
# testing extension on list of files
for name  in $(ls -l) ;              do
                for exten in .pdf .txt; do
                        echo "Trying $name$exten"
              done
done
```

We changed two characters in the `ls` command by adding `-l` to make it output in a long format. If we now run it, this is not even remotely what we expected:

```
demo@cli1:~/variables$ bash forexpand.sh
Trying total.pdf
Trying total.txt
Trying 36.pdf
Trying 36.txt
Trying -rw-rw-r--.pdf
Trying -rw-rw-r--.txt
Trying 1.pdf
Trying 1.txt
Trying demo.pdf
Trying demo.txt
Trying demo.pdf
Trying demo.txt
Trying 494.pdf
Trying 494.txt
```

We stopped the output here.

How it works...

This way of getting information from one command is probably one of the simplest things to understand in entire bash scripting. What the shell does is execute the command, get its output, and then behave as if it is a long list of separate strings using a space as a separator.

This is also the reason why we have to pay special attention to what is going to be the output of the application. The shell is unable to *understand* what we want out of it; it simply parses whatever it sees and treats spaces as separators. What will happen then rests entirely on you—the command that you embedded this expression in can treat the end result completely differently.

See also

- `https://tldp.org/HOWTO/Bash-Prompt-HOWTO/x279.html`
- `http://www.compciv.org/topics/bash/variables-and-substitution/`

12
Using Arguments and Functions

Whenever we are trying to program any kind of application or a script in any programming language, we should always try to make our code modular and easily maintainable. The thing that is going to help us a lot in this aspect of creating scripts is a concept known as a **function**.

We will cover the following recipes in this chapter:

- Using custom functions in shell script code
- Passing arguments to a function
- Local and global variables
- Working with returns from a function
- Loading an external function to a shell script
- Implementing commonly used procedures via functions

Technical requirements

For these recipes, we're going to use a Linux machine. We can use any **virtual machine** (**VM**) from our previous recipes. For example, let's say that we're going to use a cli1 VM as it's the most convenient to use, seeing that it's a **command-line interface** (**CLI**)-only machine. So, all in all, we need the following:

- A VM with Linux installed—any distribution (in our case, it's going to be **Ubuntu 20.02**).

- A bit of time to digest the complexities of using the VI(m) editor. Nano is less complex, therefore it's going to be easier to learn about that one.

So, start your VM, and let's get cracking!

Using custom functions in shell script code

Up to this point, all we did was create very simple scripts that had a few commands at most. This is going to be most of your scripts since a lot of work that is solved by scripting is the simple elimination of repetitive tasks. In this chapter, we are going to work with functions as a way of creating modules of code in your script. Their main purpose is going to be to avoid repetitive blocks of code in your scripts, further simplifying the scripts themselves.

Getting ready

When it comes to functions, Bash is a little bit strange. Things you may know about functions from other languages will look similar in bash but at the same time, completely different. We are going to start with how a function is defined. To make matters confusing from the very start, bash uses two very similar notations, one that looks more like something you would find in other languages, and another that is more in line with the rest of the bash syntax.

Before we even mention them, have in mind that there is no difference in functionality or anything else in the way functions are defined—we can use either of them with the exact same results.

The syntax of the first definition looks like something you would see in any programming language. There are no keywords—we simply specify the name of the function followed by two normal brackets, and then define a command block that makes up the function in curly brackets.

There is a big difference between `bash` and almost every programming language out there, though. Usually, brackets in any language serve to pass arguments or parameters to the function. In `bash`, they are always empty—their only purpose is to define a function. Parameters are passed in a completely different way:

```
function_name () {
<commands>
}
```

Another way to define a function is more in line with the way `bash` usually works. There is a reserved word, `function`; so, in order to define a function, we simply do this:

```
function function_name {
<commands>
}
```

This version is more likely to remind you that arguments are provided in a different way, but that is probably the only difference between the two.

A function must be defined before we can use it. This is completely logical since the shell runs every line one by one and, to understand a command, has to have it defined as either an internal command, an external command, or a function. Unlike some other languages, arguments and return values are not defined in advance—or, to be more precise, are not defined at all.

How to do it...

As always, we are going to start with a `hello world` script, but with a little twist. We are going to use our `echo` command inside a function, and the main part of the script is going to run this function. We are also going to create an alternative version of our function just to show that both ways to define a function work the same.

There are a couple of things to notice in this script—when we define a function, there is no *right* way to do it; both ways work, but they work differently. We prefer using the format that explicitly mentions the `function` keyword since it immediately draws attention to this being a definition of a function, but this is just our preference—you can use whichever format you like:

```
#!/bin/bash
# Hello World done by a function
```

```
function HelloWorld {
    echo Hello World!
}

HelloWorld_alternate () {
    echo Hello World!
}
#now we call the functions
HelloWorld
HelloWorld_alternate
```

When we run the script, we can see that both our functions behave exactly the same:

```
demo@cli1:~/scripting$ bash functions.sh
Hello World!
Hello World!
```

Now, we are going to create an example that makes much more sense. There are going to be a lot of scripts that will require you to output things to the screen or into a file. Some parts of the output are going to be repeated over and over—a task that is exactly designed for a function:

```
#!/bin/bash
function PrintHeader {
    echo ----------------------
    echo Header of some sort
    echo ----------------------
}
echo In order to show how this looks like
echo we are going to print a header

PrintHeader

echo And once again

PrintHeader
```

```
echo That was it.

demo@cli1:~/scripting$ bash function.sh
In order to show how this looks like
we are going to print a header
----------------------
Header of some sort
----------------------
And once again
----------------------
Header of some sort
----------------------
That was it.
```

What our function did is create a header for our output. When we learn to pass arguments
to functions, we are going to use this trick a lot, especially when we need to output
formatted text into logs or when we have a large block of text with a few variables that
we need to fill in.

How it works...

Functions are parts of the code that bash reproduces whenever we reference our function
inside a script. Their purpose is primarily geared toward creating scripts that are easier
to read and debug. There is another reason to use functions: avoiding errors in code. If
we need to reuse parts of the code in different parts of the script, we can always copy and
paste it, but that creates a large possibility that we will introduce bugs into the script.

See also

- https://www.shell-tips.com/bash/functions/
- https://tldp.org/LDP/abs/html/complexfunct.html

Passing arguments to a function

We started demonstrating what a function looks like by showing you a simple script, the simplest we could create. We still haven't defined how to *talk* to your function, and we still don't know how to give a function some parameters or arguments and get something in return. In this recipe, we are going to fix that.

Getting ready

Since we mentioned arguments, we need to talk a little about them. bash treats arguments in functions the same as it does in the script itself—arguments become local variables inside the function block. To return a value, we also do almost exactly the same as when we need to deal with the whole script—we simply return a value from our function block and then read it inside the main script body.

Remember when we said you can reference arguments that were given to your script when it was initially called, and that we used variables called $1, $2, $3, and so on to get the first, second, third, and other parameters that were in the command line? The exact same thing applies to functions. In this case, we use the same variable names as when referencing arguments given to our function.

How to do it...

In order to send two parameters to a simple function that will display them, we would use something like this:

```bash
#!/bin/bash
#passing arguments to a function

function output {
    echo Parameters you passed are $1 and $2
}

output First Second
```

What happens when we try to run this script is that our arguments get passed in a way that we expect, one after the other, and then our function outputs them:

```
demo@cli1:~/scripting$ bash functionarg.sh
Parameters you passed are First and Second
```

You may wonder how our scripts are going to handle arguments that are given to the script, compared to arguments we pass to the function. The short answer is that variables named $1 and so on have a value that is local to the function and is defined by arguments we passed to the function. Outside of the function code block, these variables have the value of the arguments passed to the script. The long version of the answer is going to be in the next recipe and is called local and global variables. Using arguments is nothing but a special case of declaring a local variable; arguments that we pass simply become a local variable in the function:

```
#!/bin/bash
#passing arguments to a function
function output {
    echo Parameters you passed are $1 and $2
}
#we are going to take input arguments of the script itself and
#reverse them
output $2 $1
```

The reason we are changing the order of the arguments is to show the order in which arguments are passed to the function and to make sure that we are not using the arguments we passed to the script in the function since they have the same name. What this script will do is get two arguments from the command line, reverse them, and then give them in reversed order as arguments to our function. The function is simply going to output them:

```
demo@cli1:~/scripting$ bash functionarg2.sh First Second
Parameters you passed are Second and First
```

What happened here is also what we expected. Now, we are going to check one more thing that can be confusing to some people. Is the function even aware that some arguments are passed to the script or are the arguments strictly local? In order to check that, we are going to ignore whatever was in the script command line, and we are going to pass a pair of hardcoded strings to the function. If bash is behaving like we think it is, our script will output the hardcoded values. If the variables named $1 and $2 are set to values from the command line and they persist in the function, we should see that value in our echo statement. What we are going to do is create a functionarg3.sh file containing the following code:

```
#!/bin/bash
#passing arguments to a function
```

```
function output {
    echo Parameters you passed are $1 and $2
}
#we are going to ignore input parameters
output Hardcoded Variables
```

Now, we are going to run it and check what happened:

```
demo@cli1:~/scripting$ bash functionarg3sh First Second
Parameters you passed are Hardcoded and Variables
```

We can see that our assumption was correct and that the arguments given to the function always take precedence.

The next thing that we are going to do is show you how to handle simple operations using functions. Operations that can be done on variables is something we covered elsewhere in this book, but here, we are going to use an example we haven't used yet. We are simply going to add two arguments from the command line together.

In order to do that, we are passing arguments from the command line into our function and then using echo to output the result of the calculation. Part of the function used to get the result is also very interesting since it reminds us that we have to explicitly use a function to add two numbers in order to do that. If we try to add variables together, we are going to end up creating a string—something like this:

```
demo@cli1:~/scripting$ a=1
demo@cli1:~/scripting$ b=2
demo@cli1:~/scripting$ echo $a+$b
1+2
demo@cli1:~/scripting$ echo $(($a+$b))
3
```

This is the final version incorporated into our script:

```
#!/bin/bash
#Doing some maths
function simplemath {
add=$(($1+$2))
echo $add is the result of addition
```

```
}
```

```
#we are going to take input arguments and pass them all the way
simplemath $1 $2
```

Note that in this example, we are using a new variable inside a function to add the numbers and then outputting the value of this variable as the result. This is a better way to do this than directly doing the operation in the output itself—code that uses these temporary variables is always easier to read and understand than trying to find and understand variables embedded into output strings.

How it works...

The next thing we want to show is a nifty little feature that is not so common in most programming languages. Since bash treats arguments in the function the same way as it treats arguments to the script and uses the same logic to turn these arguments into variables inside the function, we can actually send multiple arguments to the function without defining their number in advance. Of course, our function needs to be able to understand something such as this.

See also

- https://linuxize.com/post/bash-functions/
- https://linuxhint.com/create-bash-functions-arguments/

Local and global variables

When it comes to declaring any variable in a script—or for that matter, anywhere at all—one crucial attribute for that variable is its scope. By scope, we mean *where the variable has the value we declared*. Scope is very important since not understanding how it works means that we can get unexpected results in some cases.

Getting ready

Defining a global scope to our variables is something bash does by default, without any interaction with us. All variables that are defined are global variables; their value is the same in the entire script. If we change the variable value by reassigning it (remember that operations on the value do not change the value itself), this value changes globally, and the old value is lost.

There is another thing we can do when declaring variables, and that is to declare them locally. In simple terms, this means that we are explicitly telling bash that we will use this variable in some limited part of the code and that it needs to keep the value just there, not globally in the entire script.

What are the reasons to declare a local variable? There are a couple of them, the most important one being to make sure that we don't change the value of any global variable. If a variable is declared locally with the same name as a global one, bash will create another instance of the variable with the same name and will keep track of both values, the global and the local one.

Global and local variables and how they work are something that is best explained by using an example.

How to do it...

The script that we are going to use to show you how this works is something that you will find in almost every example on the internet and in any book covering the subject. The idea is to create a global variable and then create a local variable in the function that is going to have the same name as the global one. The value that the global variable has should be different than the local value, and once we display that value, we should see that the value changes depending on if we are referencing a global or local variable:

```
#!/bin/bash
# First we define global variable
# Value of this variable should be visible in the entire script
VAR1="Global variable"
Function func {
# Now we define local variable with the same name
# as the global one.
local VAR1="Local variable"
#we then output the value inside the function
echo Inside the function variable has the value of: $VAR1 \
}
echo In the main script before function is executed variable \
has the value of: $VAR1
echo Now calling the function
func
# Value of the global variable shouldn't change
```

```
echo returned from function
echo In the main script after function is executed value is: \
$VAR1
```

If we execute this script, we are going to see exactly how variables interact:

```
demo@cli1:~/scripting$ bash funcglobal.sh
In the main script before function is executed variable has the
value of: Global variable
Now calling the function
Inside the function variable has the value of: Local variable
returned from function
In the main script after function is executed value is: Global
variable
```

This is completely expected—if there are a global variable and a local variable with the same name, the local variable will have its own values in the block it is defined in; otherwise, a global value will be used.

We said scripts such as this are common as an example, but what happens if we define just the local value? bash is different from most other languages since, by default, it will not show an error if we mistakenly try to reference a variable that is undefined. When debugging scripts, this can be a big problem since an undefined variable and a defined variable with no value will, at first, look exactly the same when we try to reference them.

To show this, we are going to make a small modification to our script and just remove the first variable definition. This will make our global value undefined—only the local value will have an actual value:

```
#!/bin/bash
# We are not defining the value for our variable in the global
#block
function func {
# Now we define local variable that is not defined globally
# as the global one.
local VAR1="Local variable"
#we then output the value inside the function
echo Inside the function variable has the value of: $VAR1
}
```

```
echo In the main script before function is executed undefined \
variable has the value of: $VAR1
echo Now calling the function
func
# Value of the global variable shouldn't change
echo returned from function
echo In the main script after function is executed undefined \
value is actually: $VAR1
```

In any strict programming language, something such as this would create an error. In bash, things are different:

```
demo@cli1:~/scripting$ bash funcglobal1.sh
In the main script before function is executed undefined
variable has the value of:
Now calling the function
Inside the function variable has the value of: Local variable
returned from function
In the main script after function is executed undefined value
is actually:
```

We can see that instead of errors, the script just ignores the variable value and replaces it with nothing. As we mentioned, even though we are expecting this behavior, keep in mind that this can lead to unexpected consequences. Another important thing in this script is the local value. We can see that the local variable *exists* only in the block of code in which it is defined; defining it will not create a global variable, and the value will be lost as soon as the function or block of code is executed.

How it works...

Using global variables in scripts can be useful for one more thing—forwarding values between functions. This feature of variables is something that can be useful, but at the same time, it is something that is dependent on your personal style of programming. Using global variables this way is easy—what you do is just declare a variable at the start of the script and then change its value whenever you need to. Usually, you assign a value before executing a particular function and then read the same variable after the function is done. This way, your function only needs to change the variable to give you the value that you expect.

However, there is a big problem in this otherwise perfectly logical way of using global variables. Since you have no way of knowing if the function behaved correctly and got to the point where it had to change the value of the variable, you do not have any idea if the value itself is what you are expecting. If a function fails for any reason, your variable will have the same value you sent to the function, leaving you with something that could be wrong.

What we are trying to say is that using global variables in this way is to be avoided, even though you can do it—the right way to work with functions and passing values is by using arguments and returning values by a mechanism we will look at in the next recipe.

See also

- `https://www.thegeekstuff.com/2010/05/bash-variables/`
- `https://tldp.org/LDP/abs/html/localvar.html`

Working with returns from a function

We mentioned that it is possible to use global variables to pass values to the functions inside a script and to get results back. This is the worst possible way to do it. If we need to pass some value to a function, using arguments is the way it should be done. The problem that we still have is how to get the results back when the function finishes. We are going to solve that in this recipe.

Getting ready

If nothing else, `bash` is logical and consistent in the syntax it uses. The reason we are mentioning this is that when functions return a value, they use the exact same mechanism that scripts use when returning a variable—the `return` command. Using this command, it is possible for a function to return a value when called, but the value can be in the range of numbers between 0 and 255. There is also a possibility to set a global variable just to return a function value—for example, if we need to return a string—but try to avoid that since it creates code that is difficult to debug. When you are browsing the internet for function `return` statements, you may also run into a third solution that uses something called *reference passing* or `nameref`. This is a more complex solution that you should be aware of, but we are deliberately avoiding it in this recipe since it works only on the most recent versions of `bash` (from 4.3 up), and that breaks the compatibility and usability of our scripts.

How to do it...

We'll show you both ways to return a function, starting with the one we consider wrong. The reason that we are even showing you a wrong solution is that you will often run into this in different scripts downloaded from the internet, and if you are unaware of this method, you will probably be a little bit puzzled at first because the variable is usually first defined in the function itself and does not exist before the function is first called:

```bash
#!/bin/bash
#Doing some string adding inside a function and returning
#values
#function takes two strings and returns them concatenated
function concatenate {
RESULT=$1$2
}


# calling the function with hardcoded strings


concatenate "First " "and second"
echo $RESULT
```

What we did is just pass two strings to a function that returned them concatenated. Of course, this is silly—we could do that by simply using the expression we used in the function. This example is so basic that it doesn't even use any operators.

What's important is the way we returned our value. By just assigning a new value and therefore creating a global variable named RESULT, we got our string, and we were able to use echo to write it to the screen. Why is this a problem?

We have already explained this. What we are doing here is dangerous since we have no way of knowing if the function has done what it had to do. The only thing we have is the variable called RESULT that probably contains the value we expect. In this trivial example, we could check the outcome, but that would defeat the purpose of having a dedicated function. In order to reduce the uncertainty a little bit, there is a small trick that we can do.

Consider this change to the script:

```bash
#!/bin/bash
#Doing some string adding inside a function and returning \
values
#function takes two strings and returns them concatenated
function concatenate {
```

```
RESULT=$1$2
}
```

```
concatenate "First " "and second"
[ $? -eq 0 ] && echo $RESULT || echo Function did not finish!
```

What we did is create a conditional output. The format of the condition itself should be familiar to you by now—we are using logical functions to either print out the result of the function or to print out that the function did not work correctly. As a reminder to when we introduced logical operators, what our script does in the last line is check for the value of a variable called `$?`. If the variable value equals `0`, we print out the result of the function. If the value is not zero, we output the error message since we know that our function had an error somewhere inside its command block.

The reason we can do this is simple—we already said that functions have the same way of communicating to the script as the script itself does to the rest of the operating system. This includes passing arguments and being able to use a `return` statement to return values, but also it means that `bash` sets a variable named `?` when the function is finished. When we use it to understand what happened to the script (which we already explained), if we check this variable and it has a value of `0`, this means that the function finished correctly, or at least that the last command in it finished correctly.

This is a simple solution to a problem that we shouldn't create in the first place; whenever possible, we should use `return` to get our values. Here's an example:

```
#!/bin/bash
#simple adding of two numbers
#function takes two numbers and returns result of addition
function simpleadd {
    local RESULT=$(($1+$2))
    return $RESULT
}
#we are going to hardcode two numbers
simpleadd 4 5
echo $?
```

This is a much better way if we are sure that our numbers fit into the range from `0` to `255`. We are outputting the result of the function, and this is as easy as referencing the right variable. We could also check if the value of the variable after the execution of the function is `0`, meaning that the function behaved correctly, and then output the result.

Another thing you should know is that a function can use the `exit` command. By using it, you are telling `bash` to immediately stop what the function is doing and exit the function command block. The value that is going to be returned in this instance is going to be the error level of the last command that was executed before the `exit` command was invoked.

Here's an example:

```
#!/bin/bash
#exiting from a function before function finishes
function never {
echo This function has two statements, one will never be \
printed.
exit
echo This is the message that will never print
}
#here we run the function
never
```

What is going to get printed is just the first line of output; since we used the `exit` statement, the second part of the output will never run:

```
demo@cli1:~/scripting$ bash funcreturn3.sh
This function has two statements, one will never be printed.
```

How it works...

The main reason why all of this even exists is to enable you to more tightly control how functions and, more generally, the order of command execution works in your script. `bash` is very basic in the way it approaches this topic, and that at the same time makes it versatile. In order to use functions, you only need to know how arguments work in scripting—all the variable names and logic behind it are the same when applied to functions.

See also

- https://www.assertnotmagic.com/2020/06/19/bash-return-multiple/
- https://www.linuxjournal.com/content/return-values-bash-functions

Loading an external function to a shell script

A problem that will often pop up when you need to create more complex shell scripts is going to be how to include other code into your script. Once you start scripting, you will often create a couple of common functions that you always use—things such as opening connections to servers, getting some operations done, and other things like that.

Sometimes, your scripts will have to use a lot of preset variables that are defined by the user before they even run the script in order to avoid having to type them in each time a script is called.

Of course, the solution to both of these problems can be to simply copy and paste the relevant code into your script and to make the user edit the script before running it. The reason we should never do this is that each time we copy and paste something, we are creating a new version of our code. If we notice an error in the code, we need to fix it in all the scripts that reuse it. Luckily, there is a better way to solve this problem, and that is to split the script into different files and then include them when we need them.

Getting ready

This recipe is going to be useful in two scenarios that are not necessarily mutually exclusive. We already mentioned both of them briefly.

The first one is when using external functions. Normally, when creating a script, everything is going to be in one file. All the functions, definitions, variables, and commands are going to be in one place. This is usually completely fine if we are creating something that is specially written to accomplish a particular task.

More often than not, we will need to solve something that we already worked on before in some other solution. In this case, we usually already have some functions ready that can be considered part of the solution.

In complex scripting solutions, you might even use some common things such as menus, interfaces, headers, footers, logs, and other things that are exactly the same across every script that you make.

Another very common problem is settings that require some setup by the user. Large scripts can have server names, ports, filenames, users, and many different things that are required for the script to function. You can always put this information as arguments into the command line, but that will look bad and will make your script prone to errors since the user will have to type a lot of things by hand each time scripts are executed.

A common practice in these circumstances is to put everything in one file as variables, and then have the user edit this file as part of the installation process for the script. Of course, you can put everything together with the script itself, but that will almost certainly mean some user will change something they shouldn't have.

As always, there is a solution for that.

How to do it...

bash has built-in functionality that enables including different files into a script. The idea is pretty simple—there is a *master script file* that gets executed as the script itself. In that file are commands that tell bash to include different files and scripts.

As with everything else, even though this is a pretty easy thing to do, there are some things you need to know. The command we are going to use first is source. Before we explain everything, we are going to create two scripts. The first one is going to be the script that the user is going to run, and it is going to look like this. Name the file main.sh:

```
#!/bin/bash
#first we are going to output some environment variables and
#define a few of our own
echo Shell level before we include $SHLVL
echo PWD value before include $PWD
TESTVAR='main'

echo Shell level after include $SHLVL
echo PWD value after include $PWD
echo Variable value after include $TESTVAR
```

We are going to run it just to see how the script behaves:

```
demo@cli1:~/includes$ bash main.sh
Shell level before include 2
PWD value before include /home/demo/includes
Shell level after include 2
PWD value after include /home/demo/includes
Variable value after include main
```

The results are what we expected—our current directory is the same as the one we ran the script in, and $SHLVL is 2 since we ran our script in a separate shell (lvl2) from the command line (lvl1). Our variable is defined as main and it hasn't changed.

Now, we are going to create our second script and name it auxscript.sh:

```
echo Inside included file Shell level is $SHLVL
echo Inside included PWD is $PWD
echo Before we changed it variable had a value of: $TESTVAR
TESTVAR='AUX'
echo After we changed it variable has a value of: $TESTVAR
```

The biggest thing here is that we are not using the usual #!/bin/bash notation at the start of the script. This is intentional, as this file is meant to be included in other scripts, not run by itself.

After that, we are doing more or less the same things as in the main script, outputting some text and values, and working with variables.

The reason we are changing the variable is to show what actually happens inside this included part of the file and how it interacts with the main script body.

Now, we are going to change the main.sh script and add just one line:

```
#!/bin/bash
#first we are going to output some environment variables and
#define a few of our own
echo Shell level before we include $SHLVL
echo PWD value before include $PWD
TESTVAR='main'

source auxscript.sh

echo Shell level after include $SHLVL
echo PWD value after include $PWD
echo Variable value after include $TESTVAR
```

The main thing now is to run the main.sh script again:

```
demo@cli1:~/includes$ bash main.sh
Shell level before include 2
```

```
PWD value before include /home/demo/includes
Inside included file Shell level is 2
Inside included PWD is /home/demo/includes
Before we changed it variable had a value of: main
After we changed it variable has a value of: AUX
Shell level after include 2
PWD value after include /home/demo/includes
Variable value after include AUX
```

Some interesting things happened here. What we can see is that our environment variables haven't changed but the test variable did.

We are going to explain that, but we are going to do one more thing before that—we are going to use another command instead of source. A lot of people new to scripting tend to confuse the source command that we just showed you with executing a script. After all, we are including a script inside a script, so those things do look similar. We are going to try to do it in our example.

We are going to change a single line inside the main script, but our aux script is going to stay the same. There are multiple ways in which we can do it, but we intentionally chose to run bash and run our second script explicitly. The reason is simple—other methods require our script to have the executive bit set (something we haven't done) or depend on less readable versions of the same thing as just running a command called exec:

```
#!/bin/bash
#first we are going to output some environment variables and
#define a few of our own
echo Shell level before include $SHLVL
echo PWD value before include $PWD
TESTVAR='main'

bash auxscript.sh

echo Shell level after include $SHLVL
echo PWD value after include $PWD
echo Variable value after include $TESTVAR
```

The only thing we have changed is that we are not including the script—we are executing it:

```
demo@clil:~/includes$ bash mainexec.sh
Shell level before include 2
PWD value before include /home/demo/includes
Inside included file Shell level is 3
Inside included PWD is /home/demo/includes
Before we changed it variable had a value of:
After we changed it variable has a value of: AUX
Shell level after include 2
PWD value after include /home/demo/includes
Variable value after include main
```

We can see, however, that this small change created a huge difference in the way our script works.

How it works...

The last example we did requires quite a lot of explaining, and we need to start with how bash works.

Using the source command tells bash to go find a file and use its contents at the place where we sourced the file. What bash does is straightforward—it just replaces this line with the entire file we pointed to. All the lines are inserted and then executed as if we copy-pasted the entire file into our original script.

This is the reason why in our first example nothing changed. Our script started running from the main file, continued running commands from the auxiliary file, and then returned to the main file to finish the commands that followed.

When we changed our source for bash, we created a completely different scenario. By using the bash command inside the script, we are telling the shell to start another instance and execute the script we are referring to. This means that the entire environment is created, and unless we explicitly specify that we need some variables in the new environment, they are not going to get exported.

This is also the reason that our $SHLVL variable incremented—since we called another shell inside, the shell level had to go up.

Our test variable vanished because we didn't export it, so it had no value before being set, and since our environment was created just to run these couple of lines, the same variable simply disappeared when the script we called ended.

Remember that executing a script and sourcing it are completely different things, and when in doubt, think about what you are trying to do. If you want to execute something inside a script such as a regular command, use `bash` or `exec`. If you would otherwise copy-paste code from another script, use `source`.

Before we finish with this recipe, we also need to mention functions. Including functions is exactly the same as including any other part of any other script, with one important difference. In order for your code to work, you *must* include functions at the start of the script or immediately before you try to use said functions. If you don't do that, the resulting error is going to be the same as if you hadn't defined your function at all.

See also

- `https://bash.cyberciti.biz/guide/Source_command`
- `https://tldp.org/LDP/Bash-Beginners-Guide/html/sect_02_01.html`

Implementing commonly used procedures via functions

By this point, we have created a lot of different and very simple scripts that more or less used `echo` and a few commands just to show how a particular thing in `bash` works. In this recipe, we are going to give you a couple of ideas on how to use what we have learned so far.

Getting ready

We are going to create a small script that is going to show you how to easily automate the most mundane tasks on any system. The idea here is not to show you every task possible, but instead to show you how to tackle the most common problems.

How to do it...

Before we even start with the script, we need to go back to the recipes where we were explaining how to start writing scripts. What we are talking about are the prerequisites and presumptions we are going to make when we create and run this script.

Every script will have its own prerequisites. These are usually a list of things that your script needs to run—either it requires different packages or it needs some other condition that has to be met in order for the script to work, such as a working database or a working web server.

For this one, we are presuming that you have installed a package called `curl` and that you are connected to the internet.

Now, for the presumptions that we are depending on, this script has some commands that affect users and groups on the system. This means that in order for that part of the script to work, we absolutely need the script to be run either by the `root` user or another user who has administrative privileges.

The script also presumes a lot about users and checks only if we have enough parameters, not the quality of arguments that were provided. This means that a user can give the script a number instead of a string, and the script will happily use this as a valid parameter. We will explain how to deal with that when we start dissecting the script.

As a person responsible for writing scripts, part of your job is to be aware of these preconditions and to make sure to address them. There are two ways you can do that—the first is by stating what your script expects in some form of document that will follow your script.

The other thing you can do (and we highly recommend this) is to check for every possible condition that you can think of, and if something is wrong, either print an error message and stop your script or, if you know what the problem is, try to rectify it.

Examples of things you can solve inside the script are administrator privileges—your script can test if it can run, and ask the user to elevate privileges if the permissions are too low. You can also test if a particular package is present on the system if you see that some command that is not standard fails.

In the end, how you solve problems in your script is going to be up to you and your skill level, but before you do anything, remember that when it comes to scripting, you need to test everything.

Now, here's the actual script:

```bash
#!/bin/bash
#shell script that automates common tasks

function rsyn {
rsync -avzh $1 $2
```

```
}

function usage {

echo In order to use this script you can:
echo "$0 copy <source> <destination> to copy files from source \
to destination"
echo "$0 newuser <name> to createuser with the username \
<username>"
echo "$0 group <username> <group> to add user to a group"
echo "$0 weather to check local weather"
echo "$0 weather <city> to check weather in some city on earth"
echo "$0 help for this help"


}

if [ "$1" != "" ]
            Then
    case $1 in
        help)
            Usage
            Exit
            ;;
        copy)
                if [ "$2" != "" && "$3" != "" ]
                then
            rsyn $2 $3
        fi
            ;;


            group)
        if [ "$2" != "" && "$3" != "" ]
                then
                    usermod -a -G $3 $2
        fi
                                    ;;
```

```
            newuser)
                if [ "$2" != "" ]
                then
                            useradd $2
                    fi
                    ;;


            weather)
                if [ "$2" != "" ]
                        then
                            curl wttr.in/$2
                        else
                            curl wttr.in
                    fi
                    ;;



                *)
                echo "ERROR: unknown parameter $1\""
                usage
                exit 1
                ;;
        esac
                    else
                    Usage
    fi
```

How it works...

This script requires some explaining, and we intentionally did not comment on any of it for two reasons. One was that comments would make the script so long it would require too many print pages, and the other one was to be able to go through it block by block in this explanation without breaking your flow with short comments. Having said that, always comment on your scripts!

So, our script starts with a function. Considering this function has only one line, you may be surprised that we decided to break it into function, but we had a point to make.

Some commands, such as `rsync` or `tar`, for example, have a complicated list of switches that are often used. When creating a script, it is sometimes easier to put some of those commands into a function to be able to call the function without having to remember all the switches every time. This also goes for commands that need a lot of parameters that are predetermined when the script is configured. Put all of them into a function and then call the function with only the bare minimum of arguments.

Another thing we put into a function is *usage*, a block of text that helps the user to run the script, giving them enough information so they don't need any other type of help.

If you can, please write more verbose help pages for your script. You can even create a **manual** (**man**) or info page, but at the same time, always provide help directly inside your script. Make your script display help when there is anything wrong with the command line. There is nothing more annoying than when a script just fails without any meaningful message, or when a script simply states `read the help page for more information`.

In this function, we are using the `$0` positional argument in order to output the name of the script. Use this way of giving the user help when you are giving examples on script usage. Avoid hardcoding the script name because you don't know if the user changed the filename of the script, and hardcoded names can then completely puzzle them.

Also, if you are using any special character in your text, use quotation marks; otherwise, you may run into errors or, worse, completely unexplainable errors.

The next part of our script deals with each individual command. When creating a command-line utility such as this, decide in advance whether you are going to create a tool that will use *commands* such as this one, *switches* such as -h or −something, or some sort of simple textual interface. There are pros and cons for all of these, but in essence, the format that we chose is mostly used for scripts that can do multiple tasks one at a time. Switches enable you to introduce many parameters to a task, and **user interfaces** (**UIs**) are targeted toward inexperienced users. Also, remember that your script may be used from inside other scripts, so avoid interfaces that will block that.

In the `case` statement, we are checking for a couple of things. First, we are testing if the first argument is a valid command. After that, we are checking if there are enough arguments for a given command to make sure that we can run it without errors. Even with this, we are not doing nearly enough testing for the validity of arguments. When reading this, try to add a few more sanity checks such as *are the parameters actually valid*, *did the user input a valid parameter containing spaces that got divided into multiple strings*, and so on.

We are not going into too much detail for individual commands; we are only going to mention the one that, in all fairness, looks completely out of place. We are, of course, talking about the `weather` command that gives you a weather report for your city:

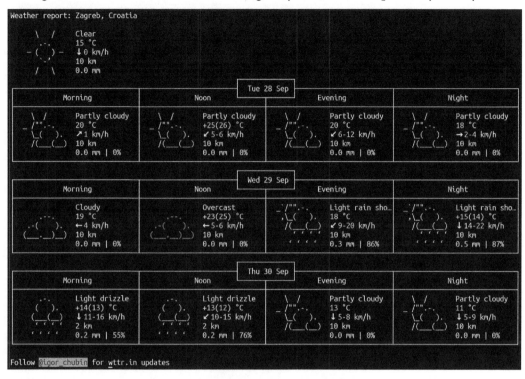

Figure 12.1 – wttr.in is one of many interesting services available online

Internet is full of useful services, and `wttr.in` is definitely one of those. If you go to `wttr.in` or run `curl wttr.in`, you are going to get a weather report for the city that the system thinks you are living in. There is some deep magic involved here—the system is going to try to guess where you are based on your **Internet Protocol** (**IP**) address, and even while having to do that, it is going to give you a pretty accurate forecast almost immediately.

We chose to show you this example on purpose—if you add a city name to the `wttr.in` link, the system is going to show you the weather in that city, while even trying to guess the exact city name. There are a couple of really useful online services such as this accessible from the command line, and using some of them means you can extend your script in the most unusual ways.

At the end of the recipe, note that we are checking for different errors in how the script is invoked in three different ways. Always try to anticipate errors such as this.

See also

The following web page is a must-see if you do anything in the command line:

- `https://stackify.com/top-command-line-tools/`

13
Using Arrays

In one of the recipes in a previous chapter, we mentioned arrays as one of the possible compound data types that bash supports. We said that what bash has is two different data types (strings and numbers), but that there are ways that we can use more than that if we need to. Arrays are just that—something that we need to be able to use since we need something a little bit more complex than single-value variables to solve some problems.

In this chapter, we are going to cover two basic recipes connected to arrays:

- Basic array manipulation
- Advanced array manipulation

You can already see that we are being intentionally broad here; arrays are like that—simple on the surface, but with quite a few small tricks if we need to use them.

Technical requirements

As with all the chapters that cover scripting, we are using any machine that works, and things as fundamental as arrays are going to work on all machines that run bash. You will therefore need the following:

A **virtual machine** (**VM**) with any distributuon of Linux (in our case, it's going to be **Ubuntu 20.10**)

So, start your VM!

Basic array manipulation

The thing with bash and variables in bash is that they look deceptively simple. There are no formal declarations of type, or basically declarations of any kind. Typing is done by the shell itself, and we can do a lot of things implicitly. This is especially true for *regular* variables. Arrays are a little bit more complex, and they offer a few syntactic peculiarities when used, but they are an extremely useful tool. You may wonder why we are even mentioning them in any context since they are nothing more than one value under the same variable name. Well, the main reason is that we often actually need exactly this. A lot of times, we must store multiple values that belong to some set of data. Typically, that will be something such as an unordered list of values in case it is something that we do not care about having in a particular order, or an ordered list of values if we do.

Getting ready

Usually, arrays are defined as *one-dimensional indexed arrays*, since they have an explicit order embedded in the very definition of an array. In essence, when we have any multi-value variable, it can be *ordered* and *unordered*. The difference is in whether we can define in which order the values are declared in the variable. If we can store values but are unable to get the order in which they are stored, that is called an unordered set. In bash, we only have ordered lists, which we call arrays. This means that every value in the variable has not only the value itself but a defined index or place. Not only can we add or remove values from an array, but we can also directly read any of them and we can reorder them if we need to.

We have two types of arrays at our disposal. Both are ordered but one is *indexed*, which means that we have a numerical value that defines a particular element of the array. We also have something named associative arrays, sometimes also called *hash tables*. This type of array is useful because it doesn't use a numeric value but instead uses a string key to define a particular array element. We will talk a lot about both.

How to do it...

The first thing we need to do is to declare a variable:

```
demo@ubuntu:~/includes$ TEST=(first second third fourth fifth)
demo@ubuntu:~/includes$ echo $TEST
first
```

We are off to a bad start. Obviously, we declared our variable correctly since there were no errors, but as soon as we tried to print it, we ran into problems. We had those before when we tried to print arrays. The solution is to use a special character to denote the index and tell bash that we want to print not only the first value but all values in the array:

```
demo@ubuntu:~/includes$ echo ${TEST[@]}
first second third fourth fifth
```

This works as intended since bash understands it needs to do a quick loop and go through every index in the array, printing all the values. An alternative to that is to use the following:

```
demo@ubuntu:~/includes$ echo ${TEST[*]}
first second third fourth fifth
```

This has the exact same output as the previous command.

This should get you thinking: *Is there another way to use indexing?* Of course there is—we can use it to directly access a single value in the array:

```
demo@ubuntu:~/includes$ echo ${TEST[2]}
third
demo@ubuntu:~/includes$ echo ${TEST[4]}
fifth
```

Make a note we are *off by one* in this example since the first element in an array has an index of 0. There is one more very interesting property rarely mentioned in terms of arrays. Before we show it, we need to explain another way of declaring an array.

Declaring a variable using simple brackets is the most common way to do this, but we can also do it by directly specifying elements in an array. The interesting thing is that we can do it in any order we want to:

```
demo@ubuntu:~/includes$ ORDER[2]=second
demo@ubuntu:~/includes$ ORDER[3]=third
```

If we now try to print our array, the result is going to be more dangerous than surprising:

```
demo@ubuntu:~/includes$ echo ${ORDER[*]}
second third
```

Our array has two values stored under indexes. We intentionally made an error in indexing—the value of `second` is stored under index number 2, making it the third array element. What we didn't do is declare the first element of our array. The reason we said that the result of the previous command is dangerous is that from its output, you cannot see the index of a particular element, so you have no idea what the actual index of a particular value is. This makes it easy to confuse the values—or, to make it more obvious, something such as this will not return a value, although we might think it should:

```
demo@ubuntu:~/includes$ echo ${ORDER[0]}
demo@ubuntu:~/includes$ echo ${ORDER[1]}
```

Things such as these are a common source of errors that we need to troubleshoot, and they are especially complicated to spot if we use the direct syntax of specifying both a value and its index.

What we are trying to say is to never use this way of declaring an array, unless you have a particular reason why you need it. Otherwise, things can become confusing later.

There is a third way to declare an array that is the least commonly used method. Using the `declare` statement and the `-a` switch, you explicitly declare that a particular variable is going to hold an array of values. The reason we rarely see this in code is that when we use either of the already mentioned implicit declarations, our variable will become appropriately typed, so there is simply no reason to do this twice unless you want to do so to make your code readable.

All of this was just different ways to create and print normal, indexed arrays. We mentioned that there is another type of array called an associative array, also known as a hash table, dictionary, or key-value paired array. This type was introduced in `bash` 4.0 and is still not available on some platforms; most notably, some versions of OS X require you to upgrade `bash` to be able to use this type.

Real life contains a lot of things that can be considered pairs of values. Things such as username/password, name/address, name/telephone number are naturally created pairs of data that are often used in scripts. Obviously, we could use a normal array to store this, but to do it in a way that enables us to understand which values are a part of a given pair, we would need not one but two separate arrays and a little bit of focus on how we declare indexes for them so that we can use the same index to get the first and the second value:

```
#!/bin/bash
#we are declaring two arrays, one for the names, one for the
#phone numbers
NAMES=(John Luke "Ivan from work" Ida "That guy")
NUMBERS=(12345 12344 113312 11111 122222)
```

```
#now we need to pair them up:
for i in {0..4}
do
                echo Name:${NAMES[i]} number:${NUMBERS[i]}
done
```

This is going to give us an output that looks like this:

```
demo@ubuntu:~/variable$ bash pairs.sh
Name:John number:12345
Name:Luke number:12344
Name:Ivan from work number:113312
Name:Ida number:11111
Name:That guy number:122222
```

This looks and works fine if we need to print out all of the data. Imagine now that we have information that we need to search for—imagine we are looking for a telephone number of a particular person. If we needed to do this using regular arrays, something like this could be done:

```
#!/bin/bash
#we are declaring two arrays, one for the names, one for the
#phone numbers
NAMES=(John Luke "Ivan from work" Ida "That guy")
NUMBERS=(12345 12344 113312 11111 122222)
#now we need to pair them up:
for i in {0..4}
do
            if [ "${NAMES[i]}" == "$1" ]
                        then
                                    echo Name:${NAMES[i]}
number:${NUMBERS[i]}
            fi
done
```

In order to check all the values, we need to go through every element of one array and then print pairs if we find a match. First, we are going to test this:

```
demo@ubuntu:~/variable$ bash search.sh John
Name:John number:12345
```

```
demo@ubuntu:~/variable$ bash search.sh
demo@ubuntu:~/variable$
```

Although this is working, it is not the right way to do it. Some of the downsides of this approach are obvious:

- An array is indexed, so it can hold the same content at different indexes.

- In order to find something, we need to go through all values.

- If we mess up any of the arrays, we can create invalid data.

- The script is very complicated for a simple task.

That other *associative* array type we mentioned is the solution to this and many more problems. While in normal arrays, the values we are using are indexed by numbers. In this particular array type, we use any value as a *key* to reference a specific value in an array.

Doing this requires an explicit declaration of the array type that has to be done using the declare statement and the -A switch. Be very careful about this switch since it uses a capital letter *A*, while *normal* arrays are declared using the same letter in lowercase. While you may implicitly declare an indexed array in multiple ways, an associative array can be declared only by using this method and must be declared explicitly. Before we can do a script to fully demonstrate that, we need to show you how to declare an array of this type. Besides the fact that we must use the declare statement, we also need to declare the indexes since they can be any string, not just a number. The output looks something like this:

```
demo@ubuntu:~/variable$ declare -A NAMES
demo@ubuntu:~/variable$ NAMES["John"]=12345
demo@ubuntu:~/variable$ NAMES["Luke"]=12344
demo@ubuntu:~/variable$ NAMES["Ivan from work"]=113312
```

There is also a way to declare these arrays in one line, by specifying pairs:

```
demo@ubuntu:~/variable$ NAMES=([Ida]=11111 ["That guy"]=122222)
```

Now that we have defined this array in two ways, let's do a small follow-up with other operations that we can carry out. We should be able to write out the array values. First, we will try the method we used with regular arrays:

```
demo@ubuntu:~/variable$ echo ${NAMES[*]}
11111 122222
```

This is, of course, a problem, but we were expecting it. This syntax is designed for arrays that have a numerical index and is translated into something a human can understand, which means: *print all the values from an array called NAMES by using all possible indices.*

The key here is that we are not printing the index of a value, just the value itself. Since we are using both the key and the value in our arrays, we need to be able to see what the key for a particular value is. This can be done using a `for` loop, but before we do that, we have a point to make—by design, arrays have multiple values, so not only can we redefine the whole array, but we can also add and remove elements from it. We already showed an example in which we added elements to a variable, and we did this by creating a new value under a new index.

All this applies not only to associative arrays but to arrays in general; however, we are going to use associative arrays just to make you more familiar with this type. We are going to repeat the example from before, but with a twist:

```
demo@ubuntu:~/variable$ declare -A NAMES
demo@ubuntu:~/variable$ NAMES["John"]=12345
demo@ubuntu:~/variable$ NAMES["Luke"]=12344
demo@ubuntu:~/variable$ NAMES["Ivan from work"]=113312
demo@ubuntu:~/variable$ echo ${NAMES[*]}
113312 12344 12345
demo@ubuntu:~/variable$ NAMES=([Ida]=11111 ["That guy"]=122222)
demo@ubuntu:~/variable$ echo ${NAMES[*]}
11111 122222
```

What happened? First, we defined an array that consists of three value pairs. We did this by declaring every single pair by itself. After that, we used the alternative way of array declaration, but since we basically redeclared the array, the values we used completely replaced values that the array had before. What we should have done is *add* those values. There are two ways to do this—one is to just define the right values for the pairs again:

```
demo@ubuntu:~/variable$ NAMES["John"]=12345
demo@ubuntu:~/variable$ NAMES["Ivan from work"]=113312
demo@ubuntu:~/variable$ NAMES["Luke"]=12344
demo@ubuntu:~/variable$ echo ${NAMES[*]}
113312 11111 12344 12345 122222
```

We can now see our array has the expected number of values, although we still don't know how to get them printed.

Another way to do this is by using addition. We are going to change only one character in our example to do this:

```
demo@ubuntu:~/variable$ declare -A NAMES
demo@ubuntu:~/variable$ NAMES["John"]=12345
demo@ubuntu:~/variable$ NAMES["Luke"]=12344
demo@ubuntu:~/variable$ NAMES["Ivan from work"]=113312
demo@ubuntu:~/variable$ NAMES+=([Ida]=11111 ["That
guy"]=122222)
demo@ubuntu:~/variable$ echo ${NAMES[*]}
113312 11111 12344 12345 122222
```

Can you even notice the change? What we did is use the plus operator in front of the equals sign to tell bash that we want these pairs added to our array. This notation is completely the same as if we had used NAMES=NAMES+([Ida]=11111 ["That guy"]=122222) —it is just a little shorter.

The last thing we need to know is how to remove a value from an array.

The solution to this is simple—there is a command called unset that simply removes the value associated with a particular index or a key. More often than not, this is used on key-value pairs since it makes much more sense there, but you can also do it on a regular array:

```
demo@ubuntu:~/variable$ echo ${NAMES[*]}
113312 11111 12344 12345 122222
demo@ubuntu:~/variable$ unset NAMES["John"]
demo@ubuntu:~/variable$ echo ${NAMES[*]}
113312 11111 12344 122222
```

Now, we are going to tackle the big problem and rewrite our script from before, using our only associative array.

The idea of how to do this is based on the way bash uses sets of objects. We are going to create one such set out of all the keys, and then print both the key and the value it points to. To access keys directly, we can use an exclamation point:

```
#!/bin/bash
#we are declaring one associative array for pairs of values:
declare -A PAIRS
PAIRS=(["John"]=12345 ["Luke"]=12344 ["Ivan from work"] =113312 \
["Ida"]=11111 ["That guy"]=122222)
```

```
#now we need to get them printed
for name in "${!PAIRS[@]}"
do
                echo Name:"$name" number:${PAIRS["$name"]}
done
```

There are some small things you need to notice in this script. For example, quotes are important since our keys contain spaces. The general rule here is that as soon you are using any string as a value of anything, it should be enclosed in quotes to get the space parsed correctly.

The for loop is going to go through keys one by one, and keys are going to be used as an entire key value, including a space. Some manuals will state that a key must be a single word, but officially it can be anything. The usage scenario for this type of variable is, however, that we will be using a word or two in most cases:

```
demo@ubuntu:~/variable$ bash associative.sh
Name:Ivan from work number:113312
Name:Ida number:11111
Name:Luke number:12344
Name:John number:12345
Name:That guy number:122222
```

How it works...

We have demonstrated the way arrays work in practice, how to create them, how to read from them, and how to delete individual elements inside an array. What makes arrays a little bit different from regular variables is that variables hold one value, so there are no operations that can modify that value. If we need to change it, we simply redeclare the entire value. When dealing with arrays, we are dealing with multiple values in one array, and that still means we must redeclare any that we need to change, but we are then changing just a single value out of many, not the entire array. This is the main reason we have operations that add and remove elements.

Some of you already familiar with different programming languages will probably be a little bit confused by the relaxed way some things are defined, especially when we are talking about regular arrays and the way they address individual values and print them. Probably the most confusing part is how you can completely skip a range of indexes and still get a valid array. We are going to talk a little bit more about this in the next recipe, but bash is inconsistently vague in some ways, and this is one of them.

Associative arrays are going to be something you will need from time to time, especially when you need to deal with objects that have some properties. It is not possible to store more than one property per key, but even this makes for a nice environment since this single value can, for example, be an indexed value that then points to different arrays containing everything else that we need to store about a particular object.

See also

Arrays are complicated and, at the same time, very simple in their syntax. Check out these links for many more examples:

- `https://www.shell-tips.com/bash/arrays/`

- `https://opensource.com/article/18/5/you-dont-know-bash-intro-bash-arrays`

Advanced array manipulation

Now we have finished dealing with the basics, we need to add much more to your knowledge of arrays. What we first need to do is to give you some ideas on how to make the arrays you create in your scripts more persistent, so we need to deal with storing and restoring them.

The reason why this is important is that arrays can be quite large, depending on the script you are creating. Dumping and reusing variables in scripts is easy since we can use the *source* to declare variables we stored in a file. Arrays make for more complicated work since they can contain multiple variables, and sometimes we even need to create them from another data source.

Getting ready

The thing with scripts is that they sometimes need to deal with a lot of data. In a lot of cases, we can use files to both store data in and load data from. There are, however, some cases where arrays—especially associative arrays—are a necessity for working with large amounts of data, and then we need to know how to save that data to the disk and reuse it. We are going to show you how to solve those problems and advise you when to do it.

How to do it...

When talking about all of this, we always need an example where a particular feature makes sense. Some of the bash features we dealt with were so generic that our examples also had to be generic. Associative arrays are not like that. Although they can be used for a number of things, some scenariós are so common that you will automatically start to declare an array even before you think how and why you are doing it.

The most common scenario is saving user settings.

Any larger script that deals with any task will have to be configurable. We mentioned that when we said that you can include files, and the most common ones are going to be the ones containing variables, storing values for different settings.

Most of the time, all the settings in a script look something like this:

```
USER=demo
CMD=testing
HOSTNAME=demounit
```

These settings are just an example, but most scripts have a block of these either in a separate file or at the start of the script.

Notice that all of them have a format of KEY=VALUE and look like they are created to be used in an associative array. Having said that, we need to make a point about using any feature of any programming language—do what is best for the performance and clarity of a given solution and don't use a feature just because you *know* it is used in a particular situation.

This is where your experience is required. If a script has a set of settings that never change after we initially set them while loading and saving them from the disk, this makes no sense. The same goes for if a script only has a few variables defined when it starts—there is no point in using arrays here.

But if a script has more than a few things that change between different script executions, and if you are using them to store some important runtime operations that are needed not only when the script starts but also during its run, arrays may be a solution.

Our example is going to be a small script that will have a few settings it needs to remember, and we are going to use arrays to load them from disk, use them, and store them later back to disk. Working on this, we are also going to perform some tasks on a given array to demonstrate how to manipulate pairs in a script.

But before we do any of that, we need to go through a few advanced examples to show you how some things we previously glanced over work.

First, we are going to deal with the difference between using * and @ operators to read indices and keys in arrays. We said that for a given array, those two operators are different ways of going through all indices. Here's an example to illustrate this:

```
demo@ubuntu:~/variable$ SOMEARRAY=(0 1 2 3 4 5)
demo@ubuntu:~/variable$ echo ${!SOMEARRAY[*]}
0 1 2 3 4 5
demo@ubuntu:~/variable$ echo ${!SOMEARRAY[@]}
0 1 2 3 4 5
```

The same goes for associative arrays:

```
demo@ubuntu:~/variable$ declare -A PAIRS
demo@ubuntu:~/variable$ PAIRS=(["John"]=12345 ["Luke"]=12344
["Ivan from work"]=113312 ["Ida"]=11111 ["That guy"]=122222)
demo@ubuntu:~/variable$ echo ${!PAIRS[@]}
Ivan from work Ida Luke John That guy
demo@ubuntu:~/variable$ echo ${!PAIRS[*]}
Ivan from work Ida Luke John That guy
```

So far, the results are at first glance identical. This is one of those things in bash that will sometimes make you pull your hair out in despair because when we use them for loops, there will be a big difference that we are going to show by creating a small example script:

```
#!/bin/bash
declare -A PAIRS
PAIRS=(["John"]=12345 ["Luke"]=12344 ["Ivan from work"] =113312 \
["Ida"]=11111 ["That guy"]=122222)
#now we are going to print keys, once using @ and then using \
#*
echo "for first example we are using @ sign"
for name in "${!PAIRS[@]}"
do
        echo Number: ${PAIRS["$name"]}
        echo Name: "$name"
        echo
done
```

```
echo ------------------------
echo "then we are using * sign"
for name in "${!PAIRS[*]}"
do
        echo Number: ${PAIRS["$name"]}
              echo Name: "$name"
              echo
done
```

If we run this script and both of those expressions return the same set of values, the output should be identical. But if we actually run it, we get this:

```
demo@ubuntu:~/variable$ bash arrayops.sh
for first example we are using @ sign
Number: 113312
Name: Ivan from work
Number: 11111
Name: Ida
Number: 12344
Name: Luke
Number: 12345
Name: John
Number: 122222
Name: That guy
------------------------
then we are using * sign
Number:
Name: Ivan from work Ida Luke John That guy
```

We see that when we use @, we get what we expected, but as soon as we replace it with *, we can see the keys (or indices) but get no values in return.

The reason behind this is, as always, the way bash works. Using the @ sign signals to bash that we are trying to get each index or key separately. Using the * sign, on the other hand, makes bash return all the indices or keys as a single string divided by spaces. So, our script in one case reads each element one by one, and the other loop is run only once. Since the value that we are trying to look up is different from any single key value in our array, this one run gets no results. In the end, those two expressions are not identical, but the simple printout is—don't let that fool you.

Now, we are going to create our master script for manipulating arrays and then add a few elements to it.

The obvious starting point is to create a file containing our array. There are multiple ways to do that, some more complicated than others, but almost all of them depend on using some loop to go through the array and either read it or write it to a file.

We are going to do it the canonical Linux way by using the `declare` statement. If we do the `-p` switch, we are telling it to print a particular variable with both its definition and the values stored in it:

```
demo@ubuntu:~/variable$ declare -p PAIRS
declare -A PAIRS=(["Ivan from work"]="113312" [Ida]="11111"
[Luke]="12344" [John]="12345" ["That guy"]="122222" )
```

Obviously, this is great since this is the only thing we need to remember to have everything that is stored in the variable itself. To save it, we just redirect it to a file on disk:

```
demo@ubuntu:~/variable$ declare -p PAIRS > PAIRS.save
```

To show that this worked, we are going to unset the variable and verify it was deleted so that the values only exist in a file:

```
demo@ubuntu:~/variable$ unset PAIRS
demo@ubuntu:~/variable$ echo ${!PAIRS[@]}
```

Now, let's look at how to reload the script from a disk. Notice that the output of the `declare` statement was an actual `declare` statement defining the variable. If we load it from disk and execute it, everything should be fine:

```
demo@ubuntu:~/variable$ source PAIRS.save
demo@ubuntu:~/variable$ echo ${!PAIRS[@]}
Ivan from work Ida Luke John That guy
```

The internet is full of more complicated solutions, but for an array of a reasonable size, this should work great. At the same time, this is easy to read in a script, and the file that contains the data is in a universal format readable by any other `bash` version installed on any system (if it is `bash` 4.0 since that is when these types of arrays were introduced).

We now know how to read and write an array to disk, but what else can we do with it? In our example, we are going to switch to a regular array to show you some things you can do:

```
demo@ubuntu:~/variable$ REGULAR=( zero one two three five four \
)
demo@ubuntu:~/variable$ echo ${REGULAR[@]}
zero one two three five four
```

So, we have created an array and we made an error. Since the array is already in the wrong order, we are going to reshuffle it even more (the `shuf` command randomizes the array):

```
demo@ubuntu:~/variable$ shuf -e "${REGULAR[@]}"
three
one
zero
five
two
four
```

Although you could create your own solution for the randomization of values, using an external command is the simplest solution.

Shuffling is easy but it is not permanent. What the command is doing is taking our array as its input, shuffling the values, and then printing the result while the original array stays the same.

We reassigned the array to the result of the command. The main reason we also created another variable and printed it is that we had to show that shuffling happens in real time, and the results are different each time the `shuf` command is started.

Sorting an array is going to be more of a problem since it requires some sorting mechanism. Either you will create one yourself or you can, with a little care, use the `sort` command already included in `bash`.

The next thing we are going to do is show you how to work with ranges of indices. We are going to reset our array and then show you some examples. When we declare ranges, in reality, we are using a mechanism built into `bash` that enables us to define a range of numbers. We already used that to create loops and iterators in them, so this will not be too much of a surprise to you:

```
demo@ubuntu:~/variable$ REGULAR=( zero one two three four five \
)
demo@ubuntu:~/variable$ echo ${REGULAR[*]}
```

```
zero one two three four five
demo@ubuntu:~/variable$ echo ${REGULAR[*]:2:3}
two three four
demo@ubuntu:~/variable$ echo ${REGULAR[*]:0:3}
zero one two
demo@ubuntu:~/variable$ echo ${REGULAR[*]:0:}
demo@ubuntu:~/variable$ echo ${REGULAR[*]:0:2}
zero one
demo@ubuntu:~/variable$ echo ${REGULAR[*]:3}
three four five
demo@ubuntu:~/variable$ echo ${REGULAR[*]:2}
two three four five
demo@ubuntu:~/variable$ echo ${REGULAR[*]:2:-2}
bash: -2: substring expression < 0
```

We are using standard notation in bash. The first number after the variable is the starting index we want to print, and the optional number after it is the number of values we need. Notice that negative numbers do not work here unlike in some other places; we cannot go back from the end of the array this way.

The next thing we can do is concatenate two arrays.

Depending on what you want to do, the result of that operation will not get you what you might have been expecting:

```
demo@ubuntu:~/variable$ ANOTHER=(sixth seventh eighth ninth)
demo@ubuntu:~/variable$ NEW="${REGULAR[*]} ${ANOTHER[*]}"
demo@ubuntu:~/variable$ echo ${NEW[*]}
zero one two three four five sixth seventh eighth ninth
demo@ubuntu:~/variable$ echo ${NEW[@]}
zero one two three four five sixth seventh eighth ninth
```

Another operation we can do is count the number of values in an array. We also already did this before, so let's check if our arrays were merged correctly:

```
demo@ubuntu:~/variable$ echo ${#REGULAR[@]}
6
demo@ubuntu:~/variable$ echo ${#ANOTHER[@]}
```

```
4
demo@ubuntu:~/variable$ echo ${#NEW[@]}
1
```

What happened here?

In our concatenation, we made a huge error. What we wanted to do is create a new array that will hold the values from both arrays. What we did is create a `string` variable that contains one huge string created from all values in the arrays. We need to fix this by using brackets:

```
demo@ubuntu:~/variable$ NEW=(${REGULAR[*]} ${ANOTHER[*]})
demo@ubuntu:~/variable$ echo ${#NEW[@]}
10
demo@ubuntu:~/variable$ NEW=(${REGULAR[@]} ${ANOTHER[@]})
demo@ubuntu:~/variable$ echo ${#NEW[@]}
10
demo@ubuntu:~/variable$ declare -p NEW
declare -a NEW=([0]="zero" [1]="one" [2]="two" [3]="three"
[4]="four" [5]="five" [6]="sixth" [7]="seventh" [8]="eighth"
[9]="ninth")
```

So, we created our array and checked how many values it holds. Since we wanted to be completely sure, we used the `declare` statement to show all the index/value pairs.

Before we move on, we need to make a small mistake by using a pair of quotation marks:

```
demo@ubuntu:~/variable$ NEW=("${REGULAR[@]} ${ANOTHER[@]}")
demo@ubuntu:~/variable$ echo ${#NEW[@]}
9
demo@ubuntu:~/variable$ declare -p NEW
declare -a NEW=([0]="zero" [1]="one" [2]="two" [3]="three"
[4]="four" [5]="five sixth" [6]="seventh" [7]="eighth"
[8]="ninth")
```

What we did is create something extremely similar to our first example, but at the same time, completely wrong. One of the values is a combination of two strings, not two separate values. Try all different combinations of quotation marks to try to see how they work and if using either @ or * makes a difference to the resulting array.

How it works...

We have dealt with different things you can do to an array in detail. What's left is to see what else is there to know about arrays and how to check if an array contains a value. The length—or, more precisely, the number of values is something we just looked at, and if you need to know if an array already exists, you should check if the length of it is longer than 0. This will tell you that either the array you are testing is not defined or it contains no elements. If you explicitly want to check if the variable is defined, use the declare statement and count the results. In our example, we have a variable called TEST1 and an undefined name, TEST2:

```
demo@ubuntu:~/variable$ TEST1=()
demo@ubuntu:~/variable$ declare -p TEST1
declare -a TEST1=()
demo@ubuntu:~/variable$ declare -p TEST2
bash: declare: TEST2: not found
demo@ubuntu:~/variable$ echo ${#TEST1[@]}
0
demo@ubuntu:~/variable$ echo ${#TEST2[@]}
0
```

For most intents and purposes, just checking for the value count is enough, but sometimes you need to know if a variable is even defined.

Another common thing to do is try to find if there is a particular value inside an array. You can do this by creating your own loop to check for the values, or you can use the built-in test that bash already has. For example, you could do something like this:

```
demo@ubuntu:~/variable$ [[ ${REGULAR[*]} =~ "one" ]] && echo \
yes || echo no
yes
demo@ubuntu:~/variable$ [[ ${REGULAR[*]} =~ "something" ]] && \
echo yes || echo no
no
```

Once again, we used a one-line logical expression to quickly see the result.

Now, here's a small script that will show some of the things we learned in this recipe:

```
#!/bin/bash
#check if settings exist
```

```
                        function checkfile {
                if [ -f setting.list ]
                        then
                                            return 0
                        else.
                                        return 1
                                        fi
}
function assign_settings {
            echo assigning settings
            SETTINGS=(["USER"]=John ["LOCALDIR"]=$PWD \
["HOSTNAME"]=hostname)


}
declare -A SETTINGS
SETTINGS=()
if checkfile
then
            source setting.list
else
            assign_settings
fi
echo Settings are:
for name in "${!SETTINGS[@]}"
do
            echo "$name"=${SETTINGS["$name"]}
done
declare -p SETTINGS > setting.list
```

We already know most of this, but we will go through the script.

The first function in that script tests if a file exists. We could have done the same thing by using the if statement later in the code, but we wanted to remind you how to use functions and logical checks. The function returns 0 if the file is there, and 1 if it is not.

Next, we have the function we called assign_settings that is used if the file is not found. What it does is simply create a new associative array that contains some data.

Then, we are in the main body of code in our script, and first, we are declaring our arrays since they cannot be declared implicitly. Then, we decide if we have our file saved and if we should load our array from there or whether we need to reassign the defaults.

After that, we are just printing out the values and then saving them to disk.

In a normal script, this would be a part of the script that does the importing and saving of important settings. The rest of the script would be right before the line that saves the variables.

We are going to start the script two times in a row. The result should be that it will detect we have no configuration and make it for us:

```
demo@ubuntu:~/variable$ bash settings.sh
assigning settings
Settings are:
USER=John
HOSTNAME=hostname
LOCALDIR=/home/demo/variable
demo@ubuntu:~/variable$ bash settings.sh
Settings are:
USER=John
HOSTNAME=hostname
LOCALDIR=/home/demo/variable
```

When you are doing something such as this, we must also warn you that there may be big problems with local and global variables. Be very careful if you are declaring any variable that is supposed to be global in a function or—even worse—if you are sourcing it from a function since the scope will limit your values from propagating throughout the script.

This is where we will leave arrays and go on to more interesting stuff—starting to create some interfaces.

See also

- *How to use bash array in a shell script*: https://linuxconfig.org/how-to-use-arrays-in-bash-script

- *The Ultimate Bash Array Tutorial with 15 Examples*: https://www.thegeekstuff.com/2010/06/bash-array-tutorial/

14
Interacting with Shell Scripts

We are almost done with explaining the basic concepts of scripting, but before we can say we are completely done with them, we need to learn how to interact with shell scripts. This isn't always necessary in shell scripting, but it may apply to most situations. For example, it's one thing to create a script that does one job and one job only. It's completely different to create a script that requires us to make some choices as it gets executed. If nothing else, this second type is a prime candidate for shell script interaction. In this chapter, we are going to cover three different ways to deal with shell script interaction.

In this chapter, we will cover the following recipes:

- Creating text-based interactive scripts
- Using expect to automate repetitive tasks based on text output
- Using dialog for menu-driven interactive scripts

Technical requirements

As with all the chapters thus far, we are going to use the same virtual machine running the Bash shell. So, we need a virtual machine with Linux installed – any distribution is fine (in our case, it's going to be *Ubuntu 20.10*).

Now, start your virtual machine!

Creating text-based interactive scripts

The one thing that we haven't done so far is put any interaction in our scripts. The reason for this is simple – at this point, we've only discussed how to output information and not how to get it from the user or any other source. In the real world, interaction is something that we need to deal with because it is at the core of creating any script. We could say that there are two kinds of interaction. First, our script can interact with the system itself. This means using different variables and other information that we can get from the system – for example, free space in memory or on mounted disks. You could say that this is not real interaction but instead just reading real-time data from the system. But, still, it's a very useful way of making sure that a script does what it needs to do.

Another thing that we can do is interact with the user starting it. If the script is run by the system, it isn't something that is going to interact with the user in any way, but user interaction is extremely important when we are creating scripts for day-to-day jobs. Consider this question – why would we create a script that backs up *a folder* (one folder only) when we can create a script that can be told to back up *one or more directories*? Isn't that way of designing a script much more usable?

Getting ready

When we start creating our scripts, we must decide on the kind of interaction that we need in them. Depending on the type of script that's required, we may use interactive prompts, menus, some sort of pre-configuration, or even some graphical interface. For now, we are going to stay away from using GUIs for our scripts. However, we can use them if we need to with the help of some appropriate tools. Remember, scripts are barely more than some execution control that dictates how different commands and applications interact, so those commands are what matters most in the first place.

For starters, our first recipe will be a simple interactive script asking the user for input and then acting on it.

How to do it...

The main commands that we are going to use in this recipe are going to be `read` and `echo`. Before we do anything else, we need to learn a few tricks regarding these commands. In theory, `read` is simple to understand – it waits for user input and then stores that input in some variable. But to show you the different things that are made possible by this simple command, we need to show you a few examples.

`read`, in its basic form, accepts one argument – the variable – and then takes whatever the user types in and puts it into this variable so that we can use it later. Let's consider the following example:

```
#!/bin/bash
echo "Input a value: "
read Value1
echo "Your input was: $Value1"
```

If we quickly test this script, this is what we will get as a result:

```
demo@cli1:~/interactive$ bash singlevar.sh
Input a value:
test
Your input was: test
demo@cli1:~/interactive$ bash singlevar.sh
Input a value:
test value
Your input was: test value
```

Sometimes, this is not enough. Sometimes, we need to get more than one value into our script. The problem here is the way users type in the values. Shell uses space characters as separators, so a space is going to be what separates the value in the `read` command. If we need to get a value that contains spaces, we will have to deal with it differently. As we saw in the previous example, this will only become a problem if we use more than one return variable.

If, however, we use values that do not contain spaces, we can simply use the following code and save it in the `doublevar.sh` file:

```bash
#!/bin/bash
echo -e "Input two numbers "
read num1 num2
echo "Two numbers are $num1 and $num2"
```

Now, let's try it and see if it works the way we are expecting it to:

```
demo@cli1:~/interactive$ bash doublevar.sh
Input two numbers
2 3
Two numbers are 2 and 3
```

We must stop here and do another test to clear a few things up. Bash performs no checks on what the type of the variable is. In our script, we presumed that the user is going to input numbers, but nothing stops them from using any string. Another thing would be how multiple values separated by spaces are going to be handled – the first value is going to be assigned to the first variable; everything after that is going to be assigned to the second one. If we use more than two variables to store values, the result is always going to be that each variable in the sequence will get one variable assigned and the last one will get whatever was left in the input line:

```
demo@cli1:~/interactive$ bash doublevar.sh
Input two numbers
First second
Two numbers are First and second
demo@cli1:~/interactive$ bash doublevar.sh
Input two numbers
first second third
Two numbers are first and second third
```

Another way we can use `read` is to get the values into a predefined variable called `$REPLY`. If we simply omit the variable name, everything you type in is going to be in that variable, which can then be used in your script:

```bash
#!/bin/bash
echo "Input a value: "
Read
echo "Your input was: $REPLY"
```

A simple test proves that this behaves exactly as if we gave the command a proper variable name:

```
demo@cli1:~/interactive$ bash novar.sh
Input a value:
test value no variable
Your input was: test value no variable
```

Another way of using `read` is by using the `-a` switch. By using this, we are saying that we want to store all the values it got as an array. After this switch, we need to state the name of the variable we are going to use to store the values, or we can simply use the default `$REPLY` variable. What we should not do is use more than one variable name. This is because we are storing multiple values in one variable, so it makes no sense to try and reference more than one variable in the first place:

```
#!/bin/bash
echo "Input multiple values: "
read REGULAR
echo "Your input was: $REGULAR"
echo "This will not work: ${REGULAR[0]}"

echo "Now input multiple values again:"
read -a REGULAR
echo "This will work: ${REGULAR[0]}"
```

Here, we are reading multiple values into a single variable. Since we didn't ask Bash to create an array, it is going to store everything into this variable, but there will be no way of referencing the elements inside this variable. Bash treats all the values in the variable as a single, first element, so if we try to print it out, we will get everything.

The second time we do this, we are getting the values from the user. Here, we are using an array. Everything looks the same but if we reference the first element of the variable, we will only print the first element:

```
demo@cli1:~/interactive$ bash array.sh
Input multiple values:
first second third
Your input was: first second third
```

```
This will not work: first second third
Now input multiple values again:
first second third
This will work: first
```

Now, we need to experiment a bit with the echo command. In the entire scripting part of this book, we have been using this command in its most basic form to, well, output text to screen or, to be more precise, to standard output. This will work for the majority of cases, but there are some scripts where we need more control over the output. The problem with the way echo works is that it always terminates the string it outputs with a newline character, forcing the output into a new line once it's printed whatever was given to it as a parameter. While this is alright for printing out information, when we try to interact with the user in our scripts, it will look strange if we always force the user to enter the values we are looking for in a new line. So, echo offers one additional option that changes the default behavior (-n). Let's consider this example:

```
#!/bin/bash
echo -n "Can you please input a word?: "
read  word
echo "I got: $word"
```

What we told echo to do here is that it should print the text inside quotation marks, but after that, it should stay in the same line. Since our read command naturally continues wherever the cursor was placed by the previous command, the result will be that the value we type in will appear most logically on screen:

```
demo@cli1:~/interactive$ bash echoline.sh
Can you please input a word?: singleword
I got: singleword
```

There is another way to do this that looks more complicated but behaves the same:

```
#!/bin/bash
echo -e "Can you please input a word?:  \c "
read  word
echo "I got: $word"
```

The reason we are showing you this is because this example uses special characters to denote the end of the line, but at the same time, there are more characters we can use for even finer control over the output. By default, the most commonly used are as follows:

- `\\`   backslash: When we need to output the actual \ character.

- `\a`   alert (BEL): When we want to warn the user by using a loud sound.

- `\b`   backspace: We use this when we need to provide a backspace character, deleting whatever is on the same line under the cursor.

- `\c`   produce no further output: This is used to tell echo to simply stop the output.

- `\t`   horizontal tab: This is used to provide tab and align the output.

Now that we know about some of the basic read and echo syntaxes, let's try to put that to good use in a script.

How it works...

Now that we know how to deal with everything in the Bash script that can be used for interaction, we can create a script that will show it all. Here, we are creating a simple menu:

```bash
#!/bin/bash
echo "Your favourite scripting language?"
echo "1) bash"
echo "2) perl"
echo "3) python"
echo "4) c++"
echo "5) Dunno!"
echo -n "Your choice is: "
read choice;
# we do a simple case structure
case $choice in
    1) echo "You chose bash";;
    2) echo "You chose perl";;
    3) echo "You chose python";;
    4) echo "You chose c++";;
    5) exit
esac
```

Try it out!

There is another way to create an interactive menu that we can use in our script – using the `select` command. This command is often used to create simple menus, like this:

```bash
#!/bin/bash
PS3='Please choose an option: '
options=("Option 1" "Option 2" "Option 3" "Quit")
select opt in "${options[@]}"
do
    case $opt in
        "Option 1")
            echo "you chose Option 1"
            ;;
        "Option 2")
            echo "you chose Option 2"
            ;;
        "Option 3")
            echo "you chose Option $REPLY which is $opt"
            ;;
        "Quit")
            break
            ;;
        *) echo "invalid option $REPLY"
            ;;
    esac
done
```

`select` looks a lot like some sort of loop; it requires us to set a few variables in advance. `$PS3` contains the question that the user will see, while `$options` contains an array of strings that represent options. When a user runs this script, it will be presented with a list of numbered options, and they can input any of them as a number. `select` is going to then substitute the string from our options list based on the number that the user selected and run the appropriate command:

```
demo@cli1:~/interactive$ bash select.sh
1) Option 1
2) Option 2
3) Option 3
```

```
4) Quit
Please choose an option: 1
you chose Option 1
Please choose an option: 2
you chose Option 2
Please choose an option: 3
you chose Option 3 which is Option 3
Please choose an option: 4
demo@cli1:~/interactive$
```

This is a great way to quickly create a script with multiple choices.

See also

A lot more can be found about the `echo` and `read` commands at the following links:

- `https://linuxhint.com/bash_read_command/`
- `https://www.javatpoint.com/bash-read-user-input`

Using expect to automate repetitive tasks based on text output

Bash is a formidable tool but sometimes, we need to do a particular thing that needs additional tools. In this recipe, we are going to be working with just such a tool called `expect`. Before we start, we must note that `expect` is not part of Bash scripting – it is a whole separate scripting language, written for a particular purpose, to enable interaction between your scripts and users and other systems. The idea behind it is to enable your scripts to not only execute `normal` commands that provide information when executed (command output) but to also be able to interact with any application that has a **command-line interface** (**CLI**) and get information from it.

Getting ready

In a simplified way, `expect` acts as a virtual keyboard that can type in some text and read what is on the screen. This is a powerful thing that is often needed because several applications and scripts are created by people who either had no reason to enable scripting support or just didn't want to do it. This means that without a tool such as `expect`, we will not be able to interact with those applications. This sometimes means that we will not be able to do what we want from inside our scripts.

In this recipe, we will learn how to use expect to interact with another shell on another computer, how to type in a password and log in, and how to type commands and get a response from the other side.

But before we even do that, we need to install expect if it isn't installed on the system since it is not a standard part of the system.

Use the following command and wait for it to finish:

```
sudo apt install expect
```

We need to use sudo here. Only an administrator can install packages, so expect to have to input your user password.

How to do it...

Since expect is not Bash, we must tell our script to use it. The syntax is the same as when we're creating a script using Bash – we need to make running expect the first line of our script. Note that this immediately means that our script no longer uses any of the commands from Bash, but we get many new things we can do.

We are going to start with the simple hello world script:

```
#!/usr/bin/expect
expect "hello"
send "world"
```

What the script does is exactly what the commands sound like they do; when we start it, the script is going to look for a hello string and after it receives it, it is going to reply with the world string.

Strings are case-sensitive, so nothing else than the exact match will work. Also, expect has a built-in period during which it expects to get the string. If nothing is matched during that period, the script is going to continue from the next command. In our example, even if we give it no input, we are going to get the world string printed.

When we start our script, we must use the expect command; we cannot start this script using Bash since it is written specifically for expect:

```
demo@cli1:~/interactive$ expect expect1.exp
HEllo
helo
hello
world
```

We tried three different ways of spelling the string here, and only the exact match worked.

Let's do something more interesting and explain what we did along the way:

```
#!/usr/bin/expect
set timeout 20
set host [lindex $argv 0]
set user [lindex $argv 1]
set password [lindex $argv 2]
spawn ssh "$user\@$host"
expect "Password:"
send "$password\r";
interact
```

This script is intended to enable you to quickly connect to another host using the ssh protocol. When we run it, we need to provide three things: the name or IP address of the host, the username to use, and the cleartext password for the user that will log in. We are aware that using a password in this way is not normal at all, but we are providing an example here.

At the start of the script, we are setting the timeout for the prompts. As we mentioned earlier, if the expect command doesn't detect any input, it will continue the script after the time specified in this timeout value. The next three lines deal with the arguments we passed to the script. We are assigning each of them to a variable so that we can use them later.

After that, we are using the spawn command to call the ssh command in a separate process. We are using the standard ssh client and giving it usernames and hostnames so that we can start the login process.

After this, our script waits until it detects that we need to type in the password. When it detects the password: part of the prompt, it sends our password in cleartext.

The last command in the script is interact and it hands over control to us so that we can use our freshly logged-in session to do what we intend to do. This is what it looks like when it's run:

```
demo@cli1:~/interactive$ expect sshlogin.exp localhost demo
demo
spawn ssh demo@localhost
demo@localhost's password:
Welcome to Ubuntu 20.10 (GNU/Linux 5.8.0-63-generic x86_64)
```

```
   Documentation:    https://help.ubuntu.com

 * Management:        https://landscape.canonical.com

 * Support:           https://ubuntu.com/advantage

 0 updates can be installed immediately.
 0 of these updates are security updates.
 demo@cli1:~$
```

This is a good start if we need to work on a remote system. But how do we run other commands on the other system and what can we do with that?

There are two ways to continue here; one is to simply wait for the prompt to show and then send commands, while another one is to script this *blindly* and just wait for a predetermined period and then send the commands that we need. So, let's make an example script for both concepts. Our goal is to have a scenario in which one part of the script does its job as an *answer* to command output (in our script, this is the ssh part and its output). The second part is related to the concept of waiting for a predetermined period (sleep 5 means waiting for 5 seconds) and then doing something. Let's check out our example:

```
#!/usr/bin/expect
set timeout 20
set host [lindex $argv 0]
set user [lindex $argv 1]
set password [lindex $argv 2]
spawn ssh "$user\@$host"
expect "password:"
send "$password\r";
sleep 5
send "clear\r";
send "ip link\r";
expect "$"
puts $expect_out(buffer);
send "exit\r";
```

We added two things that we haven't mentioned previously. The first one is that we are using the `puts` command to print information to the screen. It behaves similar to `echo` does in Bash.

The `$expect_out (buffer)` variable holds data that the script got from running commands between two matches. So, in our script, this is going to hold information that was provided by the `ip add` command. If you are wondering where the login information disappeared, it is not visible since we issued the `clear` command to clear the screen, which, in turn, cleared the buffer.

How it works...

`expect` is an amazing tool that has even more things up its sleeve. One of its main purposes is to automate administration tasks using scripts. One of the most common usages was what we did in the previous example – running commands remotely. The reason we may use it often is not only to automate logins but also to automate testing. After running a particular command, we can do whatever we want and then get the results into our scripts.

In this way, `expect` is usually used as part of another script. When we need some sort of data that only `expect` can provide, we call it and then continue processing the data in our main script. However, automating scripts can mean one more thing – getting them to accept information, which enables us to write tests to check if our script works. For this reason, `expect` has a tool that is used to create an `expect` script out of running any script that contains interaction with the user.

The following is a quick example to demonstrate what we mean, but first, we are going to create a very simple script where we will use different ways of calling `echo` and `read` in Bash:

```bash
#!/bin/bash
#echo -e "Can you please input a word?:  \c "
echo -n "Can you please input a word?:    "
read  word
echo "I got: $word"
echo -e "Now please input two words: "
read word1 word2
echo "I got: \"$word1\" \"$word2\""
echo -e "Any more thoughts? "
# read will by default create $REPLY variable
read
```

```
echo "$REPLY is not a bad thing "
echo -e "Can you give me three of your favorite colors? "
# read -a will read an array of words
read -a colours
echo "Amazing, I also like ${colours[0]}, ${colours[1]} and \n
${colours[2]}:-)"
```

Now, we are going to start the `autoexpect` tool to grab both the input and output for the script:

```
demo@cli1:~/interactive$ autoexpect bash simpleecho.sh
autoexpect started, file is script.exp
Can you please input a word?:    word one
I got: word one
Now please input two words:
two words
I got: "two" "words"
Any more thoughts?
none whatsoever
none whatsoever is not a bad thing
Can you give me three of your favourite colors?
blue yellow cyan
Amazing, I also like blue, yellow and cyan
autoexpect done, file is script.exp
```

Here, `autoexpect` was tracking what our script used as prompts, as well as what we gave as answers. When we finished executing our script, it created an `expect` script, which enables us to completely automate running our Bash script. We are going to omit part of this script since it contains a lot of comments that give us both the information about the tool and the disclaimer. Here's what the `autoexpect` output looks like; it is saved in a file named `script.exp`:

```
#!/usr/bin/expect -f
# This Expect script was generated by autoexpect on Sun Oct \n
10 13:35:52 2021


set force_conservative 0   ;# set to 1 to force conservative \n
mode even if

                                ;# script wasn't run \n
```

```
conservatively
          Originally
                if {$force_conservative} {
                        set send_slow {1 .1}
                        proc send {ignore arg} {
                        sleep .1
                        exp_send -s -- $arg
              }
}
set timeout -1
spawn bash simpleecho.sh
match_max 100000
expect -exact "Can you please input a word?:    "
send -- "word one\r"
expect -exact "word one\r
I got: word one\r
Now please input two words: \r
"
send -- "two rods "
expect -exact "
send -- ""
expect -exact "
send -- ""
expect -exact "
send -- ""
expect -exact "
send -- ""
expect -exact "
send -- "words\r"
expect -exact "words\r
I got: \"two\" \"words\"\r
Any more thoughts? \r
"
send -- "none whatsoever\r"
expect -exact "none whatsoever\r
none whatsoever is not a bad thing \r
```

```
Can you give me three of your favorite colors? \r
"
send -- "blue llow and "
expect -exact "
send -- ""
expect -exact "
send -- ""
expect -exact "
send -- ""
expect -exact "
send -- "cz"
expect -exact "
send -- "yan\r"
expect eof
```

This script is a good starting point for automating our work, but be careful of one fatal flaw. In the last part that checks the inputs and outputs, it logged every typo and error that we made during input, but it hasn't saved the entire process of deleting them, so this input will not work. Before we can use it, we must edit this script and sort all the errors and inputs out. After that, we can expand on it and use additional arguments to test how our script is going to behave.

See also

expect is amazing for testing. For more examples, please go to the following links:

- Expect command: https://likegeeks.com/expect-command/
- Expect man page: https://linux.die.net/man/1/expect

Using dialog for menu-driven interactive scripts

Now that we've used expect, we know how to interact with other applications. The only thing left to do is learn how to make our scripts more interactive. It will come as no surprise that this problem is already solved in a standard way. In this recipe, we will use dialog, a command that may look deceptively simple but enables you to create both complex and visually interesting interactions with end users.

Getting ready

By definition, `dialog`, as with any interaction, makes your scripts unusable in a non-interactive environment. This can be solved by either not using `dialog` at all or detecting if the script is running as a service or as an interactive script.

Like `expect`, we must install `dialog` to use it. Simply use the following command:

```
sudo apt install dialog
```

Everything that you need is going to be installed as required.

How to do it...

`dialog` is a whole application that contains not only menus but a lot of other widgets that are displayed in a GUI under text mode. It will use colors if they are available on your terminal (they probably are – terminals that are unable to show colors are long gone in our day and age) and will use cursor keys for navigation. We are going to show you a few of the most common ones, as well as how to use them.

For a start, try running this command:

```
dialog --clear --backtitle "Simple menu" --title "Available \n
options" --menu "Choose one:" 16 50 4 1 "First" 2 "Second" 3 \n
"Third"
```

If everything has gone well, you should see something like this (the size will depend on your terminal window):

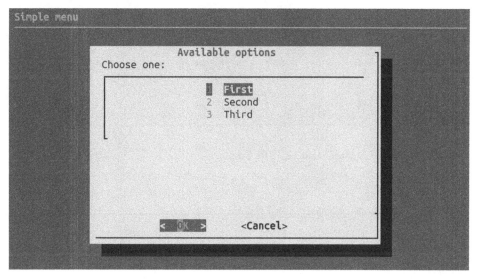

Figure 14.1 – The dialog command enables us to create menus in a single line of our script

dialog acts as a good-looking proxy between your script and the user. Your script is responsible for the logic of the process you are trying to automate; dialog is responsible for dealing with user input and output. We need to do something with this menu; if we just call it from our Bash prompt, we have no use for it. The reason we used it this way was to show you how to call a widget. A lot of people expect a complicated procedure when they first see dialog being used in a script, which is just a command in one line and a couple of arguments. To demonstrate this, let's create an actual menu and save it in a file named menu.sh:

```bash
#!/bin/bash
HEIGHT=16
WIDTH=50
CHOICE_HEIGHT=4
BACKTITLE="Simple menu"
TITLE="Available options"
MENU="Choose one:"
OPTIONS=(1 "First" 2 "Second" 3 "Third")
CHOICE=$(dialog --clear \
                --backtitle "$BACKTITLE" \
                --title "$TITLE" \
                --menu "$MENU" \
                $HEIGHT $WIDTH $CHOICE_HEIGHT \
                "${OPTIONS[@]}" \
                2>&1 >/dev/tty)
clear
case $CHOICE in
        1)
            echo "You chose First"
            ;;
        2)
            echo "You chose Second"
            ;;
        3)
            echo "You chose Third"
            ;;
esac
```

What are we doing here? `dialog` requires a couple of parameters for every widget it can display. It needs a lot of values to correctly show whether something comes from your terminal. These are values such as the width and height of the screen and how to display the output correctly. As a rule, widgets only require user-defined things – the height and width of the widget itself, titles and other strings that are used in the widget, and the choices the user has. They are different from widget to widget, depending on the way it works and is used. In our first example, we are using a menu widget that requires a list of options and the size of the menu. We needed this list of options as it helps the user who's starting this script make a correct choice. We are also providing it with the titles, although we don't need all of them for our menu to function.

An interesting thing to notice here is the way `dialog` returns the value that the user chose. To get that, we are providing indices in our list of possible options. When we run `dialog`, we are going to assign the value it got from the user directly to a variable and then act accordingly.

How it works...

`dialog` is based on displaying different widgets using different graphical characters to give the feel of a graphical interface inside the normal terminal. As somebody who uses dialog, you can change almost all the elements' appearance if you want, but usually, most scripts simply use `dialog` to quickly get the user to input the data they require.

Using something like `dialog` is a great way to enable the user to choose values for different things in their script while avoiding a lot of input errors.

For another example, let's imagine we need to ask a user to provide us with a date. We can do this using a simple entry with the `read` and `echo` commands. This will work but with a big risk of the user using the wrong format. You could solve this by explaining to your users what format you expect, but they will inevitably forget and use the wrong one.

In `dialog`, you could do something simple, like this:

```bash
#!/bin/bash
DATEPROMPT="Choose a date"
CHOSENDATE=$(dialog --stdout --calendar "$DATEPROMPT" 0 0)
echo "Chosen date is $CHOSENDATE"
```

The following is the output:

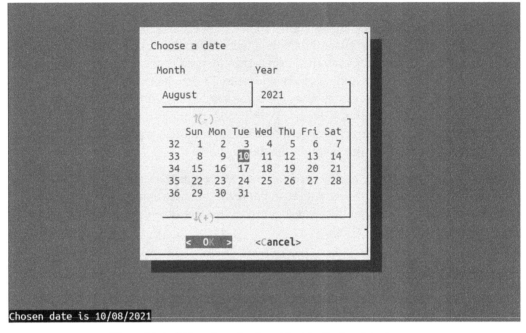

Figure 14.2 – Using the calendar widget to display dates is easy

The date we are going to get will be in the right format, depending on the regional setting of the system running the script, and the user can choose from several good-looking calendars. `dialog` helps a lot here since it will even jump to the current date if we do not specify which specific date is going to be the default. Depending on the type of terminal emulation, `dialog` may even support using a computer mouse to select data.

There are more very useful widgets that `dialog` provides, so we are going to show a few of them. We are not going to create a script for all of them since they are simple to use. The simplest way is to use the `--stdout` option to get the result of `dialog` into a variable and then work from there.

If we need to choose a directory, that can be a big problem since the user would usually forget to write the path correctly or use absolute paths instead of relative ones. By using a simple `dialog command`, such as the following, we can avoid a lot of problems:

```
dialog --dselect ~ 10 39
```

This is the expected result:

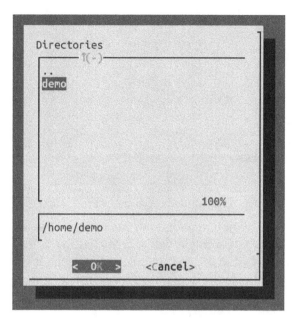

Figure 14.3 – Choosing directories using dialog avoids a lot of errors

A simple `dialog` is useful when we need to get a simple *yes* or *no* answer from the user. We can use the following code to do so:

```
dialog --yesno "Do you wish to do it?" 0 0
```

Using this `dialog` command, we should get the following output:

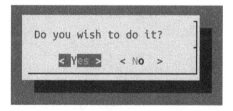

Figure 14.4 – Sometimes, you need to ask a simple question that will have a simple answer

Then, we have more informative ones. In our scripts, we are often having to present the user with some information that they need to read. Having it in a formatted box is much nicer than simply printing it in a terminal:

```
dialog --msgbox "A lot of text can be displayed here!" 10 30
```

This example will create the following output:

Figure 14.5 – Displaying text is easy and effective when you're using the right widget

There are a few widgets we will not mention as we are trying to give you a quick glimpse into what is possible. For the last example, we are going to show you a useful one – this widget shows the contents of a text file and automatically updates it with changes, enabling you, for example, to show a log to the user while they're installing something:

```
dialog --tailbox /var/log/syslog 40 80
```

This is the expected output:

```
Oct 10 15:24:55 ubuntu tracker-extract[8801]: Setting priority nice level to
Oct 10 15:24:56 ubuntu dbus-daemon[2405]: [session uid=1000 pid=2405] Succes
Oct 10 15:24:56 ubuntu systemd[2392]: Started Tracker metadata extractor.
Oct 10 15:25:12 ubuntu systemd[2392]: tracker-extract.service: Succeeded.
Oct 10 15:25:27 ubuntu dbus-daemon[2405]: [session uid=1000 pid=2405] Activa
Oct 10 15:25:27 ubuntu systemd[2392]: Starting Tracker metadata extractor...
Oct 10 15:25:27 ubuntu tracker-extract[8848]: Set scheduler policy to SCHED_
Oct 10 15:25:27 ubuntu tracker-extract[8848]: Setting priority nice level to
Oct 10 15:25:27 ubuntu dbus-daemon[2405]: [session uid=1000 pid=2405] Succes
Oct 10 15:25:27 ubuntu systemd[2392]: Started Tracker metadata extractor.
Oct 10 15:25:44 ubuntu systemd[2392]: tracker-extract.service: Succeeded.
Oct 10 15:25:46 ubuntu dbus-daemon[2405]: [session uid=1000 pid=2405] Activa
Oct 10 15:25:46 ubuntu systemd[2392]: Starting Tracker metadata extractor...
Oct 10 15:25:46 ubuntu tracker-extract[8899]: Set scheduler policy to SCHED_
Oct 10 15:25:46 ubuntu tracker-extract[8899]: Setting priority nice level to
Oct 10 15:25:46 ubuntu dbus-daemon[2405]: [session uid=1000 pid=2405] Succes
Oct 10 15:25:46 ubuntu systemd[2392]: Started Tracker metadata extractor.
Oct 10 15:25:58 ubuntu systemd[2392]: tracker-extract.service: Succeeded.
Oct 10 15:26:18 ubuntu tracker-store[8789]: OK
Oct 10 15:26:18 ubuntu systemd[2392]: tracker-store.service: Succeeded.
Oct 10 15:30:01 ubuntu CRON[8962]: (root) CMD ([ -x /etc/init.d/anacron ] &&

                              < EXIT >
```

Figure 14.6 – tailbox is the same as using tail -f on a file but is much better looking

This concludes our overview of what `dialog` can do. Make sure that you read the documentation to learn about the rest of the widgets that you can use, as well as how to use them correctly.

Note that there are other implementations of the same idea. One of those is called *whiptail* and is the same as `dialog` but uses a different way to draw objects on the screen. However, it is not as complete as `dialog`, and it lacks some widgets compared to `dialog`.

See also

- `dialog` man page: `https://linux.die.net/man/1/dialog`
- Menu box guide: `https://bash.cyberciti.biz/guide/A_menu_box`

15
Troubleshooting Shell Scripts

If you have come this far, you must have a lot of ideas about how to write a shell script, and even more questions about the ways you can make particular things in scripts work. This is completely normal. Your scripting journey has just started. No amount of reading can make up for time spent writing scripts, trying out different solutions, and understanding how different commands work.

We have some good news and some bad news for you. Being good at scripting takes a long time and, in scripting, most of that time is going to be spent trying to understand what your script should be doing and, usually, why it is doing it wrong. The good news is that scripting is never boring.

In this chapter, we will try to give you the tools needed to debug and troubleshoot scripts quickly and without a lot of confusion. The tools are going to be in the form of different methods you can use to maximize your ability to find logical and, sometimes, syntactical errors in scripts. We are going to start with the basic recipes and go on to more complex ways to work on scripts.

We will cover the following recipes in this chapter:

- Common scripting mistakes
- Simple debugging approach – echoing values during script execution
- Using the `bash -x` and `-v` options
- Using `set` to debug a part of the script

Technical requirements

In this chapter, we are going to use the same machine as in all the previous chapters on scripting in this book. Do not be alarmed that there are a couple of screenshots that are made in Windows. They are there just to illustrate a point; you don't need Windows to do anything. Just like earlier, we are using the following:

- A virtual machine with Linux installed, any distribution (in our case, it's going to be *Ubuntu 20.10*)

Now, let's dive into troubleshooting.

Common scripting mistakes

Writing a script will present many problems, including how to design it, how to find the right solutions to different problems, and how to make all of this usable in the target environment. These can be things you can easily solve in a couple of minutes, or things you will spend days or even weeks trying to solve. All this time will probably just be a small percentage of the total time you will spend debugging and troubleshooting scripts. Writing and troubleshooting scripts are two wholly different things – while you usually write your own scripts from scratch, you will not only debug and troubleshoot your own code.

Writing requires skill and deep knowledge of your environment, but it can be argued that to debug and troubleshoot, you need even more understanding of both your task and the way your script is trying to accomplish it. In this recipe, we are going to work on the skills you need to understand not only how to troubleshoot scripts you have written, but also any scripting code you run into, whether it's a part of something you've created or a separate system created by someone else that you are responsible for getting running.

Getting ready

Troubleshooting and debugging sound and look like the same task, but they are subtly different. In general, when we are debugging, we are concentrating on finding logical and other errors in our scripts. When troubleshooting, we are not only debugging but also trying gain more understanding of what should be done and how your application is trying to accomplish it. In the recipes in this chapter, we are going to use both these expressions for one thing – trying to make something that is not working correctly work as it should, or at least better.

There are a few things you can do to make this as easy as possible, and one of them is to get as much knowledge as possible about scripting under your belt. Being able to understand scripts and the specific ways things are done to solve them will enable you to quickly understand not only what the problem is but also how to solve it.

Sometimes the solution may be to simplify a part of code using a standard solution, or break code into different, more standardized modules.

When faced with a more complex script, this method of breaking code down into more manageable and understandable modules can be amazingly successful since even the greatest script-writers sometimes completely miss the point of what they are doing and complicate even the simplest tasks.

When talking about debugging, we need to talk about errors. Broadly, we can have four different outcomes from our script:

- The script works as desired.
- The script throws an error.
- The script works but not completely, making errors.
- The script works but sometimes silently breaks something either in the input or output data.

When we have the time, we can work on any of these possible outcomes and make a script behave better, even one that works correctly. Sometimes it pays to spend some time to make your script more beautiful, more commented, and more readable, even if it works alright.

Another case is scripts throwing an error. Bash has a reputation for having cryptic and generic error messages. Some of them are too vague to be of much assistance, and sometimes they make no sense at all and don't help us to understand what is actually wrong.

One common example that you will notice from time to time is not understanding ends of lines correctly. Windows and Linux treat ends of lines differently. While Windows terminate text files using both the carriage return and new line characters, Linux only uses new line characters for line termination. Bash can have a problem with that, and scripts written on Windows will sometimes break for no apparent reason. This is the same script on Windows and Linux, with an editor that shows all the characters in a file:

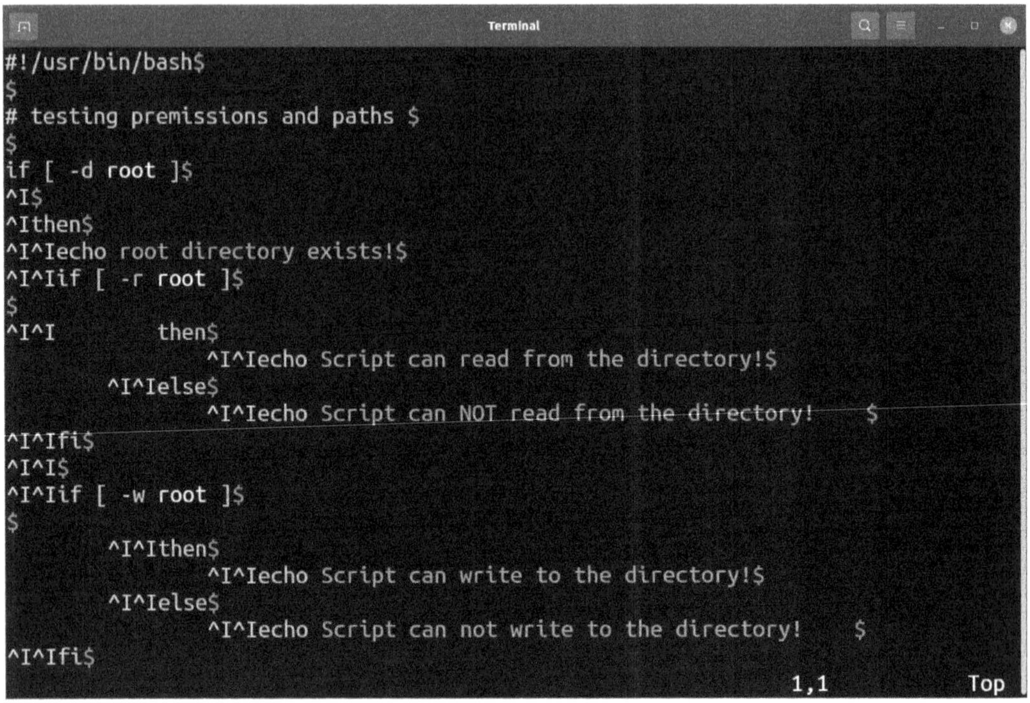

Figure 15.1 – In Linux, lines are terminated by a single character

In Windows, it looks similar, but we can see that line ends have two characters:

```
 1  #!/usr/bin/bashCRLF
 2  CRLF
 3  # testing premissions and paths CRLF
 4  CRLF
 5  if [ -d root ]CRLF
 6  CRLF
 7      thenCRLF
 8          echo root directory exists!CRLF
 9          if [ -r root ]CRLF
10  CRLF
11              thenCRLF
12                  echo Script can read from the directory!CRLF
13              elseCRLF
14                  echo Script can NOT read from the directory!    CRLF
15          fiCRLF
16  CRLF
17          if [ -w root ]CRLF
18  CRLF
19              thenCRLF
20                  echo Script can write to the directory!CRLF
21              elseCRLF
22                  echo Script can not write to the directory!    CRLF
23          fiCRLF
24  CRLF
25  CRLF
26      elseCRLF
27          echo root directory does NOT exists!CRLF
28  fi
```

Figure 15.2 – In Windows, two characters are used to terminate a line

In Linux, if we do not edit the file correctly and forget to strip out extra characters, we will end up with characters that will be invisible in a normal editor and break the code at the same time:

```
Terminal                                                    Q  ≡  _  □  ⊗
#!/usr/bin/bash^M$
^M$
# testing premissions and paths ^M$
^M$
if [ -d root ]^M$
^I^M$
^Ithen^M$
^I^Iecho root directory exists!^M$
^I^Iif [ -r root ]^M$
^M$
^I^I        then^M$
                ^I^Iecho Script can read from the directory!^M$
        ^I^Ielse^M$
                ^I^Iecho Script can NOT read from the directory!    ^M$
^I^Ifi^M$
^I^I^M$
^I^Iif [ -w root ]^M$
^M$
        ^I^Ithen^M$
                ^I^Iecho Script can write to the directory!^M$
        ^I^Ielse^M$
                ^I^Iecho Script can not write to the directory!    ^M$
^I^Ifi^M$
                                                          1,1              Top
```

Figure 15.3 – In vim, you need to turn on a couple of options to see special characters

Note that there is a utility called `dos2unix` (and `unix2dos` if you need conversion the other way around) that fixes ends of lines when transferring files. This problem is system wide and more than a couple of programs will behave strangely when they encounter text files from Windows.

If we do not deal with characters at the ends of lines, the script will break. For example, we tried running the file that came from Windows in Linux, we get completely cryptic errors mentioning commands that look like they are not even in the script. We are using `dos.sh` as the name of the script that we saved in Linux:

```
demo@ubuntu:~/Desktop/allscripts$ bash dos.sh
dos.sh: line 2: $'\r': command not found
dos.sh: line 4: $'\r': command not found
dos.sh: line 26: syntax error near unexpected token 'else'
'os.sh: line 26: '    else
```

Also, we need to make it clear that if you use copy and paste to move files between your environments, this will fix the problem directly. When you paste a line in a particular operating system, it will automatically create the right line endings. This does not cover copying and pasting the entire file; if you do that, you are transferring the entire file's content.

Another common syntax error that can be difficult to find is using the wrong quote, either by mixing them up or using a quote instead of a backtick character when executing commands inside scripts:

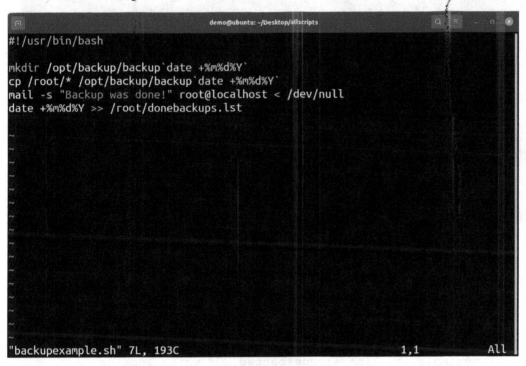

Figure 15.4 – Completely normal script, highlighted by syntax in vim

If we change one backtick to a quote, this will turn into the following:

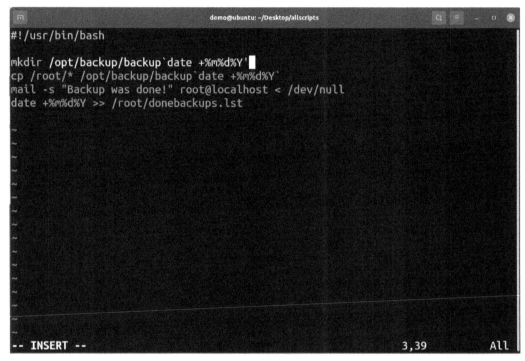

Figure 15.5 – Without proper highlighting, an error like this can cause serious problems

We are using vim here, and you will notice the change right away. The editor understands syntax and highlights the appropriate code block in a different color.

If we try to run this script, it will throw an error:

```
demo@ubuntu:~/Desktop/allscripts$ bash backupexample.sh
backupexample.sh: line 4: unexpected EOF while looking for
matching '''
backupexample.sh: line 8: syntax error: unexpected end of file
```

This error is a little confusing. Bash is telling us that it got to the end of the file while trying to find the closing quote.

One thing all of these errors have in common is using different fonts in different editors. Sometimes, the difference between characters is so minor that it is extremely hard to spot. Bash makes it even harder by reporting errors that are sometimes pointing to a completely different part of the code.

The solution to this is using a font you know is legible and using an editor that is able to pair characters such as parentheses. Quotes and backticks will probably remain a problem since most applications are unable to match them. Editors such as vim will, however, highlight comments and, as we saw in the previous example, this will make errors such as this visible.

This is an example of highlighted brackets in Notepad++ on Windows, since we mentioned the multi-platform approach:

```
 1  #!/usr/bin/bash
 2
 3  # testing premissions and paths
 4
 5  if [ -d root ]
 6
 7       then
 8             echo root directory exists!
 9             if [ -r root ]
10
11                  then
12                        echo Script can read from the directory!
13                  else
14                        echo Script can NOT read from the directory!
15             fi
16
17             if [ -w root ]
18
19                  then
20                        echo Script can write to the directory!
21                  else
22                        echo Script can not write to the directory!
23             fi
24
25
26       else
27             echo root directory does NOT exists!
28  fi
```

Figure 15.6 – Highlighting brackets in Notepad

Of course, we also have our standard run-of-the-mill syntax errors that are inevitable. A good editor will also help with these:

```
#!/usr/bin/bash
# testing premissions and paths

if [ -d root ]

        thn
                echo root directory exists!
                if  -r root

                        then
                                echo Script can read from the directory!
                        else
                                echo Script can NOT read from the directory!
                fi

                if [ -w root ]

                        then
                                echo Script can write to the directory!
                        else
                                echo Script can not write to the directory!
                fi
        else
                echo root directory does NOT exists!
fi
-- INSERT --                                                    8,7-21          All
```

Figure 15.7 – Having an editor that's capable of highlighting braces and parentheses will save you

The error is in the `then` keyword, and vim highlights that by making the keyword white instead of the yellow that it uses for regular keywords.

After dealing with syntax, it is time to see how to avoid arguably more complicated and tougher-to-spot errors in logic.

How to do it...

Mentioning logic in scripting can be deceiving. Logic can, in the very strict definition of the term, be formal logic in clauses that require logic expressions to work, or can more broadly mean any decision-making inside the script. When we say *error in logic*, we usually think of the latter; problems that are created when our script behaves like we told it to, not like we thought we told it to. Every unexpected behavior that is not a result of a syntax error falls into this category.

For example, let's presume that we want to sort a couple of numbers using the `sort` command. This may look easy but has a small flaw. `sort`, by default, sorts alphabetically, and not numerically:

```
demo@ubuntu:~/Desktop/allscripts$ du -a | sort
0               ./errorfile
0               ./settings
264        .
4               ./arrayops.sh
4               ./array.sh
4               ./associative.sh
4               ./auxscript.sh
4               ./backupexample.sh
4               ./dialogdate.sh
4               ./dos.sh
4               ./doublevar.sh
4               ./echoline1.sh
4               ./echoline.sh
```

We end up having value of `264` being larger than `0` but smaller than `4`, which is wrong. If we want to sort something as we intended to, we should be using the appropriate switch:

```
demo@ubuntu:~/Desktop/allscripts$ du -a | sort -n
0               ./errorfile
0               ./settings
4               ./arrayops.sh
4               ./array.sh
4               ./associative.sh
4               ./auxscript.sh
.
.
.
264        .
```

This is much better. Errors like this are not strictly a problem with Bash but instead happen when we are unsure of how a command is used, the result of which is that our script will misbehave.

Another thing you are going to see frequently is invalid index referencing. In arrays, indices start from `0`, but people usually count from `1`:

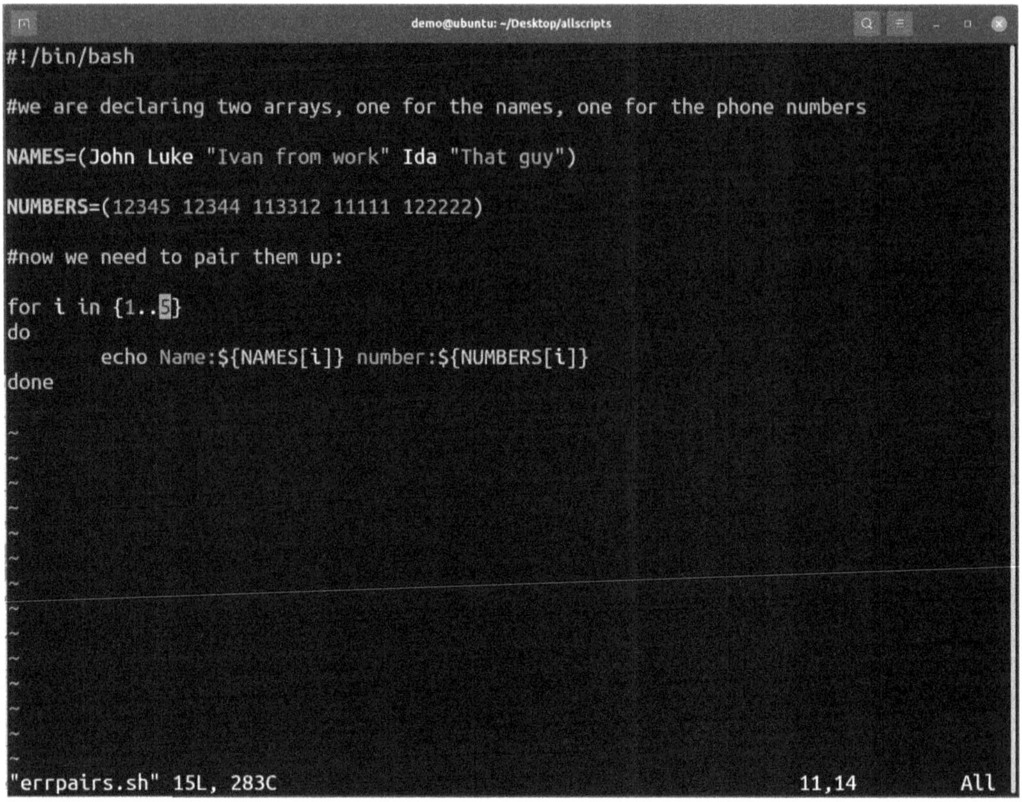
demo@ubuntu: ~/Desktop/allscripts
#!/bin/bash
#we are declaring two arrays, one for the names, one for the phone numbers
NAMES=(John Luke "Ivan from work" Ida "That guy")
NUMBERS=(12345 12344 113312 11111 122222)
#now we need to pair them up:
for i in {1..5}
do
echo Name:${NAMES[i]} number:${NUMBERS[i]}
done
"errpairs.sh" 15L, 283C
11,14
All

Figure 15.8 – Misnumbering indices is common when programming in any language

When we try and run this, we are going to lose one pair of variables in our output since we missed the first element in the array:

```
demo@ubuntu:~/Desktop/allscripts$ bash errpairs.sh
Name:Luke number:12344
Name:Ivan from work number:113312
Name:Ida number:11111
Name:That guy number:122222
Name: number:
```

The errors that doing this creates are sometimes easy to spot when the script is run, but some use cases, especially those that deal with only a part of an array, may create strange problems. The same problem can and will happen in loops using arguments, like in this example, and if we do not print the values straight away, we may not notice that we are processing only part of the array.

Fundamentally, the problem is that definition of the number we are counting from is pretty arbitrary. Usually, we use 0 as the first index, but there are some exceptions to this. If you're not completely sure, check.

All of these problems are mentioned here very broadly. You need to know them, but the way you are going to deal with them in your scripts is going to be different for every script you create. Our intention here is to make you aware that the problem exists, so you can spot it before it becomes dangerous.

The last big problem we mentioned was with scripts that work correctly most of the time, failing only in some cases and then failing only partially. This is the worst kind of problem, one that is dangerous since you cannot fully trust the output of the script, and hard to find since the output will be completely fine most of the time. The only way to deal with these problems is to carefully go through all the edge cases of your problem and test them on the script itself.

How it works...

In this recipe, we were annoyingly vague when describing possible problems, and we did it on purpose. As with all things that are directly connected with making errors while working on some problem, we would like to avoid all of them, but it is impossible to define what to avoid until we make an error. Most problems we see will be the result of a poor presumption or a false understanding of a fact. Sometimes, it is going to be a simple typing error that will go unnoticed.

There is also one more thing you can do to make things better when writing scripts. In order to avoid the most common problems with syntax and logic, primarily syntax, you can use automated tools.

There are two types of tools you can use. We have already mentioned one, although we didn't explicitly mention that it is an actual tool. We instead said that your editor is going to take care of most of your problems. Editors that are currently available usually include functionality that enables them to understand the syntax of the language you are using and to offer help if they notice something wrong. Support of this kind in editors is usually rudimentary and limits itself to being able to understand keywords and the lexical structure of a particular language. It is not uncommon for an editor to switch this functionality on as soon as it is able to identify the file and to autodetect the language you are trying to use. We have already seen examples of this. For more, please review *Chapter 2, Using Text Editors*.

There is, however, another set of tools you can use. We are talking about completely automated tools that are not only able to find errors in your scripts but are also able to find potential problems in your commands, and to even advise you how to improve your code.

You may wonder if it makes any sense to run an application on a script that will report the same errors as Bash would, and your question is valid. Bash is by itself completely capable of reporting any syntax errors, but it includes only a minimal set of messages to help you solve the problem. In essence, Bash reports only those errors that stop your code from working.

A good tool for *code analysis*, and this is the term used when talking about these applications, will find problems in your code and will give you suggestions to improve the code you have written. Things that are going to be reported may be obvious at first glance, but some of them are also errors that can lead to problems, such as missing quotes or misplaced variable assignment.

One such tool is **ShellCheck**, which is available both online and offline in the form of a package. In order to use it offline, you must install it using this command:

```
sudo apt install shellcheck
```

After that, it's only a matter of running it on your script. We will do that later when we touch upon how you can also run this tool online in a browser, and it will give you the same results as the offline version. The only difference is the interface and the simplicity of clicking on a link inside a browser. Both versions report exactly the same errors and behave the same when it comes to recommendations.

We are going to run this tool on a couple of our scripts to see what it has to say about the quality of our code. First, we are going to see what happens when we make a simple syntax error. We are using the script we used when we introduced the if statement. The script is named testif3.sh, and we have simply removed one line containing the then keyword:

```
demo@ubuntu:~/Desktop/allscripts$ shellcheck testif3.sh

In testif3.sh line 4:
if [ -d root ]
^-- SC1049: Did you forget the 'then' for this 'if'?
^-- SC1073: Couldn't parse this if expression. Fix to allow more checks.


In testif3.sh line 23:
        else
          ^-- SC1050: Expected 'then'.
              ^-- SC1072: Unexpected keyword/token. Fix any mentioned problems and t
ry again.

For more information:
  https://www.shellcheck.net/wiki/SC1049 -- Did you forget the 'then' for thi...
  https://www.shellcheck.net/wiki/SC1050 -- Expected 'then'.
  https://www.shellcheck.net/wiki/SC1072 -- Unexpected keyword/token. Fix any...
demo@ubuntu:~/Desktop/allscripts$
```

Figure 15.9 – ShellCheck provides much better warnings about syntax errors than Bash does

We can see that the tool has found the problem immediately and has not only reported it, but has also given us a suggestion about what to do next. The interesting thing is that it has marked the if statement that has the error in it, while Bash gave us an error that was comparatively misdirected, pointing to a piece of code that comes in much later in the script.

If we fix the error, we can rerun the tool, as shown here:

```
demo@ubuntu:~/Desktop/allscripts$ shellcheck testif3.sh
demo@ubuntu:~/Desktop/allscripts$
```

If there is no output from the tool, this means that no errors were detected.

Now, let's do it on a more complex script. In this case, we are working with a script named `funcglobal.sh` from *Chapter 12, Using Arguments and Functions*:

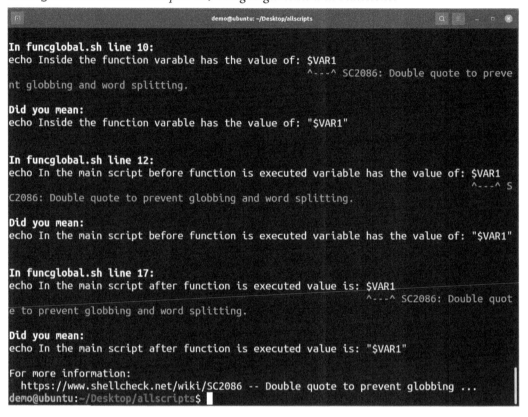

Figure 15.10 – Using variables in this way is not an error as such but it can lead to problems

The output does not look pretty because of the large font size in the terminal, but it gives us an idea of what to do better in our script. As we mentioned earlier, spaces are a big problem and so by using double quotes, we will prevent a space character completely messing up our script.

We are going to do one more example, a modified version of a script we used earlier and saved under the name `functions.sh`:

```
#!/bin/bash
#shell script that automates common tasks
function rsyn {
```

```
    rsync -avzh $1 $2
}


function usage {
echo In order to use this script you can:
echo "$0 copy <source> <destination> to copy files from source\
to destination"
echo "$0 newuser <name> to createuser with the username\
<username>"
echo "$0 group <username> <group> to add user to a group"
echo "$0 weather to check local weather"
echo "$0 weather <city> to check weather in some city on earth"
echo "$0 help for this help"
}
if [ "$1" != "" ]
            then
    case $1 in
        help)
            usage
            exit
            ;;
        copy)
            if [ "$2" != "" && "$3" != "" ]
            then
            rsyn $2 $3
            fi
            ;;


        group)
            if [ "$2" != "" && "$3" != "" ]
                then
                    usermod -a -G $3 $2
            fi
```

```
                              ;;

            newuser)
                  if [ "$2" != "" ]
                  then
                              useradd $2
                  fi
                  ;;

            weather)
                  if [ "$2" != "" ]
                              then
                                    curl wttr.in/$2
                              else
                                    curl wttr.in
                  fi
                  ;;


      *)
            echo "ERROR: unknown parameter $1\""
            usage
            exit 1
            ;;
      esac
            else
            usage
fi
```

If we run ShellCheck on this, we are going to end up with a long output, part of which looks like the following:

```
rsync -avzh "$1" "$2"


In functions.sh line 29:
        if [ "$2" != "" && "$3" != "" ]
                        ^-- SC2107: Instead of [ a && b ], use [ a ] && [ b ].


In functions.sh line 31:
        rsyn $2 $3
             ^-- SC2086: Double quote to prevent globbing and word splitting.
                ^-- SC2086: Double quote to prevent globbing and word splitting.

Did you mean:
        rsyn "$2" "$3"


In functions.sh line 36:
        if [ "$2" != "" && "$3" != "" ]
                        ^-- SC2107: Instead of [ a && b ], use [ a ] && [ b ].


In functions.sh line 38:
        usermod -a -G $3 $2
                         ^-- SC2086: Double quote to prevent globbing and word sp
litting.
                            ^-- SC2086: Double quote to prevent globbing and word
 splitting.
```

Figure 15.11 – Output of ShellCheck is going to warn you if it sees logical errors

If we click on the link that ShellCheck provides as the last line of the output, we are taken to a detailed explanation of why this is a problem, as can be seen in this screenshot:

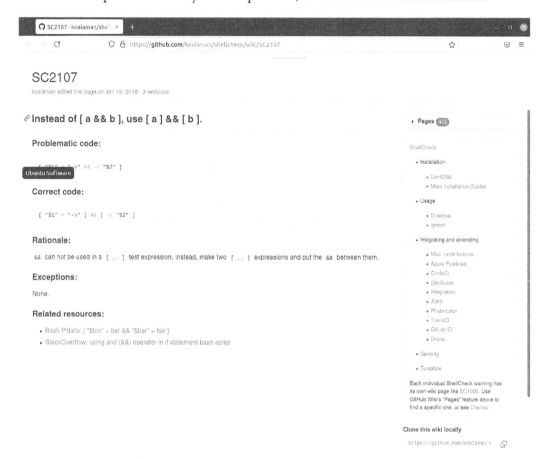

Figure 15.12 – Links that ShellCheck provides give you detailed information about the error

This explanation is not only useful, but it also contains more links for when we want to understand what actually went wrong, why it went wrong, and what is the reason for this being an issue to look at in the first place. Sometimes, the problems that the tool detects are going to have limited scope and will be solved in some versions of the Bash interpreter yet misbehave in another.

See also...

Troubleshooting is complicated since we are unable to anticipate all the possible problems. Some of them as follows are, however, common:

- `https://mywiki.wooledge.org/BashPitfalls`
- `https://mywiki.wooledge.org/BashGuide`
- `https://www.shellcheck.net/`

Simple debugging approach – echoing values during script execution

The first thing you will learn when using Bash is how to regularly use the `echo` command when running any script. This approach is simple as it gives us an opportunity to follow the workflow of the script and to print the values of the variables as they are in different points of the script. Being able to understand both those things is going to help us to follow all the inputs to our script and to see how they transform into outputs that we expect.

Getting ready

In this recipe, we are going to deal with simple ways we can make our script help us understand what is happening during its run. There are three ways we can use this simple method.

The first thing we can do is use the echo command in every place in the script that we think is helpful. As an example, take a look at one of the scripts from previous chapters (funcglobal.sh) that is already pretty verbose:

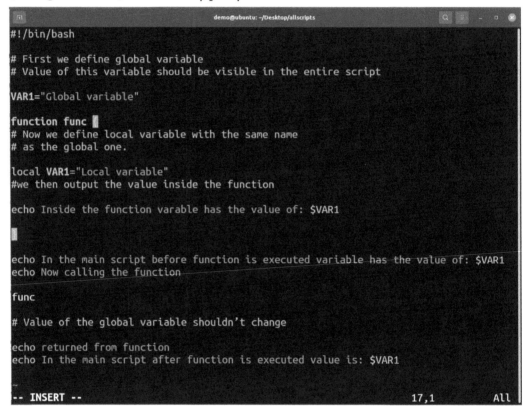

Figure 15.13 – Using echo to debug program flow is useful when dealing with functions

We are going to add even more `echo` statements here to enable us to see exactly what is happening and in what order:

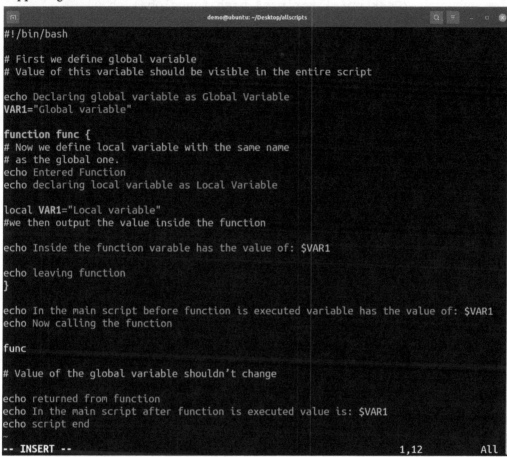

```bash
#!/bin/bash

# First we define global variable
# Value of this variable should be visible in the entire script

echo Declaring global variable as Global Variable
VAR1="Global variable"

function func {
# Now we define local variable with the same name
# as the global one.
echo Entered Function
echo declaring local variable as Local Variable

local VAR1="Local variable"
#we then output the value inside the function

echo Inside the function varable has the value of: $VAR1

echo leaving function
}

echo In the main script before function is executed variable has the value of: $VAR1
echo Now calling the function

func

# Value of the global variable shouldn't change

echo returned from function
echo In the main script after function is executed value is: $VAR1
echo script end
~
-- INSERT --                                                    1,12          All
```

Figure 15.14 – There are never enough echo commands when debugging

If we now run our modified script, we will be able to precisely follow the flow of the script:

```
demo@ubuntu:~/Desktop/allscripts$ bash funcglobal.sh
Declaring global variable as Global Variable
In the main script before function is executed variable has the
value of: Global variable
Now calling the function
Entered Function
declaring local variable as Local Variable
Inside the function variable has the value of: Local variable
leaving function
returned from function
In the main script after function is executed value is: Global
variable
script end
```

This is particularly useful when we're dealing with a lot of code blocks, functions, and conditional statements. The rough idea here is that we can use echo to announce entering and leaving each code block so that we can see if our script ran correctly.

Another thing we are going to do is to print the values of variables during the script execution. While doing this, we suggest you always mention the place this particular command is printing variables from in the code of the script. When debugging this way, the values of the variables are going to be printed to the output in the order they are assigned, helping us follow the flow of the script. Our previous example already does that.

How to do it...

There is one more thing you can do to make your scripts provide more information when you are debugging them. There is a command built into bash called trap. The main reason it is there is to help you react to interrupts and to ensure that your script works even if something unexpected happens. The syntax it uses is simple – we need to tell it what to do and under what circumstances to do it. By circumstances, we mean any interrupt signal possible under Linux. The most common ones are SIGHUP, SIGKILL, and SIGQUIT, but a lot of others are used.

For example, we can create a script like this:

```
#!/bin/bash
#trapping end of script

trap "echo Script was interrupted; exit" SIGINT


echo -n "Can you please input a word?: "
read  word
echo "I got: $word"
echo Script cleanly finished executing
```

```
"echoline.sh" 11L, 199C                              4,39              All
```

Figure 15.15 – Using trap inside the script to stop Ctrl + C

What this first line does is that it establishes something considered to be an interrupt routine. If at any point in our script someone uses *Ctrl + C* to interrupt it, our script will detect that and execute two commands inside quotes:

```
demo@ubuntu:~/Desktop/allscripts$ bash echoline.sh
Can you please input a word?: dsasd^CScript was interrupted
demo@ubuntu:~/Desktop/allscripts$ bash echoline.sh
Can you please input a word?: nointerrupt
I got: nointerrupt
Script cleanly finished executing
```

When we first tried, we pressed the interrupt key and our *routine* did what we told it to, which was to exit the script right away and give us a warning about it. This command can also be very useful to block attempts to stop the script since it will execute whatever we tell it to and then just continue running the script.

Another thing that you can do is to use EXIT as the keyword in the trap command like this:

```
trap "some command" EXIT
```

This keyword covers any possible way to exit from a script, meaning that this trap is going to be executed no matter what happens to the script, and it will run right before the control returns to whatever process ran our script.

When used this way, trap is useful not only for debugging but also for cleaning up after your script, since it will run as the last command, enabling you to do whatever needs to be done to close the files and clean things up after your script.

How it works...

No matter which way you choose to debug your scripts, it all boils down to heavily modifying it. Using echo is useful but at the same time requires adding a lot of commands to the script we are debugging. Having said that, this is probably going to be the first thing you will try when debugging any script since it enables you not only to understand how the values inside the scripts change but also how the entire script works because you have the exact information of where your command is executed from, enabling you to understand both the variables and the workflow of the script.

Using trap is a slightly more nuanced way of debugging and can be very useful to gain knowledge of what is happening when we go outside of the program flow we imagined. If a script breaks or gets interrupted in any other way, trap will give you information on what happened and where.

There isn't a particular way of debugging to recommend here as all of them work in a particular scenario. What we can say is that you should try using all of them and see which fits a particular scenario.

See also

- https://www.linuxjournal.com/content/bash-trap-command
- https://tldp.org/LDP/Bash-Beginners-Guide/html/sect_12_02.html

Using the bash -x and -v options

Up to this point, we have tried debugging using different methods that involved commands inserted into our scripts. Regardless of the command we used, this approach has one drawback – whatever we do, using commands inside the script is either very localized to a particular part of a given script or too global since it has to cover a good chunk of code. We are not saying that this is not a valid way of solving problems in scripts, but we still need more ways to debug.

Getting ready

The only thing we need to know before we start using this is that we will be running scripts by invoking them as parameters of the interpreter, so something like this:

```
bash -options <scriptname>
```

This is important since we can not use any of these options if we invoke scripts in any other way.

How to do it...

Bash is in this respect pretty complicated since it offers little, in fact almost nothing, in terms of support for any reasonable systematic debugging, and we have already mentioned that. There is a glimmer of hope, however, and that comes in the form of two switches, -x and -v. The first one turns on the printing of every command that is run in the script and it also prints all the command arguments used. This makes understanding the workflow of commands easy.

Using -v is arguably less useful. It simply prints all the script lines as they are read.

In order to understand these options, we are going to create a small example using one of the scripts we used in a different recipe, but this time we are going to use different switches when running it.

First, we are going to use -v:

```
demo@ubuntu:~/Desktop/allscripts$ bash -v testif3.sh
#!/usr/bin/bash
# testing premissions and paths
if [ -d root ]
            then
                    echo root directory exists!
                    if [ -r root ]
```

```
                                            then
                            echo Script can read from the directory!
                                            else
                        echo Script can NOT read from the directory!
                            fi
                            if [ -w root ]
                            then
                            echo Script can write to the directory!
                            else
                                        echo Script can not
    write to the directory!
                            fi
                else
                        echo root directory does NOT exists!
    fi
    root directory exists!
    Script can read from the directory!
    Script can write to the directory!
```

Now, we are going to use -x to run the script:

```
demo@ubuntu:~/Desktop/allscripts$ bash -x testif3.sh
+ '[' -d root ']'
+ echo root directory 'exists!'
root directory exists!
+ '[' -r root ']'
+ echo Script can read from the 'directory!'
Script can read from the directory!
+ '[' -w root ']'
+ echo Script can write to the 'directory!'
Script can write to the directory!
```

Both switches have their place in debugging. When we said that -v is less useful than
-x, we meant that it only gives us an insight into how Bash interpreted your script, but
nothing more than that.

Using -x shows us how Bash executed the script and what commands it ran during the execution. What you must understand is that this is not going to be the list of all commands in the script but only those that actually ran. If a particular part of the script was not used, for example, if it belonged to a block of commands that are in the code only for a specific condition, this way of running a script will not show that.

How it works...

The most common thing is using both switches together, since it enables us to quickly understand what the script looks like and what Bash does when executing it. In a large script, this is going to generate a lot of output, but this is usually what we actually want to do. Then, we can go through the script step by step and understand the logic it implements.

On the other hand, we cannot consider this as a universal solution to anything. Although it gives us a lot of information about the command it runs, it is very limited by what the variable values are and what is actually going on when processing data. Take, for example, this loop included in the file forloop1.sh, available as part of the files included with the book:

```
demo@ubuntu:~/Desktop/allscripts$ bash -x forloop1.sh
+ for name in {user1,user2,user3,user4}
+ for server in {srv1,srv2,srv3,srv4}
+ echo 'Trying to ssh user1@srv1'
Trying to ssh user1@srv1
+ for server in {srv1,srv2,srv3,srv4}
+ echo 'Trying to ssh user1@srv2'
Trying to ssh user1@srv2
+ for server in {srv1,srv2,srv3,srv4}
+ echo 'Trying to ssh user1@srv3'
Trying to ssh user1@srv3
+ for server in {srv1,srv2,srv3,srv4}
+ echo 'Trying to ssh user1@srv4'
```

We are not going to copy the entire output since it has no other useful information. Here, we can see that we are looping in a for loop, and we can see possible values but the actual value of a particular variable is not seen unless we print it. This means that we will have to combine this way of debugging with the other ways we presented in this chapter.

See also...

- `https://linux.die.net/man/1/bash`
- `https://tldp.org/LDP/Bash-Beginners-Guide/html/sect_02_03.html`

Using set to debug a part of the script

In the previous recipe, we dealt with globally using two options to tell Bash to include a lot of useful information in its output. We mentioned that this offers another way to deal with debugging and troubleshooting how your scripts work. At the same time, we mentioned that this approach is in stark contrast with using commands in the script itself since we can deal with things globally without too many changes to the scripts when debugging.

In this recipe, we are going to cover another way to debug, one that shares a lot of similarities with the ones we introduced before, while also being different.

Getting ready

One very interesting built-in command in Bash is `set`. What it does is give us the ability to change the options Bash uses. A lot of things can be changed by using `set`, and by a lot we mean almost every option Bash has. In this recipe, we are using only two of them, but you can turn all of them on or off.

`set` enables us to set a particular option on in just a small block of code, instead of using it globally. You also need to know that `set` can turn an option both on and off. If we use `set` with a – sign, we turn the option on. For example, we could use this:

```
set -x
```

This is telling Bash to start showing us commands as they are executed.

A slightly confusing way is if we turn off any option that is currently used. To do that, we have to use the + sign, something that is a little bit counterintuitive since *adding* is usually used to turn something on, not off. For example, look at this command:

```
set +x
```

This will turn off the output of commands in Bash.

We are going to see a couple of examples of this just to make you comfortable.

How to do it...

Using set is simple. In any script we wish to debug, we are going to insert the set statement right before the place we are starting our trace from. After we no longer need to trace our script, we simply unset the option and we are done. We can do this as many times as we need to.

Here's an example:

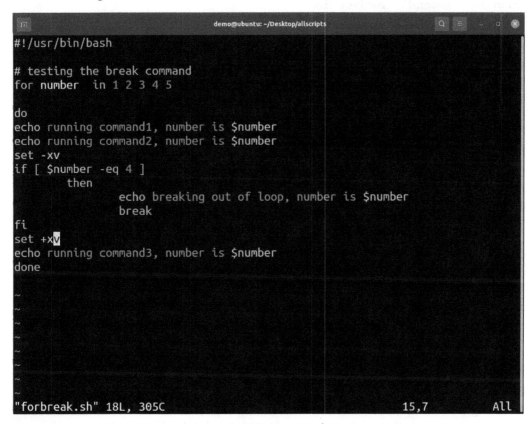

```
#!/usr/bin/bash

# testing the break command
for number  in 1 2 3 4 5

do
echo running command1, number is $number
echo running command2, number is $number
set -xv
if [ $number -eq 4 ]
        then
                echo breaking out of loop, number is $number
                break
fi
set +xv
echo running command3, number is $number
done
```

```
"forbreak.sh" 18L, 305C                              15,7            All
```

Figure 15.16 – How to set and unset options while running the script

In this example, we are starting our trace just before we do the test of our variable. This means that we are going to start tracing right before we do a test and stop tracing right before we continue the loop. The result is this:

```
demo@ubuntu:~/Desktop/allscripts$ bash forbreak.sh
running command1, number is 1
running command2, number is 1
+ '[' 1 -eq 4 ']'
```

```
+ set +xv
running command3, number is 1
running command1, number is 2
running command2, number is 2
+ '[' 2 -eq 4 ']'
+ set +xv
running command3, number is 2
running command1, number is 3
running command2, number is 3
+ '[' 3 -eq 4 ']'
+ set +xv
running command3, number is 3
running command1, number is 4
running command2, number is 4
+ '[' 4 -eq 4 ']'
+ echo breaking out of loop, number is 4
breaking out of loop, number is 4
+ break
```

We can even consider this way of debugging as a special case of using options globally. We are basically doing the same thing as in previous examples where we used `set` globally, but this time we are limiting the scope to make the output more readable. Sometimes that can make working a lot easier since we are not creating much irrelevant output.

How it works...

The `set` command can be used for one more thing, and that is to force Bash to do some things differently than are usually done. For example, we can make scripts fail if any command inside them fails, we can make scripts fail if they reference a value of a variable that is not set, or we can even change the way Bash expands characters in the command line.

All these things can be useful when working but at the same time, can be a little too much to grasp all at once when you first begin scripting, so we decided to not include them in these recipes.

See also...

- `https://linuxhint.com/debug-bash-script/`
- `https://www.gnu.org/software/bash/manual/html_node/The-Set-Builtin.html`

16

Shell Script Examples for Server Management, Network Configuration, and Backups

Now that we've covered everything that we wanted to cover in terms of various concepts and structures in Bash shell scripting, let's dig into some examples. This will allow us to put these last few chapters to good use as shell scripts are the most commonly used tools in system engineers' daily jobs. Due to this, we are going to go through some shell scripts to emphasize the point of scripting – to make our lives a lot easier.

In this chapter, we're going to cover the following recipes:

- Creating a file and folder inventory
- Checking if you're running as root
- Displaying server stats
- Finding files by name, ownership, or content type and copying them to a specified directory
- Parsing date and time data
- Configuring the most common firewall settings interactively (`firewalld` and `ufw`)
- Configuring network settings interactively (`nmcli`)
- Backing up the current directory with shell script arguments and variables
- Creating a current backup based on the user input for the backup source and destination

Technical requirements

Let's continue using our Ubuntu machine, specifically the `cli1` machine. If you've not started it up yet, please start it now so that we can go through our examples. We will use the `cli2` CentOS machine for a few recipes as well, so make sure that you start that one when the time comes.

Creating a file and folder inventory

Let's start with something basic – a script that reports in terms of folder and file inventory. As simple as this is, this type of script can use a variety of different tools, including commands, built-in CLI applications, loops – there are a lot of choices to be made. We're going to do this in the simplest way possible – by taking advantage of our knowledge of commands and CLI applications. We're going to create a couple of different versions of this script as it can be used in a variety of different ways – for example, as input into future shell scripts or as plain text reporting tools.

How to do it...

Let's start by creating a script that's just going to tell us the following:

- The number of folders in the current folder and their sizes, sorted by size in descending order

- The number of files in the current folder and their size, sorted by size in descending order

Here's the first version of our script – we saved it as a file called `sscript1.sh`:

```bash
#!/bin/bash
# V1.0 / Jasmin Redzepagic / 01/11/2021 Initial script version
# Distribution allowed under GNU Licence V2.0
echo "Number of directories in current directory is:"
find . -type d | wc -l
echo "Directory usage, sorted in descending order, is as
follows:"
find . -type d | du | sort -nr
echo "Number of files in current directory is:"
find . -type f | wc -l
echo "File usage, sorted in descending order, is as follows:"
find . -type f -exec ls -al {} \; | sort -k 5 -nr | sed 's/
\+/\t/g' | cut -f5,9
```

As we can see, we only used basic commands here, without going into a lot of looping, actual programming, and so on. Let's treat this as a reporting script and go from there.

On our `cli1` machine, this was the result:

```
student@cli1:~$ ./sscript1.sh
Number of directories in current directory is:
6
Directory usage, sorted in descending order, is as follows:
112      .
16       ./.ssh
8        ./.config
4        ./copylocation
4        ./.config/procps
4        ./.cache
Number of files in current directory is:
19
File usage, sorted in descending order, is as follows:
12887    ./.viminfo
12285    ./.bash_history
4073     ./fw.sh
4073     ./ubuntu.sh
3771     ./.bashrc
2602     ./.ssh/id_rsa
1579     ./sscript5.sh
888      ./.ssh/known_hosts
807      ./.profile
566      ./.ssh/id_rsa.pub
528      ./stats.sh
419      ./sscript1-2.sh
402      ./sscript1.sh
220      ./.bash_logout
190      ./root.sh
75       ./.selected_editor
34       ./.lesshst
0        ./.sudo_as_admin_successful
0        ./.cache/motd.legal-displayed
student@cli1:~$
```

Figure 16.1 – First version of our folder and file inventory script

This is working and can be used for reports – yep, that's all good. But what happens if we want some more functionality? What if we were to use the first version of this script to generate a `.txt` file that contains lists of files in the current directory, modify it a little bit, and then use this file for something else, such as to copy those files to a pre-configured location?

We would need to make some adjustments, as follows:

```bash
#!/bin/bash
# V1.0 / Jasmin Redzepagic / 01/11/2021 Initial script version
# Distribution allowed under GNU Licence V2.0
# First, let's find out if the destination directory exists by
# using test function. If it does, go on with the script. If it
# doesn't, create that destination directory. In our example,
# destination directory is called copylocation.
if [ -d "./copylocation" ]
then
        echo "Directory ./copylocation exists."
else
        echo "Error: Directory ./copylocation does not
exist."
        mkdir ./copylocation
fi
# next step, let's create a friendly file list with all of the
files
# in current folder
find . -type f -exec ls -al {} \; | sed 's/  */ /g' | cut -f9
-d" " > filelist.txt
# Last step, let's load this file into variable so that we can
loop
# over it and copy every file from it to our destination folder
file_list='cat filelist.txt'
for current_file in $file_list
do
        echo "Copying $current_file to destination"
        cp "$current_file" ./copylocation
done
```

The next script we'll look at is simple but very, very useful – it's about checking if we're running the script as root. The reasoning is simple – there will be scripts that we don't want to run as root for fear of messing something up, in which case we'd use some accessible resources. Let's see how that would work.

See also

If you need more information about sort, find, wc, cut or sed, we suggest that you visit these links:

- sort command man page: https://man7.org/linux/man-pages/man1/sort.1.html

- find command man page: https://man7.org/linux/man-pages/man1/find.1.html

- wc command man page: https://man7.org/linux/man-pages/man1/wc.1.html

- cut command man page: https://man7.org/linux/man-pages/man1/cut.1.html

- sed command man page: https://man7.org/linux/man-pages/man1/sed.1.html

Checking if you're running as root

There are different ways of checking if we're running a script as root. We can use environment variables, just as we can use the whoami or id commands to check if it equals root/number 0 or not.

Getting ready

We'll continue using the cli1 machine for this recipe, so make sure that it's powered on.

How to do it...

Let's create a short snippet of Bash shell script code that's going to help us find out whether we're running a script as root or not. It's a rather simple thing to do in Linux, considering that we have easy access to an environment variable called EUID, and reading its value is enough to determine whether we're running as root (EUID=0) or not (EUID value > 1):

```
#!/bin/bash
# V1.0 / Jasmin Redzepagic / 01/11/2021 Initial script version
# Distribution allowed under GNU Licence V2.0
# First, we need to check if our environment variable UID is
set to
```

```
# 0 or not and branch that out to either yes or no with
appropriate
# status messages
if [ "$EUID" -eq 0 ]
            then
                        echo "You are running as root user.
Please be careful!"
            else
                        echo "You are not root. It's all
sunshine and roses, you can't do much damage!"
fi
exit 0
```

The next example that we're going to cover is about displaying server stats. We're going to use the sar command to do so. Let's go!

See also

If you need more information about internal variables, we suggest that you go to https://tldp.org/LDP/abs/html/internalvariables.html.

Displaying server stats

Let's say that we have to write a shell script that's going to display the following pieces of information:

- Current hostname
- Current date
- Current kernel version
- Current CPU usage
- Current memory usage
- Current swap space usage
- Current disk I/O
- Current network bandwidth

This is more of an exercise in filtering data and using commands, but there are some interesting concepts in terms of how to format data to look *nice* and *readable*. This is something we consider to be very important.

Getting ready

We need to leave the `cli1` machine running. Also, for this script to work, we need to deploy the `sysstat` package, and then enable the necessary service. We can do this by using the following command for Ubuntu:

```
sudo apt-get -y install sysstat
```

We can use the following command for CentOS:

```
sudo yum -y install sysstat
```

After that, we need to start the `sysstat` service:

```
sudo systemctl enable --now sysstat
```

Now, we can start working on our script.

How to do it...

We are going to use the `sar` command to get a lot of information about our Linux machine. We are also going to filter out some of the unnecessary details. Our script should look like this:

```
#!/bin/bash
# V1.0 / Jasmin Redzepagic / 01/11/2021 Initial script version
# Distribution allowed under GNU Licence V2.0
echo "Hostname: $(hostname)"
echo "Current date: $(date)"
echo "Current kernel version and CPU architecture: $(uname -rp)"
# sar command has a default first line output telling us that it's
# running on Linux, and which kernel we are using. It's pointless
```

```
# to get this information four or five times, so let's filter
that
# out from the get-go (grep -v "Linux" part of every command)
echo "Current CPU usage:"
sar -u 1 1| grep -v "Linux"
echo ""
echo "Current memory usage:"
sar -r 1 1| grep -v "Linux"
echo ""
echo "Current swap space usage:"
sar -S 1 1| grep -v "Linux"
echo ""
# When sar displays disk I/O info, it displays that info per
# device, which isn't all that important. What's important for
# us are sd* and vd* devices, as well as the status line
telling
# us which specific metrics are shown in the column (DEV).
echo "Current disk I/O:"
sar -d 1 1| grep -E "(DEV|sd|vd)" | grep -v "Linux"
echo ""
# When sar displays network information, it shows it per
device.
# Having in mind that we have a loopback network device (lo)
and
# that its statistics isn't important, let's just filter that
out
# so that we can see network bandwidth info per real network
device
echo "Current network bandwidth usage:"
sar -n DEV 1 1| grep -v lo | grep -v "Linux"
```

We used `echo  " "` multiple times here so that our output looks clean and readable. The output should look like this:

```
student@cli1:~$ ./stats.sh
Hostname: cli1
Current date: Mon Jan 10 23:14:50 UTC 2022
Current kernel version and CPU architecture: 5.8.0-63-generic x86_64
Current CPU usage:

23:14:50        CPU     %user      %nice     %system    %iowait    %steal     %idle
23:14:51        all      0.00       0.00       0.50       0.00       0.00      99.50
Average:        all      0.00       0.00       0.50       0.00       0.00      99.50

Current memory usage:

23:14:51     kbmemfree    kbavail kbmemused  %memused kbbuffers   kbcached   kbcommit    %c
ommit  kbactive    kbinact    kbdirty
23:14:52      1211776    1480208    254560     12.72     26716     351408     697200
17.01    246680     233852        248
Average:      1211776    1480208    254560     12.72     26716     351408     697200
17.01    246680     233852        248

Current swap space usage:

23:14:52     kbswpfree kbswpused  %swpused  kbswpcad   %swpcad
23:14:53      2097148         0      0.00         0      0.00
Average:      2097148         0      0.00         0      0.00

Current disk I/O:
23:14:53         DEV       tps      rkB/s      wkB/s      dkB/s    areq-sz    aqu-sz
await      %util
23:14:54         sda      4.00       0.00       0.00       0.00       0.00      0.00
 0.00       0.40
Average:         DEV       tps      rkB/s      wkB/s      dkB/s    areq-sz    aqu-sz
await      %util
Average:         sda      4.00       0.00       0.00       0.00       0.00      0.00
 0.00       0.40

Current network bandwidth usage:

23:14:54       IFACE    rxpck/s    txpck/s     rxkB/s     txkB/s    rxcmp/s    txcmp/s   rxm
cst/s   %ifutil
23:14:55       ens33      4.00       0.00       0.71       0.00       0.00      0.00
 0.00       0.00

Average:       IFACE    rxpck/s    txpck/s     rxkB/s     txkB/s    rxcmp/s    txcmp/s   rxm
cst/s   %ifutil
Average:       ens33      4.00       0.00       0.71       0.00       0.00      0.00
 0.00       0.00
student@cli1:~$ 
```

Figure 16.2 – Displaying the server stats from our script

The next recipe is about finding content – by name, ownership, or extension – so that we can copy the content we find to a specific location. Let's get started!

There's more...

If you need to learn more about the `sar` command, take a look at the `sar` command's man page at `https://man7.org/linux/man-pages/man1/sar.1.html`.

Finding files by name, ownership, or content type and copying them to a specified directory

Managing files can be a bit of a burden. Usually, we have thousands of them, and if it's an enterprise-level company that we're discussing, there might be millions. What happens if we need to find files that follow specific criteria?

We'll start with something simpler – finding by name. Then, we'll move on to ownership-based searches, and then, the most involved – content type-based searches.

Getting ready

Before you start this recipe, you need to make sure that our `cli1` virtual machine is up and running.

How to do it...

This is a perfect script to do a bit more interaction, so case loops are in store for us. We're making a conscious effort to use case a lot, with a lot of status/debugging code that can guide us through script usage.

We want to slice this script into three parts as it's going to do three different things. Here's what the script will look like:

```
#!/usr/bin/bash
# V1.0 / Jasmin Redzepagic / 01/11/2021 Initial script version
# Distribution allowed under GNU Licence V2.0
read -p "Enter directory to move file to: " DESTDIR
echo -e "\n"
# Let's first establish a destination directory with a loop
that can test if that directory exists or not
if [ "$DESTDIR" == "" ];
then
        echo "You must specify a directory."
```

```
else
        if [ ! -d "$DESTDIR" ]
        then
                echo "Directory $DESTDIR must exist. Exiting!"

                exit
        fi
fi


# Directory is ready, let's go to the main part of the script. First
# step is selecting which type of search we want to use.
echo "Enter number denoting criteria for search: "
echo "1 = by name "
echo "2 = by ownership "
echo "3 = by content extension "
echo -e "\n"
read CRIT


# Let's start our case loop against CRIT variable.
case $CRIT in
        1)
                read -p "Enter name to search for: " NAME
                echo -e "\n"
                if [ ! -z   $NAME="" ]
                        then
                                find / -name "$NAME" -exec cp
{} $DESTDIR \; 2> /dev/null
                        else
                                echo You have to enter the
name!
                fi
                ;;
```

```
    2)
            read -p "Enter owner to search for: " OWNER
            echo -e "\n"
            if [ ! -z $OWNER="" ]
                then
                            find / -user $OWNER -exec cp {}
$DESTDIR \;  2> /dev/null
                else
                            echo You have to input an
owner!
            fi
            ;;


    3)
            read -p "Enter content extension: " CEXT
            echo -e "\n"
            if [ ! -z $CEXT="" ]
                then
                            read -p "Where are we looking
for files, in which directory?" LOOKUP
                            find "$LOOKUP" -type f -name
"$CEXT" -exec cp {} $DESTDIR \; 2> /dev/null
                else
                            echo You have to enter the
content type!
            fi
             ;;


        *)      echo please make a choice, either 1, 2 or 3!
esac
```

Note that when we're asked about the extension, we have to type something like `*.txt` for this script to work. Here's what the script execution looks like with that extension in mind:

```
student@cli1:~$ ./sscript5.sh
Enter directory to move file to: temp


Enter number denoting criteria for search:
1 = by name
2 = by ownership
3 = by content extension


3
Enter content extension: *.txt


Where are we looking for files, in which directory?.
student@cli1:~$ ls -al temp
total 12
drwxrwxr-x 2 student student 4096 Jan 10 23:21 .
drwxr-xr-x 7 student student 4096 Jan 10 23:20 ..
-rw-rw-r-- 1 student student  302 Jan 10 23:21 filelist.txt
student@cli1:~$
```

Figure 16.3 – Script execution with a file extension as the criteria

In the next recipe, we'll learn how to work with date and time-based data, a concept that's often used in shell scripting for indexing purposes. While easy to use and understand, we need to learn how to use this concept programmatically, via variables. So, let's do that next!

There's more…

If you need more information about the `sar` command, we recommend that you check out the following link to learn more: `https://www.howtogeek.com/662422/how-to-use-linuxs-screen-command/`.

Parsing date and time data

Working with time-based data is often less than fun, especially when you're working with a lot of time-based content. But for our use cases, we often use date/time information for indexing; that is, to name our backup files and similar purposes. So, learning how to get information from the date command and putting that information into variables so that our code can be as modular as possible is very important. Let's create a shell script that we are going to be able to use in future scripts as a snippet of code for a lot of our shell scripts – at least bits and pieces of it.

Getting ready

We don't need any special utilities to be installed, just our Linux machine to be alive and ready for action.

How to do it...

We are going to go back to the basics and use the date command to extract all of the date and time pieces that we'll ever need:

- Information about the current time in terms of hours, minutes, and seconds

- Information about today's date

- Information about what day it is today

Let's type the following in our text editor and execute our script:

```bash
#!/bin/bash
# V1.0 / Jasmin Redzepagic / 01/11/2021 Initial script version
# Distribution allowed under GNU Licence V2.0
# This part of our script is just plain using date command to assign
# values to "obviously named variables". This further shows two
# things - how to assign a variable value from external command,
# and how to use that principle on date and time data.
hour=$(date +%H)
minute=$(date +%M)
second=$(date +%S)
day=$(date +%d)
month=$(date +%m)
year=$(date +%Y)
```

```
# Let's print that out
echo "Current time is: $hour:$minute:$second"
echo "Current date is: $day-$month-$year"
```

Here's an example of the sample output. We called this script `sscript2.sh`:

```
student@cli1:~$ bash sscript2.sh     student@cli1:~$ ./sscript2.sh
Current time is: 23:25:45             Current time is: 13:51:14
Current date is: 10-01-2022           Current date is: 26-11-2021
student@cli1:~$                       student@cli1:~$ _
```

Figure 16.4 – Sample output from our date and time script

This can be very useful for backup scripts – for example, when we're indexing backup files (`.tar.gz` or something else) by dates. This is a concept that we're going to use later in this chapter. For now, let's learn how to configure firewall settings via shell scripts.

Configuring the most common firewall settings interactively

Firewall configuration is just one of those things – we often need to do it, but we don't necessarily know all of the commands off the top of our heads. Let's do this via shell scripts, for both CentOS (`firewalld`) and Ubuntu (`ufw`).

Getting ready

Before you start this recipe, you need to make sure that you have `firewalld` on your CentOS machine and `ufw` on your Ubuntu machine up. So, first, you need to use the following command:

```
systemctl status firewalld
```

Use the following command for CentOS and Ubuntu:

```
systemctl status ufw
```

If they're disabled, we need to turn them on, like so:

```
systemctl enable --now firewalld
```

On CentOS and Ubuntu, you can use the following command:

```
systemctl enable --now ufw
```

Now, we're ready to get started. Of course, you need to be logged in as an administrator to be able to change your firewall configuration, so make sure that you're either logged in as root (or a user with similar capabilities) or use the sudo configuration to change your firewall configuration.

Furthermore, with `firewalld`, a lot of people have trouble remembering the service names that it uses. That's not a problem – we just need to use the following command:

```
firewall-cmd --get-services
```

For `ufw`, we just need to go and look at `/etc/service`, since all of the service names are listed there, and `ufw` uses them for configuration purposes.

How to do it...

First, let's make a CentOS-based script for `firewalld`. We'll include eight standard operations – manipulating service configuration, TCP and UDP ports and rich rules, both adding and removing them, as well as the capability to list current configuration. Here's what the script should look like:

```bash
#!/bin/bash
# V1.0 / Jasmin Redzepagic / 01/11/2021 Initial script version
# Distribution allowed under GNU Licence V2.0
echo "1 = firewalld (CentOS) - manipulate service configuration
- add"
echo "2 = firewalld (CentOS) - manipulate service configuration
- remove"
echo "3 = firewalld (CentOS) - manipulate TCP ports - add"
echo "4 = firewalld (CentOS) - manipulate TCP ports - remove"
echo "5 = firewalld (CentOS) - manipulate UDP ports - add"
echo "6 = firewalld (CentOS) - manipulate UDP ports - remove"
echo "7 = firewalld (CentOS) - manipulate rich rules - add"
echo "8 = firewalld (CentOS) - manipulate rich rules - remove"
echo "9 = firewalld (CentOS) - list current configuration"
echo -e "Your choice:"
read CRIT
# Let's start our case loop against CRIT variable.
case $CRIT in
        1)
                echo "Enter service names, using space as
```

```
separator."
                echo "Hint: ssh http https etc. Get list from
firewall-cmd --get-services"
                echo "Your input:"
                read -a FW1
                for svcs1 in ${FW1[@]}
                do
                        firewall-cmd --permanent
--add-service=$svcs1
                done
                firewall-cmd --reload
                ;;
        2)
                echo "Enter service names, using space as
separator."
                echo "Hint: ssh http https etc. Get list from
firewall-cmd --get-services"
                echo "Your input:"
                read -a FW2
                for svcs2 in ${FW2[@]}
                do
                        firewall-cmd --permanent --remove-
service=$svcs2
                done
                firewall-cmd --reload
                ;;
        3)
                echo "Enter TCP port numbers, using space as
separator."
                echo "Hint: 22 80 443 etc."
                echo "Your input:"
                read -a FW3
                for svcs3 in ${FW3[@]}
                do
                        firewall-cmd --permanent
--add-port=$svcs3/tcp
                done
```

```
                    firewall-cmd --reload
                    ;;
        4)
                    echo "Enter TCP port numbers, using space as
separator."
                    echo "Hint: 22 80 443 etc."
                    echo "Your input:"
                    read -a FW4
                    for svcs4 in ${FW4[@]}
                    do
                            firewall-cmd --permanent --remove-
port=$svcs4/tcp
                    done
                    firewall-cmd --reload
                    ;;
        5)
                    echo "Enter UDP port numbers, using space as
separator."
                    echo "Hint: 22 80 443 etc."
                    echo "Your input:"
                    read -a FW5
                    for svcs5 in ${FW5[@]}
                    do
                            firewall-cmd --permanent
--add-port=$svcs5/udp
                    done
                    firewall-cmd --reload
                    ;;
        6)
                    echo "Enter UDP port numbers, using space as
separator."
                    echo "Hint: 22 80 443 etc."
                    echo "Your input:"
                    read -a FW6
                    for svcs6 in ${FW6[@]}
                    do
                            firewall-cmd --permanent --remove-
```

```
port=$svcs6/udp
                done
                firewall-cmd --reload
                ;;
     7)
                echo "Let's manipulate rich rules - to add
specific IPs access to specific port."
                echo "Hint: first, we need an endpoint IP
address, like 45.67.98.43                        "
                echo "Your input (IP address):"
                read -a FW71
                echo "To which TCP port you want to allow
access?"
                echo "Your input (TCP port number):"
                echo "Your input:"
                read -a FW72
                for svcs71 in ${FW71[@]}
                do
                        for svcs72 in ${FW72[@]}
                        do
                                firewall-cmd
--permanent --add-rich-rule='rule family="ipv4" source
address="'$svcs71'/32" port protocol="tcp" port="'$svcs72'"
accept'
                        done
                done
                firewall-cmd --reload
                ;;
     8)
                echo "Let's manipulate rich rules - to add
specific IPs access to specific port."
                echo "Hint: first, we need an endpoint IP
address, like 45.67.98.43"
                echo "Your input (IP address):"
                read -a FW81
                echo "To which TCP port you want to allow
access?"
                echo "Your input (TCP port number):"
```

```
                    echo "Your input:"
                    read -a FW82
                    for svcs81 in ${FW81[@]}
                    do
                              for svcs82 in ${FW82[@]}
                              do
                                    firewall-cmd --permanent
--remove-rich-rule='rule family="ipv4" source
address="'$svcs81'/32" port protocol="tcp" port="'$svcs82'"
accept'
                              done
                    done
                    firewall-cmd --reload
                    ;;
        9)
                    echo "Let's just list the firewalld settings
first:"
                    firewall-cmd --list-all
                    echo "Let's list all the rich rules, if any:"
                    firewall-cmd --list-rich-rules
                    ;;
        *)          echo "Please make a correct choice, available
choices are 1-9!"
esac
```

This is a lot of code, but it makes it so much more readable (since we're using a case loop). We could've done this in a couple of different ways, but this is the easiest code to debug, and, most importantly, it works well.

Now, let's look at Ubuntu's `ufw` script, which is going to be very similar – we just need to get the `ufw` commands correct. We're also going to look at two different ways of deleting rules (by an index number and by rule), just so that we know how to get on with both:

```
#!/bin/bash
echo "1 = ufw (Ubuntu) - manipulate service configuration -
add"
echo "2 = ufw (Ubuntu) - manipulate service configuration -
remove"
echo "3 = ufw (Ubuntu) - manipulate TCP ports - add"
```

```
echo "4 = ufw (Ubuntu) - manipulate TCP ports - remove"
echo "5 = ufw (Ubuntu) - manipulate UDP ports - add"
echo "6 = ufw (Ubuntu) - manipulate UDP ports - remove"
echo "7 = ufw (Ubuntu) - manipulate whitelist IP/port
configuration - add"
echo "8 = ufw (Ubuntu) - manipulate whitelist IP/port
configuration - remove"
echo "9 = ufw (Ubuntu) - list current configuration"
echo -e "Your choice:"
read CRIT


# Let's start our case loop against CRIT variable.
case $CRIT in
        1)
                echo "Enter service names, using space as
separator."
                echo "Hint: ssh http https etc. Get list from /
etc/services"
                echo "Your input:"
                read -a FW1
                for svcs1 in ${FW1[@]}
                do
                        ufw allow $svcs1
                done
                ;;
        2)
                echo "Enter rule numbers from the list:"
                ufw status numbered
                echo "Your input, single number or multiple
numbers separated by space:"
                echo "Hint: Best way to do it would be
backwards - from top rule number to bottom rule number!"
                read -a FW2
                for svcs2 in ${FW2[@]}
                do
                        echo "y" | ufw delete $svcs2
                done
```

```
                ;;
        3)
                echo "Enter TCP port numbers, using space as
separator."
                echo "Hint: 22 80 443 etc."
                echo "Your input:"
                read -a FW3
                for svcs3 in ${FW3[@]}
                do
                        ufw allow $svcs3/tcp
                done
                ;;
        4)
                echo "Enter TCP port numbers, using space as
separator."
                echo "Hint: 22 80 443 etc."
                echo "Your input:"
                read -a FW4
                for svcs4 in ${FW4[@]}
                do
                        ufw delete allow $svcs4/tcp
                done
                ;;
        5)
                echo "Enter UDP port numbers, using space as
separator."
                echo "Hint: 22 80 443 etc."
                echo "Your input:"
                read -a FW5
                for svcs5 in ${FW5[@]}
                do
                        ufw allow $svcs5/udp
                done
                ;;
        6)
                echo "Enter UDP port numbers, using space as
separator."
```

```
                echo "Hint: 22 80 443 etc."
                echo "Your input:"
                read -a FW6
                for svcs6 in ${FW6[@]}
                do
                        ufw delete allow $svcs6/udp
                done
                ;;
        7)
                echo "Let's manipulate whitelist rules - to add
specific IPs access to specific port."
                echo "Hint: first, we need an endpoint IP
address, like 45.67.98.43"
                echo "Your input (IP address):"
                read -a FW71
                echo "To which port you want to allow access?"
                echo "Your input (port number):"
                echo "Your input:"
                read -a FW72
                for svcs71 in ${FW71[@]}
                do
                        for svcs72 in ${FW72[@]}
                        do
                                ufw allow from $svcs71 to any
port $svcs72
                        done
                done
                ;;
        8)
                echo "Let's manipulate whitelist rules - to
remove specific IPs access to specific port."
                echo "Hint: first, we need an endpoint IP
address, like 45.67.98.43"
                echo "Your input (IP address):"
                read -a FW81
                echo "To which port you want to allow access?"
                echo "Your input (port number):"
```

```
                    echo "Your input:"
                    read -a FW82
                    for svcs81 in ${FW81[@]}
                    do
                                for svcs82 in ${FW82[@]}
                                do
                                        ufw delete allow from $svcs81
to any port $svcs82
                                done
                    done
                    ;;
        9)
                    echo "Let's list the ufw settings:"
                    ufw status
                    ;;


        *)          echo "Please make a correct choice, available
choices are 1-9!"
esac
```

There we go – that's another long script done. This should help us when we're using Ubuntu a lot. Next, we will be going in a different direction – using nmcli in interactive, scripted mode to configure network settings on CentOS.

There's more...

For additional information about the firewall-cmd and ufw command-line options, we suggest that you visit the following links:

- firewall-cmd man page: https://firewalld.org/documentation/man-pages/firewall-cmd.html

- Configuring complex firewall rules with the *rich language* syntax: https://access.redhat.com/documentation/en-us/red_hat_enterprise_linux/7/html/security_guide/configuring_complex_firewall_rules_with_the_rich-language_syntax

- ufw cheatsheet: https://blog.rtsp.us/ufw-uncomplicated-firewall-cheat-sheet-a9fe61933330

Configuring network settings interactively

Often, we don't have access to GUIs and GUI-based configuration tools. If we need to configure network settings, this can lead to a bunch of problems. Either we need to learn the syntax of `/etc/sysconfig/network-script` files (not user-friendly), or we need to use the tools that are at our disposal to configure network settings from the CLI. Let's learn how to use `nmcli` for that purpose.

Getting ready

Before you start this recipe, you need to make sure that you are using our `cli2` CentOS machine as Ubuntu doesn't use `nmcli` by default. Once you've done that, you're all set!

How to do it...

Configuring network settings via `nmcli` isn't difficult, but at the same time, it's far from super user-friendly. There's quite a bit of syntax involved and sometimes, that can get a bit overwhelming. So, let's create a script that's going to do three things for us:

- Configure network settings via `nmcli` so that we use static IP network configuration.
- Configure network settings via `nmcli` so that we use DHCP network configuration.
- Check/output the current network settings

Our script should look like this:

```
#!/bin/bash
# V1.0 / Jasmin Redzepagic / 01/11/2021 Initial script version
# Distribution allowed under GNU Licence V2.0
echo "1 = nmcli - static IP address configuration for existing
interface"
echo "2 = nmcli - reconfigure a static IP-based configuration
to DHCP"
echo "3 = nmcli - list current device and connection status"
echo -e "Your choice:"
read CRIT


# Let's start our case loop against CRIT variable.
case $CRIT in
        1)
                echo "Let's first check current connection
configuration:"
                nmcli con show
```

```
                echo "Which interface do you want to configure
from this list?"
                echo "HINT: We need to use an entry from NAME
field"
                echo "Type in the interface name: "
                read -a interface1
                echo "Type in the IP address/prefix: "
                read -a address1
                echo "Type in the default gateway IP address: "
                read -a gateway1
                echo "Type in DNS servers, use space to
separate entries: "
                read -a dns1
                echo
                nmcli con mod $interface1 ipv4.address
"$address1" ipv4.gateway "$gateway1"
                nmcli con mod $interface1 ipv4.method manual
                for dnsservers in ${dns1[@]}
                do
                        nmcli con mod $interface1 ipv4.dns
$dnsservers
                done
                systemctl restart NetworkManager
                ;;
        2)
                echo "Let's first check current connection
configuration:"
                nmcli con show
                echo "Which interface do you want to configure
from this list?"
                echo "HINT: We need to use an entry from NAME
field"
                echo "Type in the interface name: "
                read -a interface1
                nmcli con mod $interface1 ipv4.method auto
                systemctl restart NetworkManager
                ;;
```

```
        3)
                echo "Current status of network devices: "
                nmcli dev show
                echo "Current status of network connections: "
                nmcli con show
                ;;
        *)      echo "Please make a correct choice, available
choices are 1-3!"
esac
```

Here's what the output will look if we use this script:

```
[root@cli2 ~]# ./nmcli.sh
1 = nmcli - static IP address configuration for existing interface
2 = nmcli - reconfigure a static IP-based configuration to DHCP
3 = nmcli - list current device and connection status
Your choice:
1
Let's first check current connection configuration:
NAME    UUID                                TYPE      DEVICE
ens33   c613f601-9658-4ac4-8d92-7ee9a6cdcbfe  ethernet  ens33
Which interface do you want to configure from this list?
HINT: We need to use an entry from NAME field
Type in the interface name:
ens33
Type in the IP address/prefix:
192.168.0.20/24
Type in the default gateway IP address:
192.168.0.254
Type in DNS servers, use space to separate entries:
8.8.8.8 8.8.4.4

[root@cli2 ~]# cat /etc/sysconfig/network-scripts/ifcfg-ens33
TYPE=Ethernet
PROXY_METHOD=none
BROWSER_ONLY=no
BOOTPROTO=none
DEFROUTE=yes
IPV4_FAILURE_FATAL=no
IPV6INIT=yes
IPV6_AUTOCONF=yes
IPV6_DEFROUTE=yes
IPV6_FAILURE_FATAL=no
NAME=ens33
UUID=c613f601-9658-4ac4-8d92-7ee9a6cdcbfe
DEVICE=ens33
ONBOOT=yes
IPADDR=192.168.0.20
PREFIX=24
GATEWAY=192.168.0.254
DNS1=8.8.4.4
```

Figure 16.5 – Configuring a network interface from a shell script to make it a static IP configuration

As we can see, all of the network settings get applied. Also, for the second use case – which is to revert to using DHCP from an existing configuration – the output will look like this:

```
[root@cli2 ~]# ./nmcli.sh
1 = nmcli - static IP address configuration for existing interface
2 = nmcli - reconfigure a static IP-based configuration to DHCP
3 = nmcli - list current device and connection status
Your choice:
2
Let's first check current connection configuration:
NAME    UUID                                    TYPE        DEVICE
ens33   c613f601-9658-4ac4-8d92-7ee9a6cdcbfe    ethernet    ens33
Which interface do you want to configure from this list?
HINT: We need to use an entry from NAME field
Type in the interface name:
ens33
[root@cli2 ~]# cat /etc/sysconfig/network-scripts/ifcfg-ens33
TYPE=Ethernet
PROXY_METHOD=none
BROWSER_ONLY=no
BOOTPROTO=dhcp
DEFROUTE=yes
IPV4_FAILURE_FATAL=no
IPV6INIT=yes
IPV6_AUTOCONF=yes
IPV6_DEFROUTE=yes
IPV6_FAILURE_FATAL=no
NAME=ens33
UUID=c613f601-9658-4ac4-8d92-7ee9a6cdcbfe
DEVICE=ens33
ONBOOT=yes
IPADDR=192.168.0.20
PREFIX=24
GATEWAY=192.168.0.254
DNS1=8.8.4.4
```

Figure 16.6 – Reverting to our DHCP configuration with the BOOTPROTO parameter set up correctly

This file also looks good, so we're good to go with this script as well.

The next set of scripts we'll be looking at is about backups – one will use shell script arguments and variables, while the other will use one very handy `tar` characteristic. Let's work on some backup scripts!

There's more...

`Screen` requires a bit of trial and error and getting used to. We recommend that you check out the following links to learn more:

- `nmcli` man page: `https://linux.die.net/man/1/nmcli`
- `nmcli` examples: `https://people.freedesktop.org/~lkundrak/nm-docs/nmcli-examples.html`

Backing up the current directory with shell script arguments and variables

One of the most common reasons why system engineers use Bash shell scripting is for backup purposes. There are various tools available, but for shell scripting purposes, we are going to make a couple of `tar`-based shell scripts, work with arguments and variables, and learn how to make our jobs easier by using shell scripting for backup purposes. Let's take a look!

Getting ready

Before you start this recipe, you need to make sure that you have `tar` installed on your Linux machine. For this, you need to use the following command:

```
sudo apt-get -y install tar
```

If you're using a CentOS-based machine, use the following command:

```
sudo yum -y install tar
```

Now, you're ready to get started.

How to do it...

Our premise for this first backup script, which is based on `tar`, is simple:

- We want to be able to create a backup while using an argument to set the backup's filename.
- We want to be able to easily change our shell script so that it can back up whatever number of directories we want (this is easy to do by listing the source directories in the `backup_source` variable).

Let's see how this would work:

```sh
#!/bin/sh
# V1.0 / Jasmin Redzepagic / 01/11/2021 Initial script version
# Distribution allowed under GNU Licence V2.0
# This script contains some pre-defined parameters:
# - which directories we want to backup, used as a variable
# backup_source
# - destination folder, via variable called backup_dest
# - indexing according to date, used as a variable date
#

# Also, it uses a shell script argument $1 (first argument that we
# use to call on the script) to set value for variable filename
filename=$1

# let's set the directory that we want to backup
# if we want to backup more of them, we create a space-
separated
# list
backup_source="./"

# let's set the destination folder
backup_dest="/tmp"

# let's set value of the date variable in accordance to current
date
date='date '+%d-%B-%Y''

# let's set the value of the hostname variable in accordance to
host
# name
hostname=$(hostname -s)

# let's start the backup process
echo "Starting backup"
```

```
sleep 2
tar cvpzf    $backup_dest/$filename-$hostname-$date.tar.gz
$backup_source

# let's announce the end of the backup process
echo "Backup done"
```

The process should look like this:

```
student@cli1:~$ bash backup1.sh
Starting backup
./
./fw.sh
./.config/
./.config/procps/
./stats.sh
./.selected_editor
./filelist.txt
./.bashrc
./temp/
./temp/filelist.txt
./root.sh
./.cache/
./.cache/motd.legal-displayed
./backup1.sh
./sscript1.sh
./.lesshst
./sscript2.sh
./.bash_history
./.viminfo
./.sudo_as_admin_successful
./sscript5.sh
./.bash_logout
./.ssh/
./.ssh/id_rsa.pub
./.ssh/known_hosts
./.ssh/id_rsa
./sscript1-2.sh
./.profile
./ubuntu.sh
Backup done
```

Figure 16.7 – Simple backup script with argument

The backup word that we typed as part of the script is our $1 argument in the script – the first argument that we start the script with. As we can see, the script did its job properly.

There's more...

If you need any more information about the `tar` command, we suggest that you look at the following links:

- `tar` command man page: `https://man7.org/linux/man-pages/man1/tar.1.html`

- 18 useful `tar` examples: `https://www.tecmint.com/18-tar-command-examples-in-linux/`

Creating a current backup based on the user input for the backup source and destination

After making backup scripts via `tar`, we need to make a completely interactive script that asks for all of the details from us to be inputted with the keyboard. Let's learn how to make this happen!

Getting ready

If you followed the previous recipe, then you won't need anything new – the same requirements apply.

How to do it...

Our premise has changed somewhat this time around. We want a fully functional backup script, but one that doesn't use any static variables (like the previous one). Also, we want to be able to call this script at will, which is why we're using multiple questions to set up the necessary variables. Here's what the script should look like:

```
#!/bin/bash
# V1.0 / Jasmin Redzepagic / 01/11/2021 Initial script version
# Distribution allowed under GNU Licence V2.0
# This script does a custom backup, based on our arguments
# We need to give it a couple of arguments @start:
# - backup file name
# - list of directories (or a single directory) that we
# want to backup
# We also added a bit of code to skip standard error
```

```
echo -e "Type in the backup file name, use something like file-
date.tar.gz:"
read filename
echo -e "Type in the list or a single directory that you want
to backup:"
read directories
echo "Let's do this thing!"
tar cfvz $filename $directories 2> /dev/null
```

This is simple, yet effective. Note that we used one very, very cool capability of the `tar` command, which is to use a list of directories for backup purposes, but specifically, as the list that's at *the end* of the `tar` command's syntax, which makes things a bit easier.

In the next chapter, things are going to get progressively more complicated. Make sure that you check out the second part of our shell scripting examples there.

There's more...

If you need any more information about the `tar` command, we suggest that you visit the following links:

- `tar` command man page: `https://man7.org/linux/man-pages/man1/tar.1.html`

- 18 useful `tar` examples: `https://www.tecmint.com/18-tar-command-examples-in-linux/`

17
Advanced Shell Script Examples

So far, we have done all we could to show you different ways scripts can be written, and we went through a lot of examples of how different tasks can be accomplished. In this chapter, we are going to implement all this in a much more complex way in **scripts** that can be used in real life.

The scripts we are going to show you in this chapter solve everyday problems for system administrators, from dealing with creating new users to working with **virtual machines** (**VMs**). By walking you through these examples, our aim is not only to show you how scripts should work but also what they should look like and how to approach writing them.

In this chapter, we are going to cover the following shell script examples:

- Implementing a web server service and security settings

- Creating users and groups from a standardized input file and a standardized password and forcing users to change them on the next login

- Creating users and groups from a standardized input file and a random password for every user

- Scripted VM installation on **Kernel-based Virtual Machine** (**KVM**)

- A shell script to provision **Secure Shell (SSH)** keys, create standard users, install a standardized set of packages for a **Linux, Apache, MySQL, PHP/Perl/Python (LAMP)** server, configure basic firewall settings, **Security-Enhanced Linux (SELinux)** configuration, and `sudo` configuration
- A shell script for VM administration

Technical requirements

In almost all the other chapters, we were working with a generic setup that simply required any Linux distribution, as long as it could run a Bash shell. In this chapter, we are going to change things a bit—by necessity, these scripts will have to run on Ubuntu or any other Debian-based distribution. We are going to mention the reasons for this in the following recipes when something has to be done differently in order to make it happen on any other Linux distribution. So, in order to run scripts in this chapter, you need the following:

- A VM with Linux installed—we are using *Ubuntu 20.10*, but any Debian-based distribution will work
- Understanding of all the things we did in the previous chapters since we are going to presume you understand how Bash scripting works

So, start your VM for us to start doing many useful things!

Implementing a web server service and security settings

In this particular recipe, the idea is to use a small `shell` script to help us configure an already installed web server. We are going to enable our script to change where the web pages served by the server are located, but you will quickly see that changing any other option is easily added to this script.

By using this script, all users would have to do to get the system running is this:

- Install the web server
- Run the script to change where website files are located

As always, the main problem when preparing something that will be a simple operation for the user is understanding and hiding all the complexity while making it reasonably easy for the administrator to add new features. How do we do this? Read on.

Getting ready

This is our scenario:

A user has installed an Apache web server on their Ubuntu machine. They want to change the location of files that make up their website.

Before we go into this, we must work on our presumptions for this task, as is usual with almost any script.

First, we expect the web server to be already installed before we run the script, and we expect it to be Apache. The simplest way to do it is to use the following command:

```
sudo apt install apache2 -y
```

Now, we wait for the package manager to do its job.

Our script will not work with nginx or `lighttpd` nor any other web server since the configuration is parsed directly and there is no common way to set the parameters we require. Having said that, since the parsing we are using to change the configuration is pretty basic, if you need to modify this script to work with another server, it will probably take just a few minutes.

Next, we are presuming that the user is changing the default website, one called `000-default.conf` in the configuration directory. This value is hardcoded in our script, which means that if you have multiple websites on the same server, this script will only change the one configured as default.

Sometimes, administrators just add websites directly into this part of the configuration instead of creating new files for every site, as it should be done. Our script accomplishes its task by finding and replacing any mention of the `DocumentRoot` directive in the file. If we specify multiple `DocumentRoot` directives, the script is going to change all of them to the same value.

Another thing we must think about is error checking. Inside the script itself, we are trying to catch if there was an error in the configuration, but the way we do it leaves a lot to be desired. Although our script will try to restore files to the state that it was before we changed their content, we are not trying to do any real syntax checking in the values that we are changing. This can prove to be a problem if the user makes an error when specifying the path they want to use, but there is no easy way to solve this; implementing a check that will be smart enough to scan for a valid path is too complicated for a task such as this.

How to do it...

In the recipes in this chapter, we are going to first give you our version of the script and then explain the details we think are important. All the scripts are going to have plenty of comments inside them, and we strongly advise you to do the same if possible. Comments can also be used when creating a script to define a rough outline of all the things you want to do before you type out a single command.

First, let's start with the script itself:

```
#!/bin/bash
# V1.0 / Jasmin Redzepagic / 01/11/2021
# Distribution allowed under GNU Licence V2.0


# Script configures apache DocumentRoot with a given path
# and sets firewall accordingly
# Script is interactive, no arguments are accepted


# This script has to be run as root, we need to check that

if [[ $(id -u) -ne 0 ]]
then
    echo "This script needs to be run as root!" >&2
    exit 1
fi


# If there are multiple sites configured we will show a warning

if [[ $(ls /etc/apache2/sites-enabled/ | wc -l) -gt 1 ]]
then
echo "Warning: you may have more than one site!" >&2
            exit 1
fi


# First we are going to get what the root of the site is now
# When checking for DocumentRoot we are only checking
# in default web site
```

```
HTTPDIR='grep DocumentRoot /etc/apache2/sites-available/000-\
default.conf'
HTTPDIR="/$( cut -d '/' -f 2- <<< "$HTTPDIR" )"


# We are going to print current directory
# that we read from inside the configuration file
echo "Current HTTPDIR is set as $HTTPDIR"
read -p  "Press Enter to accept current value or input absolute\
path for new DocumentRoot: " NEWDIR


# If user pressed enter we are going to
# simply use the value we already read,
# otherwise we use the new value
# Note: there is absolutely no sanity checking
# if the given value is actually a path


NEWDIR=${NEWDIR:-$HTTPDIR}
echo "Directory is going to be set to $NEWDIR"


# Since we are dealing with a path we need to
# preprocess it before we use it in sed
# otherwise this is going to break


# There is an alternative, sed allows for
# any other character in place of /
# but this is going to be a problem
# if our path contains any nonstandard character
# so we simply escape all the slashes
# we need to use the _ character in this
# case to be able to search for slash
ESCNEWDIR=$(echo $NEWDIR | sed 's_/_\\/_g')
ESCHTTPDIR=$(echo $HTTPDIR | sed 's_/_\\/_g')
# before we change the configuration
# we are going to back it up so we can restore if we need to
cp /etc/apache2/sites-available/000-default.conf /etc/apache2/1\
sites-available/000-default.conf.backedup
sed -i "s/$ESCHTTPDIR/$ESCNEWDIR/g" /etc/apache2/sites-
```

```
available/000-default.conf
# now we need to restart the service
# in order to use the new configuration.
systemctl reload apache2
# after every command we must check to see if there were any
errors.
# In this particular case, we restore from backup if there were

if [ $? -ne 0 ]
  then
      cp /etc/apache2/sites-available/000-default.conf. \
      backedup/etc/apache2/sites-available/000-default.conf


# we need to exit if we triggered this condition
# since we are finished here, nothing was changed.


# before exit we need to reload apache once more
# to make sure old configuration is used
# we are doing a start and stop here
# because reload obviously failed in the step above


      echo "Apache was not reloaded correctly, maybe there was \
      an error in the syntax"
      systemctl stop apache2
      systemctl start apache2
      return 1
fi
# if we came this far we need to get our firewall sorted out
# we are adding ports 80 and 443 as permitted.
ufw allow http
ufw allow https
# alternative to this is ufw allow "Apache Full"
# but using exact ports and aliases makes this easier to read.
# end of script
```

We need to note a few things here. Apache as a web server is right now the most used web server in all the distributions by default, but nginx is slowly becoming more and more popular. The thing to remember is that depending on the distribution package containing `apache`, this is called either `apache2` (on Debian-based distributions such as Ubuntu) or `httpd` (on Red Hat-based distributions such as **Red Hat Enterprise Linux** (**RHEL**) or CentOS). Other than the package name, there is a small difference in the placement of the configuration files for the server itself, although the syntax is exactly the same.

Another thing is the firewall. Ubuntu uses `ufw` while CentOS uses `firewalld`. The third big thing to note is `apparmor` (Ubuntu) and `SELinux` (CentOS).

Our version of the script works on Debian-based machines. Slight modifications are needed if we want to use it on, for example, CentOS.

See also

- `https://www.digitalocean.com/community/questions/which-ufw-service-to-use-for-apache2`

- `https://www.tecmint.com/setup-ufw-firewall-on-ubuntu-and-debian/`

Creating users and groups and forcing users to change them on the next login

One of the most common things you are going to do on Linux machines is create a lot of users. There is a way to avoid this by using a centralized database for user authentication, but in reality, this is used only on machines in large deployments, so local users are still prevalent in most cases.

Having a way to deploy users and assign them passwords is something every admin needs whenever deploying a new server or desktop.

Getting ready

This recipe calls for two things.

The `script` has to be used with administrative privileges since it changes users on the system. Also, we need to prepare a file containing a user list in advance.

Before we show you our script, we must also mention that there is more than one way to read and parse a file when getting values for our scripts. It makes sense to try to understand different ways of doing this in order to up your scripting game. For this exact reason, in this recipe and the next one, we decided to avoid using a `for` loop but instead opted to parse a file using arrays and delimiters.

How to do it...

As we have become accustomed to, we are starting with our script before we note things to remember:

```bash
#!/bin/bash
# V1.0 / Jasmin Redzepagic / 01/11/2021
# Distribution allowed under GNU Licence V2.0
# script creates users from a csv file, and makes the user
# change his password at next logon
# argument for the script is csv file name

# csv file is structured as follows:
# user1,password1
# user2,password2
# ....


# First thing to do is check if we have any arguments supplied

if [ $# -eq 0 ]
  then
    echo "No arguments, proper usage is $0 <CSV file>"
fi

# next we need to get our information from the file
# here we are doing it one line at a time
# and to do that we need to adjust the delimeter
# that shell uses to understand how values are
# separated.
IFS=$'\n'
```

```
# Read the lines into an array

read -d '' -ra USRCREDS < $1

# Now we are going to deal with individual lines
# since right now our array contains both the user
# and the appropriate password in one value
# separated by a , character
# we chose to do this by telling the shell
# we want to use , as a value separator
IFS=','
# Iterate over the lines

for USER in "${USRCREDS[@]}"
do
# Split values into separate variables
    read usr pass <<< "$USER"
# Create the user

    useradd -m $usr

# Set the password, we need to do this using passwd command
# alternative would be to use a hashing function
# passwd asks for password twice!
    echo "$pass"$'\n'"$pass" | passwd $usr
# then we expire the user password
    passwd --expire $usr
done
```

There are a couple of concepts that we need to mention here. The first is dealing with passwords. Having any password readable in plaintext for any amount of time is a security risk, so the idea of making the user change their password as soon as possible is wise.

When creating passwords for new users, we basically have two choices—one is to create a list of users and passwords in advance, as we did in this example, and the other is to create a list of users and then assign them random passwords, as we will do in the next recipe.

Whenever you are dealing with any password, always remember that once a single user is compromised, you have a big security problem because a lot of ways to break into the system are depending on being able to run an application locally. Minimize the time anybody other than the user knows the password for the account, and never store passwords in a plain, readable format.

See also

- https://linuxconfig.org/linux-reset-password-expiration-age-and-history

- https://www.tecmint.com/force-user-to-change-password-next-login-in-linux/

Creating users and groups from a standardized input file and a random password for each user

In the previous recipe, we dealt with a way of creating new users. In this one, we are going to expand on this using a similar script to not only create new users but also assign them groups provided with the user, giving the administrator information on new user passwords.

Getting ready

We are creating users, so this script has to be run under an administrator account. In this particular case, we also probably want to redirect the output of the script to some file since passwords for new users are created when the script is run, and passwords are not stored anywhere. If we don't save them somewhere, they are going to be lost and recreated.

How to do it...

In the previous recipe, we mentioned that passwords should never be stored anywhere, but when creating new users, this is completely inevitable. We feel that the way we deal with passwords in this recipe is better than having them ready in advance since passwords in this script are created while the script is running so that the administrator can establish more control over them from the start:

```bash
#!/bin/bash

# V1.0 / Jasmin Redzepagic / 01/11/2021
```

```
# Distribution allowed under GNU Licence V2.0


# script creates users from a csv file
# creating a group specified for each user
# and adding the user to a group
# password is generated and printed with the username

# argument for the script is csv file name

# csv file is structured as follows:
# user1,group1
# user2,group2
# ....

# output is structured as:
# user / password



# First thing to do is check if we have any arguments supplied

if [ $# -eq 0 ]
  then
    echo "No arguments, proper usage is $0 <CSV file>"
fi

# This script has to be run as root, we need to check that

if [[ $(id -u) -ne 0 ]]
            then
            echo "In order to add users and groups this \
            scripts needs to be run as root!" >&2
            exit 1
fi
```

```
# next we need to get our information from the file
# here we are doing it one line at a time
# and to do that we need to adjust the delimeter
# that shell uses to understand how values are
# separated.

IFS=$'\n'

# Read the lines into an array

read -d '' -ra USRCREDS < $1

# Now we are going to deal with individual lines
# since right now our array contains both the user
# and the appropriate password in one value
# separated by a , character
# we chose to do this by telling the shell
# we want to use , as a value separator

IFS=','

# Iterate over the lines

for USER in "${USRCREDS[@]}"
do
# Split values into separate variables

read usr grp <<< "$USER"

# Create the user
```

```
useradd -m $usr

# if the group does not exist, create it

getent group $grp || groupadd $grp

# add the user to a group

usermod -a -G $grp $usr

# now we create a random password

pass=$(cat /dev/urandom | tr -dc A-Za-z0-9 | head -c8)

# Set the password, we need to do this using passwd command
# alternative would be to use a hashing function
# passwd asks for password twice!

echo "$pass"$'\n'"$pass" | passwd $usr

# in the end we print user and password

echo $usr/$pass

done
```

One thing to note in this script is that we are relying on a lot of messages from the commands we are using instead of checking things by ourselves. For example, if the user is already created, an error message is going to be created by `useradd` instead of us:

```
demo@ubuntu:~/scripts$ sudo bash createusersrandom.sh groups.csv
useradd: user 'user1' already exists
New password: Retype new password: passwd: password updated successfully
user1/l1nDoaLV
useradd: user 'user2' already exists
New password: Retype new password: passwd: password updated successfully
user2/CIUNjHvh
useradd: user 'user3' already exists
New password: Retype new password: passwd: password updated successfully
user3/d3I8asvt
useradd: user 'user4' already exists
New password: Retype new password: passwd: password updated successfully
user4/Qv6blAX7
```

Figure 17.1 – Error messages provided by commands inside the script

See also

- https://superuser.com/questions/533126/how-to-execute-command-and-if-it-fails-execute-another-command-and-return-1

- https://linux.die.net/man/4/urandom

Scripted VM installation on KVM

Another common task done in some environments is creating new VMs from the command line. The reason we do this is usually flexibility and speed—using a **graphical user interface** (**GUI**) can be an **order of magnitude** (**OOM**) slower than using a **command-line interface** (**CLI**).

KVM provides a very simple solution for creating machines in the command line. All the user needs to know are some basic parameters.

Getting ready

We, of course, need a functioning KVM on the server we are running this script on. Other than that, our script presumes the user understands all the different options that KVM requires to be able to create a machine. Before trying to understand how the script works, be sure to go through as much information as possible about creating VMs from the command line in order to be sure what the different options do. Also, refresh your knowledge about using the `dialog` toolkit for graphical interfaces since this script relies on this for input.

How to do it...

The only big thing in this small script is the way we are assigning values using `dialog`. As always, there are a couple of ways to do it. We are using the most logical one, for us at least:

```bash
#!/bin/bash


# V1.0 / Jasmin Redzepagic / 01/11/2021
# Distribution allowed under GNU Licence V2.0


# script creates a virtual machine on the host it is run on
# asking the user for parameters of the VM


# in this script we are going to use dialog to show
# how a script can get values that way


name=$(dialog --inputbox "What is the name of the VM?" 8 25 \
--output-fd 1)
cpus=$(dialog --inputbox "How many VCPUs?" 8 25   --output-fd 1)
mem=$(dialog --inputbox "Enter the amount of memory in MB" 8 25 \
--output-fd 1)
cdrom=$(dialog --inputbox "Path to CDROM:" 8 25 --output-fd 1)
disksize=$(dialog --inputbox "Enter the disk size:" 8 25 \
--output-fd 1)
osv=$(dialog --inputbox "What is the OS variant installed?" 8 \
25 --output-fd 1)


virt-install --name=$name --vcpus=$cpus --memory=$mem \
--cdrom=$cdrom --disk size=$disksize --os-variant=$osv
```

When using `dialog`, you must handle the way input from the user is redirected. In this example, we are using `-output-fd 1` in order to tell `dialog` to get everything redirected to **standard output** (**stdout**) where we can directly assign the values to variables.

See also

- `https://linux.die.net/man/1/dialog`
- `https://www.geeksforgeeks.org/creating-dialog-boxes-with-the-dialog-tool-in-linux/`

Using a shell script to provision SSH keys

The safest way to deal with passwords is to not use them at all. Using SSH keys is a great way of avoiding passwords completely if we are able to get a public key connected to a user account that the user can log in to without using the password, and since only their private key enables login, this makes the whole transaction much safer.

This recipe deals with just such a task, installing a new machine that is going to serve as a LAMP server and that will enable users to log in using no passwords at all.

Getting ready

In reality, a script such as this will be used if we have a few servers to install and not too much time. An alternative to something like this would be to use a proper orchestration tool such as Ansible, but although it is an enormously powerful tool, Ansible is too complicated for small deployments.

In any case, this script presumes only that our server has a working internet connection to be able to get the packages that need to be installed and that we have acquired a public SSH key from the user we plan on creating.

How to do it...

We are transferring the key by using a regular plaintext file. This is completely fine since it actually contains no information that can pose a security problem—in order to use SSH to connect, a user has to have a private key that corresponds to the public key we are using.

Since this key is—or should be—controlled by only one particular user, we are not worried about safety here:

```bash
#!/bin/bash


# V1.0 / Jasmin Redzepagic / 01/11/2021
# Distribution allowed under GNU Licence V2.0


# script installs lamp, creates a user and assigns him SSH key
# key is provided in a file


# script expects filename that contains SSH key


# First thing to do is check if we have any arguments supplied

if [ $# -eq 0 ]
  then
    echo "No arguments, proper usage is $0 <file containing SSH \
key>"
fi


# This script has to be run as root, we need to check that

if [[ $(id -u) -ne 0 ]]
            then
                    echo "In order to add services this \
scripts needs to be run as root!" >&2
                    exit 1
fi


# now we follow the standard installation procedure for LAMP


# first we aquire new updates


apt update
```

```
# then we install apache server

apt install apache2 -y

# we reconfigure the firewall to allow all the traffic in

ufw allow "Apache Full"

# then we install mysql server
# we could also install mariadb as the alternative

apt install mysql-server -y

# then we install php and required modules

apt install php libapache2-mod-php php-mysql -y

# we create our user

useradd lampuser

# we create directory for the ssh keys

mkdir /home/lampuser/.ssh

# we copy the key directly, allowing login without password
# note that user has no password by default, only ssh works

cp $1 /home/lampuser/.ssh/authorized_keys

# apply permissions to files in directory

chown -R lampuser:lampuser /home/lampuser/.ssh

chmod 700 /home/lampuser/.ssh
```

```
chmod 600 /home/lampuser/.ssh/authorized_keys

# add user to sudo group enabling sudo command

usermod -a -G sudo lampuser

# finally we run the secure installation to finish setting up \
mysql

mysql_secure_installation
```

See also

- https://www.hostinger.com/tutorials/ssh-tutorial-how-does-ssh-work

- https://www.digitalocean.com/community/tutorials/how-to-install-linux-apache-mysql-php-lamp-stack-on-ubuntu-20-04

A shell script for VM administration

Some tasks are complicated to do from the command line, simply because we have a lot of commands that we have to repeat over and over again and then reuse the values that we got in one step in the step that follows it.

In this recipe, we are going to deal with such a task, doing basic maintenance on a VM. What we plan to do is create a script that will enable the user to do a couple of standard tasks on VMs running on the local server, simplifying administration tasks and removing the need to remember long commands. Our plan is to enable the user to start, stop, check status, and revert a VM running on the local server. The script is going to provide the user with the list of machines and give them the opportunity to choose any available machines or apply the command to all of them.

Let's see what is needed for this task.

Getting ready

By this point, you have become accustomed to our disclaimers and requirements that we have to enable our script to run. This one is no different. First, this script requires one important thing—the server has to have support for KVM installed on it before we even begin to do anything. In the script itself, we are using a single command to accomplish all the tasks, but in reality, all of KVM needs to be installed and configured.

Another option is that this script can be, with minor modifications, used to perform tasks on other KVM hosts, but we will leave this as an exercise for you.

So, before you start the script, do a small check if everything works, with a simple command:

```
virsh list -all
```

This should return a list of all the VMs running on your server. If there is any error, it needs to be sorted out before you even try to run the script itself since the script is based on this command working.

How to do it...

First, we are going to start with the script itself:

```
#!/bin/bash


# V1.0 / Jasmin Redzepagic / 01/11/2021
# Distribution allowed under GNU Licence V2.0


# Simple interface to virsh command
# this script enumerates all machines on this KVM host
# and enables user to perform basic commands


# script is interactive and has no command line arguments


# in this script we are going to create a simple two level menu
# that will first ask user what virtual machine he wants to \
  perform commands on.
# User has to specify the machine from a list or type ALL
# if he wants to run the command on all the machines on the
```

```
host
# we need to get the list of virtual
Machines
# notice we are not redirecting errors in order for the user
# to be able to see what actually happened

virsh list --all

# then we do some rudimentary error checking to make sure
# we are at least able to use virsh

if [ $? -ne 0 ]
  then
      echo "Something is wrong with your KVM instance, exiting!"
      return 1
fi

# if we come this far our script can talk to the user




read -p "Choose VM you want to change state of or type ALL for \
all machines:" HOSTN
echo -e "\n"

if [ $HOSTN == "ALL" ];
then
            echo "You chose all machines."
else
            echo "You chose: " $HOSTN "."
fi

echo "What do you want to do"

echo "1 = START"
```

```
echo "2 = STOP"
echo "3 = RESET"
echo "4 = STATUS"
echo -e "\n"
read CHOSENOP


if [ $HOSTN == "ALL" ]; # we are running the commands on all
the machines


# every command is run in a loop on all the machines


then
                if [ $CHOSENOP -eq 1 ];     # user chose start
                  then
                            for i in $(virsh list --name --all);
                            do
                            echo "Starting  $i"
                            virsh start $i;
                            done
                            exit 0
                elif [ $CHOSENOP -eq 2 ]; # user chose stop
                then
                            for i in $(virsh list --name --all);
              do
                            echo "Stopping  $i"
virsh shutdown $i
done
                            exit 0
          elif [ $CHOSENOP -eq 3 ]; # user chose to revert \
    to snapshot
                then
                            for i in $(virsh list --name --all);
                            do
                            echo "Reverting $i to latest
snapshot: "
                            virsh snapshot-revert $i start;
                            done
```

```
                              exit 0
                   elif [ $CHOSENOP -eq 4 ]; # user chose to \
                   display status of machines
                   then
                              for i in $(virsh list --name --all);
                              do
                              echo "Status of $i: "
                              virsh dominfo $i;
                              done
                              exit 0
                   else

# user made an invalid input

                              echo "Input was not valid!"
                              exit 0
                   fi

else

# we do everything the same way but with a particular VM


          if [ $CHOSENOP -eq 1 ];
          then
                              echo "Starting $HOSTN"
                              virsh start $HOSTN
                              exit 0
          elif [ $CHOSENOP -eq 2 ];
          then
                              echo "Stopping $HOSTN"
                              virsh shutdown $HOSTN
                              exit 0
          elif [ $CHOSENOP -eq 3 ];
```

```
        then
                        echo "Reverting $HOSTN to last \
                        snapshot"
                        virsh snapshot-revert $HOSTN
                        exit 0
        elif [ $CHOSENOP -eq 4 ];
        then
                        echo "Status of $HOSTN: "
                        virsh dominfo $HOSTN
                        exit 0
        else

# user made an invalid input

                        echo "Input was not valid!"/ \
                        exit 0
        fi


fi
```

In this particular script, the main thing that we needed to decide on is how to handle two different conditions. The first condition is: *Are we dealing with a particular VM or all of them?* The second condition is: *What operation is needed?*

There are a couple of ways we can do this—we chose this one because it looks the most logical. We first give the user a list of all the machines on the host, and after they have decided which machine they want to run the command on, we ask them to select what they want to do.

We could have easily done this the other way around and let them choose the command first and then let them select the VM they want to perform it on.

Another thing that we decided on was how to display the names of the machines. We are leaving it to the user to get the machine name right, and we are not doing any checks. One thing that could be done is to try to compare user input to the list of actual machine names. This way, if a user makes a mistake, we can catch it before the script tries to perform an operation on the invalid machine.

Another thing that can be done in this, and in pretty much any script that has a lot of logical decisions and not enough checks, is to do a `try-catch` loop for the entire script so that we can deal with any possible errors without the script breaking completely and leaving us in an unknown state.

See also

- `https://access.redhat.com/documentation/en-us/red_hat_enterprise_linux/6/html/virtualization_administration_guide/chap-virtualization_administration_guide-managing_guests_with_virsh`

- `https://help.ubuntu.com/community/KVM/Virsh`

Index

Symbols

A

B

Packt.com

Subscribe to our online digital library for full access to over 7,000 books and videos, as well as industry leading tools to help you plan your personal development and advance your career. For more information, please visit our website.

Why subscribe?

- Spend less time learning and more time coding with practical eBooks and Videos from over 4,000 industry professionals

- Improve your learning with Skill Plans built especially for you

- Get a free eBook or video every month

- Fully searchable for easy access to vital information

- Copy and paste, print, and bookmark content

Did you know that Packt offers eBook versions of every book published, with PDF and ePub files available? You can upgrade to the eBook version at packt.com and as a print book customer, you are entitled to a discount on the eBook copy. Get in touch with us at customercare@packtpub.com for more details.

At www.packt.com, you can also read a collection of free technical articles, sign up for a range of free newsletters, and receive exclusive discounts and offers on Packt books and eBooks.

Other Books You May Enjoy

If you enjoyed this book, you may be interested in these other books by Packt:

Linux Kernel Programming

Kaiwan N Billimoria

ISBN: 9781789953435

- Write high-quality modular kernel code (LKM framework) for 5.x kernels
- Configure and build a kernel from source
- Explore the Linux kernel architecture
- Get to grips with key internals regarding memory management within the kernel
- Understand and work with various dynamic kernel memory alloc/dealloc APIs
- Discover key internals aspects regarding CPU scheduling within the kernel
- Gain an understanding of kernel concurrency issues
- Find out how to work with key kernel synchronization primitives

Linux System Programming Techniques

Jack-Benny Persson

ISBN: 9781789951288

- Discover how to write programs for the Linux system using a wide variety of system calls

- Delve into the working of POSIX functions

- Understand and use key concepts such as signals, pipes, IPC, and process management

- Find out how to integrate programs with a Linux system

- Explore advanced topics such as filesystem operations, creating shared libraries, and debugging your programs

- Gain an overall understanding of how to debug your programs using Valgrind

Packt is searching for authors like you

If you're interested in becoming an author for Packt, please visit `authors.packtpub.com` and apply today. We have worked with thousands of developers and tech professionals, just like you, to help them share their insight with the global tech community. You can make a general application, apply for a specific hot topic that we are recruiting an author for, or submit your own idea.

Share Your Thoughts

Now you've finished *Linux Command Line and Shell Scripting Cookbook*, we'd love to hear your thoughts! Scan the QR code below to go straight to the Amazon review page for this book and share your feedback or leave a review on the site that you purchased it from.

`https://packt.link/r/1-800-20519-8`

Your review is important to us and the tech community and will help us make sure we're delivering excellent quality content.

www.ingramcontent.com/pod-product-compliance
Lightning Source LLC
Chambersburg PA
CBHW081451050326

40690CB00015B/2761